THE LABORATORY APPROACH
TO TEACHING CALCULUS

EDITORS

L. Carl Leinbach, Gettysburg College
Joan R. Hundhausen, Colorado School of Mines
Arnold M. Ostebee, St. Olaf College
Lester J. Senechal, Mt. Holyoke College
Donald B. Small, Colby College

MAA Notes   Number 20

THE MATHEMATICAL ASSOCIATION OF AMERICA

MAA Notes and Reports Series

The MAA Notes and Reports Series, started in 1982, addresses a broad range of topics and themes of interest to all who are involved with undergraduate mathematics. The volumes in this series are readable, informative, and useful, and help the mathematical community keep up with developments of importance to mathematics.

MAA Notes

1. Problem Solving in the Mathematics Curriculum,
   Committee on the Teaching of Undergraduate Mathematics,
   a subcommittee of the Committee on the Undergraduate Program in Mathematics, *Alan H. Schoenfeld,* Editor.

2. Recommendations on the Mathematical Preparation of Teachers,
   Committee on the Undergraduate Program in Mathematics, Panel on Teacher Training.

3. Undergraduate Mathematics Education in the People's Republic of China,
   Lynn A. Steen, Editor.

4. Notes on Primality Testing and Factoring,
   Carl Pomerance.

5. American Perspectives on the Fifth International Congress on Mathematical Education,
   Warren Page, Editor.

6. Toward a Lean and Lively Calculus,
   Ronald G. Douglas, Editor.

7. Undergraduate Programs in the Mathematical and Computer Sciences: 1985–86,
   D. J. Albers, R. D. Anderson, D. O. Loftsgaarden, Editors.

8. Calculus for a New Century,
   Lynn A. Steen, Editor.

9. Computers and Mathematics: The Use of Computers in Undergraduate Instruction,
   Committee on Computers in Mathematics Education, D. A. Smith, G. J. Porter, L. C. Leinbach, and R. H. Wenger, Editors.

10. Guidelines for the Continuing Mathematical Education of Teachers,
    Committee on the Mathematical Education of Teachers.

11. Keys to Improved Instruction by Teaching Assistants and Part-Time Instructors,
    Committee on Teaching Assistants and Part-Time Instructors, Bettye Anne Case, Editor.

12. The Use of Calculators in the Standardized Testing of Mathematics,
    John Kenelly, Editor, published jointly with The College Board.

13. Reshaping College Mathematics,
    Committee on the Undergraduate Program in Mathematics, Lynn A. Steen, Editor.

14. Mathematical Writing,
    by *Donald E. Knuth, Tracy Larrabee, and Paul M. Roberts.*

15. Discrete Mathematics in the First Two Years,
    Anthony Ralston, Editor.

16. Using Writing to Teach Mathematics,
    Andrew Sterrett, Editor.

17. Priming the Calculus Pump: Innovations and Resources,
    Committee on Calculus Reform and the First Two Years,
    a subcommittee of the Committee on the Undergraduate Program in Mathematics, *Thomas W. Tucker,* Editor.

MAA Reports

First Printing
© 1991 by the Mathematical Association of America
ISBN 0-88385-074-5
Library of Congress Catalogue Card Number 91-062171
Printed in the United States of America

The Laboratory Approach to Teaching Calculus

Editors

L. Carl Leinbach, Gettysburg College
Joan R. Hundhausen, Colorado School of Mines
Arnold M. Ostebee, St. Olaf College
Lester J. Senechal, Mt. Holyoke College
Donald B. Small, Colby College

Production of this Volume

This volume was produced from original papers and illustrations submitted by the authors. The papers were scanned into a Word Perfect version 1.03 (Word Perfect Corporation; 1555 North Technology Way; Orem, UT 84057) document using the OmniPage version 2.1 (Caere Corporation; 100 Cooper Court; Los Gatos, CA 95030) optical character recognition program residing on a MacIntosh II computer with an Apple Scanner device attached. The Word Perfect documents were then edited. The figures and illustrations were scanned using the Apple Scan program and the Apple Scanner. The articles were produced using PageMaker version 3.2 (Aldus Corporation; 411 First Ave. South; Seattle, WA 98104) and printed in their final form on an Apple LaserWriter II printer.

The titles of the articles are in 24 point Times - Bold

The author(s) and institution notation are in 18 point Times - Bold

The text is in 10 point Times - Roman, 10 point Times - Bold, 10 point Times - Italic. In some cases special terms were distinguished using 10 point Helvetica - Roman

Products Mentioned in this Volume

Throughout this volume several hardware and software products are mentioned. The following is a listing of these products and their producers.

Hardware:

Apple, Mac, MacII, MacIntosh refer to products of Apple Computer Inc.; 20525 Mariani Ave.; Cupertino, CA 95014

IBM-PC refers to personal microcomputers produced by the International Business Machines Corporation; 150 Kettletown Road; Southbury, CT 06488

NeXT refers to the product of NeXT Computer, Inc.; 900 Chesapeake Drive; Redwood City, CA 94063

Software:

EXP is produced by Simon L. Smith and distributed by Wadsworth Publishing Company; 10 Davis Drive; Belmont, CA 94002

Calculus T/L is produced by Doug Child and is distributed by Brooks-Cole, a division of Wadsworth Publishing Company; 10 Davis Drive; Belmont, CA 94002

DERIVE is a product of the Soft Warehouse; 3615 Harding Avenue, Honolulu, HA 96816

MAPLE is produced by the Symbollic Computation Group at the University of Waterloo, Waterloo, CA

MathCAD Student Edition is distributed by Addison-Wesley Publishing Company, Inc.; Reading, MA 01867

Mathematica is a product of Wolfram Research Inc.; 201 W. Springfield Ave; Champaign, IL 61826-6059

Micro Calc is a product of MathCalcEduc; 1449 Covington Drive; Ann Arbor, MI 48103

MS/DOS is the product of Microsoft Corporation; 1 Microsoft Way; Redmond, WA 98052-6399

True BASIC and *True* BASIC Calculus are the products of True BASIC, Inc.; 39 S,. Main St.; Hanover, NH 03755

Acknowledgement

The Production of this volume would not have been possible with out the support and technical expertise of the staff of Computing Services of Gettysburg College. Dr. William P. Wilson, Coordinator of Academic Computing was particularly instrumental in assisting with this project

Mrs. Janet Upton of the Department of Mathematics and Computer Science was an invaluable aid in typing material that was not able to be scanned and arranging for the duplication of draft copies of this volume.

The graphics in the article by Thomas Banchoff are the work of Curtis Hendrickson a student of Banchoff's at Brown Univesity. They were produced in PostScript format at Brown and forwarded via electronic mail.

The graphic on the back cover was produced by Howard Lewis Penn using the MPP package developed by Professor Penn, Jim Buchanan and Frank Pittelli of the Mathematics Department of the United States Naval Academy.

Table of Contents

INTRODUCTION
The Laboratory Approach to Teaching Calculus

"Laboratory Calculus" is not a new term; however, it has been heard much more frequently since the advent of the personal computer and the availability of good numerical, graphics and symbolic software suitable for use in introductory calculus classes. Generally, the term conjures up the image of a room filled with microcomputers loaded with the instructor's choice of software. Activities in this space have been varied. On one end of the spectrum, the students use the "lab" to solve or check their answers for traditional homework problems encountered in a traditional course. On the other extreme, the existence of the computer lab has been the impetus for a complete reshaping of the calculus course and a rethinking of the goals of introductory calculus. The articles in this volume reflect this diversity, but concentrate on the end of the spectrum illuminated by the latter extreme.

A preferable designation for the type of activity that is described by the articles in this volume is "The Laboratory Approach." This designation shifts the emphasis from the equipment and place to the delivery of the material. In fact, Howard Penn and Craig Bailey describe their project at the U.S. Naval Academy where every student has access to a computer in their dormitory and discovery based assignments are given virtually every class. There is no formal laboratory period, but students are engaged in laboratory activities on a daily basis.

On the other hand, Don Brown, Horatio Porta, and Jerry Uhl view the physical computer laboratory as essential to the success of their project. The conversations and discussions that take place in the lab at the University of Illinois are, in fact, the course. David Smith and Lawrence Moore meet their classes in both a laboratory and a classroom; in both settings, they stress conversation as the primary source for their students' understanding of calculus.

The Laboratory Approach has five activities related to the study of phenomena:
- 1. Observation
- 2. Identification
- 3. Exploration
- 4. Analysis
- 5. Explanation

These activities take place within the context of the "Laboratory," be it a physical location or a set of activities designed by the instructor for student investigation. For the most part they involve the use of computers and computing. However, as Donna Beers points out in her article, it is also possible to have "paper and pencil" labs. Another example of a "non-computer" lab is found in the article by Michael Hvidsten. These projects are as inviting and stimulating as those that use a computer.

The use of technology is not limited to the personal computer or the mainframe. Tom Dick of Oregon State University describes his project for using the super hand held calculators that are readily available and relatively inexpensive. Joan Hundhausen and Richard Yeatts use the same technology in their project at Colorado School of Mines.

Even when there is agreement to use the computer there is certainly no consensus on the mode of use. Ed Dubinsky and Keith Schwingendorf insist on student programming and use the ISETL language because of its highly mathematical flavor. They find this an ideal means of communicating mathematical ideas to the computer. James Hurley also insists on programming and has his students use the True BASIC language because of its easy, clear syntax for communicating algorithms. Others, such as Jean Alliman, use spreadsheets to design laboratory projects because of their ability to quickly process data and implement processes as well as to give a graphical display. Still others use course specific packages such as Micro Calc and True BASIC Calculus. A growing number of instructors are using Computer Algebra Systems in an effort to free their students from the more technical aspects of problem solving so that they can concentrate on the conceptual basis for a laboratory project.

Throughout this volume you will see three basic types of laboratory activity.
- 1. Those that anticipate material to be presented in class.
- 2. Those that deal with interesting applications and are not tethered by the need to deal with "computationally nice" data.
- 3. Those that expand upon or extend material that has been presented in class.

The material presented by Ed Dubinsky and Keith Schwingendorf shows a very careful construction of laboratories that lead the student to the "interiorization" and "encapsulation" of the objects and actions of calculus. Through this process the students develop a familiarity with these objects and actions that is based on their own discoveries and not a pronouncement of an instructor. The papers by Richard Sours and by Bruce Edwards and Patrick Stanley discuss other approaches to helping students anticipate important results from calculus. Doug Child illustrates the use of his Calculus T/L package to actively involve his students in the learning of calculus concepts that will empower them to apply calculus to problems that they have not seen before.

The use of applications to motivate the study of calculus is a very popular theme in today's laboratories. Joan Hundhausen and Richard Yeatts teach an integrated Calculus and Physics course in which the two subjects are presented in parallel. The labs for this course have a very strong physical science flavor. Other labs that deal with experimentation and observation of physical phenomena are given by Robert Decker and John Williams. The paper by Susan Hurley and Thomas Rousseau also contains the use of physical science principles to motivate calculus concepts. Motivating applications are not limited to the physical science laboratory. The paper by Elgin Johnston, Jerry Mathews, Clifford Bergman, and Alan Heckenbach includes whale harvesting strategies, subway travel, automobile test data and the design of a field irrigation device. Keith Stroyan's discussion of computer labs at the University of Iowa includes examples of topics that students choose for term projects. These projects are selected to meet the students interest and illustrate significant points in the calculus.

Every instructor has wished for more time in the course to present a favorite extension of a topic or to deal with a topic in more detail. The laboratory approach allows the instructor to guide students through an investigation of these topics. The papers by Dusa McDuff and Eugene Zaustinski, by Elton Graves and Robert Lopez, by Michael Kalleher and Michael Moore, and by Herve Lehning all contain samples of the deeper investigation that can be done by students without the use of additional in class discussion. Furthermore, the point that the instructor desires to make is made in a much more powerful way because the student is involved in the development of the material and the pace is one that the student, not the instructor sets.

This volume is organized into three distinct parts. Part I contains a discussion of the general issues of the laboratory approach to teaching calculus. These issues range from the philosophical issues and choices that an instructor faces when deciding to teach calculus to the more mundane, but very important, issues of designing a lab, choosing equipment, and selecting software.

In part II examples are given of specific programs that have been in operation and have received some national recognition in the form of funding and publicity. These programs are described in detail and, whenever possible include an evaluation of the program together with a discussion of the student reactions to their instruction. Schools represented in this section do not fit any particular pattern. Many readers may be surprised to find projects at larger institutions with a student population made up primarily of commuters. This is the environment for the project at the University of Michigan in Dearborn as described by Margaret Hoft. The largest project described in this volume is the one at Brock University described by Eric Muller. This project involves 600 students in a service calculus course with a MAPLE laboratory.

Let us then begin with our own exploration of the Laboratory Approach to Teaching Calculus. Tom Banchoff begins with a discussion the fact that "Computer laboratories magnify the instructors' idiosyncracies" and a discussion of some of the ideas expressed in the calculus reform as they relate to calculus laboratories. Don Small picks up on the reform theme and discusses the use of Computer Algebra Systems in Calculus Laboratories. David Tall reviews some of the recent developments in the use of the computer to visualize and symbolize calculus concepts. Donna Beers discusses the effects of cooperative learning in the students college experience and as alumni. Tom Dick then addresses the objections that are raised to the introduction of technology into the calculus sequence and discusses these objections within the context of using the hand-held supercalculator. I wish you enjoyable, informative, and exciting reading.

Carl Leinbach, for the Editors

Contributors
Institutions and Addresses

Atelier Logiciel de l'Enseignement
  Superieur
13 rue Letellier
75015 Paris, France
  Author: Hervé Lehning

Brock University
St Catherines
Ontario, Canada S2S 3A1
  Author: Eric R. Muller

Brown University
Providence, RI 02912
  Author: Thomas F. Banchoff

Colby College
Waterville, ME 04901
  Author: Donald B. Small

Colorado School of Mines
Golden, CO 80401
  Authors: Joan R. Hundhausen
          F. Richard Yeats

Duke University
Durham, NC 27706
  Authors: Lawrence C. Moore
          David A. Smith

Gustavus Adolphus College
St. Peter, MN 56082
  Author: Michael D. Hvidsten

Hesston College
Hesston, KS 67062
  Author: Jean M. Alliman

Iowa State University
Ames, IA 50011
  Authors: Clifford H. Bergman
          Alan J. Heckenbach
          Elgin H. Johnston
          Jerold C. Mathews

Oregon State University
Corvallis, OR 97331-4605
  Author: Thomas Dick

Purdue University
West Lafayette, IN 47907
  Authors: Ed Dubinsky
          Keith E. Schwingendorf

Rollins College
Winter Park, FL 32789
  Author: J. Douglas Child

Rose-Hulman Institute of Technology
Terre Haute, IN 47907
  Authors: G. Elton Graves
          Robert J. Lopez

Sienna College
Loudonville, NY 12211
  Authors: Susan E. Hurley
          Thomas H. Rousseau

Simmons College
300 The Fenway
Boston, MA 02115-5898
  Author: Donna L. Beers

SUNY at Stony Brook
Stony Brook, NY 11794
  Authors: Dusa McDuff
          Eugene Zaustinsky

United States Naval Academy
Chauvenet Hall
Anapolis, MD 21402-5002
  Authors: Craig K. Bailey
          Howard Lewis Penn

University of Connecticut
196 Auditorium Rd, Rm 111
Storrs, CT 06269-3009
  Author: James F. Hurley

University of Florida
201 Walker Hall
Gainesville, FL 32611
  Authors: Bruce H. Edwards
          Patrick H. Stanley

University of Hartford
West Hartford, CT 06117
  Authors: Robert J. Decker
          John K. Williams

University of Illinois
1409 W. Green Street
Urbana, IL 61801
  Authors: Don Brown
          Horatio Porta
          J. Jerry Uhl

University of Iowa
Iowa City, IA 52242
  Author: Keith D. Stroyan

University of Michigan - Dearborn
Dearborn, MI 48128
  Author: Margret H. Höft

University of Warwick
Coventry CV 4 7AL, U. K.
  Author: David Tall

Washington State University
Pullman, WA 99164
  Authors: Michael J. Kallaher
          Michael E. Moody

Wilkes University
Wilkes Barre, PA 18766
  Author: Richard E. Sours

Part I

General Issues Related to The Laboratory Approach

Computer Laboratory Magnification of Idiosyncracies

Thomas F. Banchoff
Brown University

Mathematics teachers exhibit at least three recognizable personality types: numerical, algebraic, and geometric. Preferences for one or the other of these will affect the way any course is taught, and the course in which the differences often show up most clearly is calculus. The differences in approach are particularly evident when calculus courses are augmented by computer laboratories. Such laboratories do give opportunities for new insights into calculus, but along with these benefits, there is a danger that some of the valuable aspects of older courses will be discarded too swiftly. In this fairly personal essay, I would like to discuss experiences with computers and calculus stretching back over twenty years, indicating some of the changes that have taken place, describing some successful projects as well as some dead ends, and giving a caution about the kinds of calculus courses we are developing as we respond to the technology of the computer laboratory.

With the advent of pocket calculators, the numerically minded teachers had a field day, producing a great many examples that have found their way into a variety of standard texts. Almost any computer laboratory will give students access to considerable calculating power, so mathematicians with numerical tendencies have the opportunity to introduce even more examples and topics, bringing numerical analysis techniques into the earliest calculus courses. In recent years, as a number of algebraic routines have become available on reasonably small machines, the teachers who have always stressed algebraic manipulation have recognized the chance to incorporate larger numbers of algebraic and algorithmic techniques into elementary courses. And recently sophisticated computer graphics programs have appeared for producing graphs of functions or families of functions, and for manipulating images on screens in computer laboratories. Naturally these have invited instructors with a strong geometrical bent to bring graphing technology into beginning courses.

Only in rare instances will the inclusion of new techniques shorten the amount of time given to a topic. More often, in order to make room for these innovations, some topics have to be substantially reduced or even eliminated. It is important for mathematicians of all persuasions to take part in the discussions that shape a particular calculus course, especially when it involves a computer graphics laboratory. This is the only way to prevent a course from becoming too slanted in the direction of numerical or algebraic or geometric emphasis. There might be a time in the future when a new race of teachers and students are equally at home with all three aspects, but until then it seems that we are in for an even greater divergence in the appearances of the computer laboratory component of our calculus courses.

The Fall of 1990 marks the twentieth anniversary of the first computer calculus course I ever taught. During this time we have tried several different approaches in calculus and differential geometry courses, primarily at the third semester level and above, as reported in [1] and [2] . During the 1990-91 academic year, I will teach a lecture course in second semester calculus, so in this essay I will concentrate on topics that might appear in a laboratory associated with such a course, with the hope that much of what I say will apply equally well to other levels.

Frontier Experiences

Within a year after I had arrived as a young assistant professor at Brown University in 1967, the chairman of the mathematics department asked if I would be willing to participate in a two-year experiment using computers in an introductory calculus course. For the first year of the project, the lectures were to be given by Professor Philip Davis of the Division of Applied Mathematics, and the supplementary computer lectures would be given by my colleague Charles Strauss, with whom I had already begun to work on computer graphics films in differential geometry research. In the second year of the project, I would give the mathematics lectures and Charles would once again handle the computer aspect. I agreed, and I learned quite a bit. In particular I learned the difference between pure and applied mathematics.

In the first semester, students were introduced to a number of natural topics: bisection routines for finding roots, Newton's method, numerical limits of sequences and difference quotients, and various approximations of definite integrals. As he was making up the final exam, Phil Davis included a question asking the students to estimate the integral of the square of the cosine from minus pi to pi by using three techniques: left-hand endpoint Riemann sums, the trapezoid rule, and Simpson's rule, all with error estimates. I said, "And then you'll ask them to evaluate the integral and see how close their answers are to the exact value?" "No," he explained, "because that would give them the wrong idea. The integrals that they will have to deal with in the real world won't have closed-form solutions, so they shouldn't be looking for them. They should not have to worry about techniques like trigonometric identities that they will never have to use." I tried to defend classical topics in methods of integration by pointing out that one of the ways to do that particular integral was integration by parts, a technique that any applied mathematician would have to agree was a valuable one in the real world. He agreed with me in principle, but the problem was not changed, the first year anyway. The next year, when I taught the course, in addition to the numerical techniques, students were expected to solve the problem explicitly using three different techniques of integration. But I did feel bound to tell the students that there were professors who felt the such algebraic exercises in techniques of integration have been made obsolete by the computers we were just beginning to use.

In those days the graphing programs were truly rudimentary, often consisting of a series of asterisks printed out at different heights on a page to indicate the positions of function values. In a sense, that gave a better idea of what the computer was actually doing than some of the more slick routines of today that draw very smooth curves even though they are using the same finite amount of information. As a visual person, I found myself stressing the graphical aspects of calculus, just as I had before the computer was available. I was quite struck by the way that the differences between my teaching approach and that of my colleague in applied mathematics were accentuated by the way we made use of the computer aspect.

As it happens, the two-year experiment was not sufficiently successful to be continued after the seed money ran out in the late 1960s. In those days, there were not many computers available, and it was necessary to hire student consultants for a number of hours each week to help students deal with a comparatively unwieldy programming language and somewhat unreliable terminals. Some students with computer experience found the additional instruction boring; the majority found learning elementary programming difficult and not particularly helpful in understanding calculus. The time was not yet ripe.

Recent Laboratory Developments

Much has changed in the ensuing twenty years. These days, nearly every student enters calculus with enough

exposure to word processors to be able to handle pull-down menus and command sequences immediately. Readily available software packages make it possible for the students to investigate a large number of phenomena as soon as they know what a program does, without having to go to the trouble of learning how to write programs from scratch. The logistical details that loomed so great two decades ago are for the most part forgotten.

Speed of response is much greater that it used to be, although the increase is not so noticeable in the fairly uncomplicated problems of first-year calculus (often the same problems that we used in the pioneer days). A dramatic change is the enhanced quality and speed of the graphics displays, with high resolution and/or color adding immeasurably to the effectiveness of the images that students can now explore even with relatively unsophisticated machines. Three-dimensional display techniques are now widely available and easy to use, leading to vivid presentations of topics like volumes of revolution, one of the very first and most successful examples of visual display. As it happens, some of these earliest examples are important to remember. They should not be discarded too rapidly in our rush to take up new ideas.

Revolution, Si! Volumes of Revolution, No?

Early in the summer of 1990, I participated in a site visit to review the first year of operation of a carefully conceived calculus curriculum project incorporating numerical and graphical techniques with symbolic manipulation routines. Since it would obviously take a certain amount of course time to bring in all these new concepts, I asked what topics in the traditional course would go by the boards to make room. The presenter quickly offered up as a sacrifice the subject of volumes of revolution. I was told that some

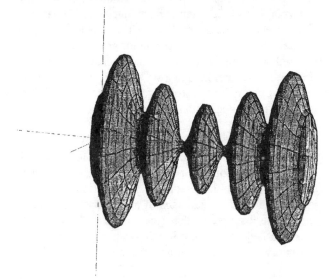

of those who discuss calculus reform generally consider this topic expendable. Someone recalled a slogan from a calculus curriculum session: "Revolution, Si; Volumes of Revolution, No!" I begged to disagree. I have always considered it a high point when it was possible to use calculus to prove the correctness of a manifestly useful formula like the volume of a cone or the volume of a sphere, something which, up to that point, had merely been a memorized fact. I have never found students unappreciative of Pappus' theorem, as they learn to calculate the volume of a figure of revolution in terms of the area of its cross-section and the distance travelled by its center of gravity. I count on the fact that students already appreciate this theorem when I generalize it in the study of tube domains in differential geometry. Anyone who works in CAD-CAM recognizes the importance of figures of revolution as primitive objects for geometric modeling. I do not appreciate the suggestion that volumes of revolution be removed to make room for symbolic manipulation packages for algebraic techniques of integration. If there are those who want to replace one of my favorite applications with something more "relevant", I am prepared to challenge them.

This strong defense of volumes of revolution exposes my prejudices if not my idiosyncrasies. I know what topics I liked when I took calculus, and I know which ones have continued to show up in my research and teaching at higher levels. These are the calculus topics I present with excitement, one of the chief characteristics of any course likely to engage students. Each professor has his or her own list of such topics. Having a computer with strong graphics capabilities encourages me to develop geometric topics even further. Having symbolic manipulation software leads some of my algebra colleagues even further away from where I think the course should go. The same can be said for the number manipulation facilities of machines, leading to more problems featuring numerical approximation. The machines will accentuate our differences before they bring us all together. There isn't time in the curriculum to take advantage of all the enhancements that could occur if we exploited fully the symbolic, the numerical, and the graphical aspects of computers in the teaching of elementary calculus. Until there is general agreement about which topics should be left aside to accommodate even an abbreviated version of these three basic thrusts, we will experience some frustrations, as will our students. We can be assured that twenty years from now the students and teachers of the future will consider our dilemmas completely irrelevant, and they will be dealing with problems undreamed of today. The one thing that is safe to predict is that there will still be problems and differences of opinion.

Various Laboratory Models

All this being said, there are a number of comments that can be made based on our experiences over the past twenty years. The major difference between the first time I taught calculus with computers and the present situation is the availability of laboratories filled with machines. In the old days, computer assignments were in reality homework problems, done individually by students who had to find the best times to avail themselves of scarce resources. Any cooperation between students was purely incidental. All that changes in the modern laboratory setting.

Over the years, we have tried a number of different models for laboratory interaction, keeping some aspects and rejecting others. A list of various approaches might be useful to others who are developing their own strategies.

1) The Science Lab--Based on traditional biology or physics laboratories, this model features setup demonstrations, followed by carefully laid out experiments to be recorded on standardized "lab report" forms. Students work singly or in pairs during a set period, and are graded on the accuracy and completeness of their reports. In this model, everyone is supposed to come up with essentially the same results, and sharing of experiences is of minor significance. Assignments are like standard homeworks, with fixed answers that the grader already knows. Success in this model depends on the quality of the materials used. The exploratory aspect is minimized, leading to greater efficiency at the expense of spontaneity.

2) The Language Lab--Here students are given interactive drill materials, with feedback to indicate how well they are mastering pronunciation and reading and listening comprehension. Students work individually, at their own pace, through standardized materials, possibly with the assistance or intervention of instructors who can listen in on what is happening. Once again, this model requires careful preparation and selection of materials. It may be most effective in remedial situations, especially when students in a class have widely differing backgrounds.

3) The Mathematical Video Arcade--The video game approach seems well suited to a certain kind of mathematical competition, solving problems very quickly. This approach treats mathematics rather like chess, so students study certain kinds of problems the way they would pour over books of openings or gambits or endgames. High performance in such specialized competitions may be related to mathematical talent, but not necessarily. Eye-hand coordination may or may not be an essential element

of the experience. Logical puzzles may be particularly well adapted for this kind of presentation, although it does not seem especially well suited to communicating the ideas of calculus.

4) <u>The Studio Art Class</u>-- After an initial presentation by the laboratory leader (possibly the instructor but more likely a student assistant), students singly or in pairs investigate examples of a particular phenomenon, formulating and testing conjectures. Near the end of the session, students are invited to present their best insights to the class as a whole. If the group is small enough, members of the class can circulate and see the images or animations prepared by each of them. A leader prepares a "critique", responding to each of the pieces, stressing the positive aspects and making suggestions for improvement. Such a laboratory works well when coordinated with standard lecture presentations, especially when the conjectures that arise from the lab investigations can be treated in a formal way in the class. Selection of topics is quite important, providing a sufficiently rich set of phenomena while at the same time providing some guidance so the class is not overwhelmed. The looser structure of exploration in a laboratory based on such a model is harder to lead, but the results can be quite dramatic when it succeeds.

Communicating Results--Saving and Hard Copying

Effective implementation of laboratory experiences demands some way for students to save their work, for transmission to the instructor, and, to a certain extent, to the rest of the class. Ideally students should be able to save not only numerical results and static images but also image sequences, even animations, showing how a given example can be modified to yield families of examples. Students should be able to attach notes to files, indicating how to interpret the images which it generates. In some cases, a printed image is sufficient to illustrate a given discovery or to motivate a conjecture. For these, there should always be some facility for hard copies of material on graphics screens, although there should be some restraints on the number of images saved in printed form. Even better is a procedure for saving an entire process, so that the instructor or another student can enter the program and continue investigating it. In this way, the explorations developed by members of the class can serve as the basis of later laboratory topics or theoretical discussions in the classroom.

Follow-Up Opportunities

Students should be polled more than once during a term about the ease of using existing programs and about "wish lists" of features desired in future versions of the programs. This is important feedback for those responsible for the maintenance and improvement of the software, and it often suggests new directions for subsequent courses. Two years ago, after I had requested such a list at the end of a third semester honors calculus course, four of the best students petitioned to continue in a semester-long group independent study so that they could implement a number of suggestions that they and other students had formulated at the end of the course. That independent study led to substantial improvements in the program we were using, and paved the way for the current version, which is coupled with another newly developed senior honors program for the study of curves and surfaces in introductory differential geometry.

It is important to give students an opportunity to follow up their accomplishments in one course with the chance to make further contributions in the future. This is one of the big advantages of computer graphics as a subject, with its encouragement of group activities and the relative accessibility of funds for the support of summer research work with computer research teams. As mathematicians, we are probably best advised to try not to woo students away from computer science, but rather to encourage them to continue to take mathematics courses along with their computer studies. There is enough evidence around that mathematical experience enhances the ability to be accepted at top-ranked programs in computer science in the United States and in other parts of the world. We do well to urge students to continue mathematical studies no matter what may be their eventual concentrations.

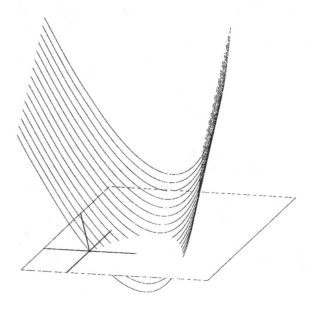

Investigating Families of Examples

Over the years, we have found that laboratory experiences are especially effective when they treat not collections of unrelated examples but rather families of examples. How many zeroes does the function $f(x) = x^2 + 2x + c$ have? The answer depends on c, and as c varies, we get an entire collection of answers. A closely related question is to find the domain of $1/(x^2 + 2x + c)$, and to sketch the graphs of the different functions in this family for different values of c. A graphics computer can be especially useful in indicating this dependence. We can also analyze the integral of $f(x)$ from -1 to 1 and consider the dependence of the answer on c. We can then go on to investigate the one-parameter family of functions $f(x) = x^3 + cx$, or the two-parameter family $f(x) = x^4 + cx^2 + bx$. Students can decide what are the crucial properties to look for, and then determine what are the relationships between c and b at the exceptional positions at which the qualitative behavior changes.

We can apply the same approach to study other types of families, for example families of polar coordinate function graphs. The equations $r(\theta) = \cos(\theta) + c$ determine very different curves depending on the nature of c. More gener-

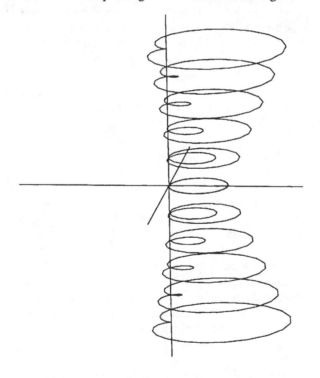

ally in the study of parametric equations, we can investigate what happens as we vary the value of c in the parametric curve $x(t) = t^2$, $y(t) = t^3 - ct$ for various values of c.

We can also use parameters in the exponent, as for example when we look at the integral of the function $f(x) = x^{(-1+a)}$ from 1 to 2. We can take the limit of this function as a approaches 0, to find an expression for the logarithm of 2. Naturally the dependence on the constant might not always lead to a good limit, as we can see by considering a family of functions which is linear from O to a and from a to 2a, with $f(x) = O$ if $x \geq 2a$, $f(0) = 0$, and $f(a) = 1/a$. We have $\lim f(x) = O$ for all x as a approaches 0, but the integral of f from O to 1 approaches1, since this integral equals 1 if $a > 0$.

Kinesthetic Response

Students who have grown up in the era of video games are quite aware that it is possible to make shapes move around on a screen in response to analogue input devices. In our programs, we use either numerical values entered from a keyboard, or selected from some preset options from pulldown menus, using a mouse or light pen, or by manipulating slider bars with buttons or dials. With a fast enough machine, we get close to real-time response, and the student is able to feel the effect of the parameter changes. Such display does have great advantages over chalkboards, or even overhead projectors. If we are studying a one-parameter family of functions, for example, we want to be able to see graphs of several members on the same coordinate system. If the graph moves from one position to another as the operator changes the parameter, all the better. We should always prefer the kinds of problems that can be done best by a given program and a given machine. For laboratory use, a program should give a response in not more than thirty seconds. This is not to say that certain projects might not take considerably longer than that, but such projects are better relegated to homework problems so they do not disrupt the flow of the class activity. In an actual research laboratory, students may sit for several minutes or longer in a nearly trancelike waiting state, and although such patience is a requirement for successful computer science, it does not seem desirable as part of a laboratory for an elementary calculus course.

Aims of the Calculus Course

Every once in a while we have to ask ourselves what we hope to accomplish by handing to the students tools that will enhance their abilities to do algebraic manipulations, numerical calculations, and graphs. It is all well and good to provide these means of augmenting what the students already have, but the results will not be very good unless the students already have a rudimentary idea of what is going on. We all know that it is deficiencies in algebra that often cause students the most difficulty in coming to grips with calculus. Any student can learn how to differentiate a polynomial, but in order to solve an elementary maximum-minimum problem, it is necessary to find the zeros of the polynomial. We know that a number of students have difficulty with such problems even when the original function is cubic, so the technique for finding the roots of the derivative is a simple application of the quadratic formula. Giving a student a symbolic manipulation routine for solving such a problem is the equivalent to providing a calculator for the multiplication of two two-digit numbers or the sum of two simple fractions. It is true that students who can handle cubics are sometimes stumped when a quartic function leads to a cubic derivative, and the ability to recognize the one obvious root carefully arranged by the poser of the question is often the faculty that distinguishes the honor student from the rest of the pack. We might ask what will distinguish our students in the new dispensation, when we provide each of them with a computer that automatically finds roots of polynomials. One might hope that a student could take the information given by such root-finding programs and use it to solve word problems or graphing problems. But if the student is incapable of doing that for cubic functions, it seems unlikely that he or she will have much better luck with more complicated examples. It is true that the presence of a computer takes some of the pressure off the persons designing the examinations, since it is no longer necessary to make sure that the answers "come out even". In fact, it might be considered better when they don't, thus making them closer to that real world out there. But I for one would hate to see the old problems disappear. I'm fond of the examples that I have put together over the years, with first and second derivatives that are relatively easy to factor so that the maxima and minima are nice numbers and the graph fits nicely on a standard grid. Even with the luxury of a computer, we instructors will still have to exercise a measure of care in choosing the coefficients in our polynomials and our rational functions, to make sure that the answers are reasonable, i.e. consistent with some sort of a priori analysis of the problem without which students can be led into the most extreme sort of errors.

It is also true that some students have a great deal of difficulty learning to graph functions. It would be a serious mistake to lead them to think that now that graphic computers are around, it is no longer necessary to be able to graph simple functions by hand. Any student who goes to a computer to graph a quadratic function is the equivalent to a clerk who need the calculator to divide a number in half. I don't mind it when someone uses a calculator to find out the dollar equivalent of 37.20 in Dutch guilders at $.5489

per guilder, but it does bother me when I see someone using a calculator to solve that problem when the rate is $.50. And I do want to see my students come up with good looking parabolas on appropriately chosen grids, with straight tangent lines at given points. Only after that, have they earned the right to use the computer to do the same problem for a quintic.

So what are the algebra problems that I expect my students to do in calculus? I often give a simple quiz, without asking students to put their names on the paper so there is no individual pressure. I ask them to expand $(x + a)^3$ and to factor $x^3 - a^3$. A discouragingly large number of students miss at least one of those problems. Telling them that it is all right not to know how to do such problems because now there is a computer to do them is a cruel hoax. Once again, it seems a good idea to require a basic mathematical facility with simple calculations before giving a student a calculator, with simple algebraic techniques before providing a symbolic manipulator, and with simple graphing facility before providing a graphics routine.

This being said, I think that the graphing programs can be invaluable in topics like Taylor series, where the student can find the first two terms easily and then can set the machine to work to graph a number of polynomial approximations, seeing how they converge where they do and how they can fail to converge near endpoints of the interval of convergence. I don't know of anyone who draws enough approximating sums to handle that problem without a graphics computer.

What about integration? It seems strange that some teachers move wholeheartedly into using symbolic integration packages who previously would never let their students use a table of integrals, even the abbreviated one on the inside cover of the calculus book. Somehow the idea of looking for an integral in the table book was something engineers did, not mathematicians. But there is an art to using those tables, since you have to recognize what form the integral is in, and often it is necessary to manipulate the integrand a bit before it resembles one of the forms listed in the tables. That at least requires a bit of finesse. It takes less than that to invoke a command on a computer to integrate some rational function in closed form.

In a recent consultation, I was surprised to see as one of the premier examples of the use of symbolic manipulation a unit on partial fractions, one of the techniques of integration which I had always liked and which I had feared doomed to the waste can when techniques were deemed expendable. It is true that the main problem with such examples was algebraic--if you set the problem up correctly, you often ended up with two equations in two unknowns, and you had to know how to solve such things. In rare instances, it is possible to use the partial fractions technique on a rational function with a cubic or quartic denominator, but the number of difficulties is great. First of all, only the better students are able to do the factorization, and even if the denominator is presented in factored form, it requires some expertise to set up and solve a three-by-three or four-by-four system of equations, after which the results would have to be reinterpreted to express the integral as a sum of rational functions, logarithms, and inverse tangents. The whole purpose of such exercises was to enable the student to exhibit some finesse. The problems themselves were not especially interesting in terms of applications (except for the sole example of the epidemiological model), and few students ever seemed to appreciate the theorem that all rational functions could be integrated in this way, modulo the ability of the integrator to factor and solve linear systems. Now the computer provides us with expert algebraic assistance, and it will be interesting to see if students once again show an interest in integrating complicated rational functions. But I would still like to believe that anyone who uses such a machine could integrate "by hand" any rational function with a quadratic denominator.

The same holds true for graphing rational functions. You earn the right to use a sophisticated graphing routine for complicated rational functions by first showing that you can graph basic examples by hand. It's a simple message, but one that bears repetition.

It seems that a great variety of things that can be done with a computer are actually merely higher-exponent versions of things that many students never learn well enough in the elementary cases. It is not at all clear that the ability to confront quintics without flinching is going to make students better at understanding the ideas of calculus, which after all should be one of the primary long-standing goals of our teaching.

All that I have written above assumes that the purpose of the calculus course is to teach calculus, as opposed to algebraic manipulation, numerical analysis, or, for that matter, mechanical drawing or computer science. Computer laboratories offer us great opportunities for improving the quality of calculus courses, and by bringing the best parts of our mathematical personalities to this challenge, we can look forward to very stimulating experiences in the future, for ourselves and for our students.

References

[1] Thomas Banchoff and Richard Schwartz, EDGE: The Educational Differential Geometry Environment, **Educational Computing in Mathematics** (Eds. Michele Emmer, Thomas Banchoff, et al.) North Holland, Amsterdam (1988) 11-31.

[2] Thomas Banchoff and Student Associates, Student Generated Interactive Software for Calculus of Surfaces in a Workstation Laboratory, **UME Trends**, Vol. 1, No. 3, (1980) 7-8.

Calculus Reform - Laboratories - CAS's

Don Small
Colby College

Introduction

Involve students in doing mathematics, instead of lecturing at them; stress conceptual understanding, not mechanics; develop meaningful problem-solving skills, not just "plug-and-chug"; have students explore patterns, not just memorize formulas; engage students in open-ended, discovery-type problems, not closed-form mechanical exercises; approach mathematics as an alive exploratory subject, not just as a description of past work: these are the "battle cries" of the calculus reform movement. Computer algebra systems (CAS's) are the tools and mathematics laboratories are the settings for implementing these cries.

Students working in pairs at a computer become involved in *doing* mathematics. They talk (often argued) about what to do. I saw a student pull her partner's hand away from the keyboard saying "Stop! Tell me what you are doing." Questions are different, "Why?" and "How come?" questions intersperse the "How to?" ones. A multimodal approach to problem solving evolves naturally when students have access to a CAS. For example, analyzing the function: g: [0,4] --> R defined by

$$g(x) = 10 \int_{\cos(x)}^{1} \frac{\log(\sin(x) + 1)}{e^x} \, dx$$

involves symbolics (computing the first and second derivatives of g to determine critical points and concavity), graphics (plotting the first and second derivatives of g), and numericics (approximating g(x) for selected x).

Some Examples

I have found that as a student's confidence in using a CAS grows, so does the willingness to experiment. Different approaches to solving a problem are often seen. The following exercise was a lab problem assigned in the fifth week of a Calculus I class that is taught with three fifty minute classes and a ninety minute lab per week.

Determine the interval over which sin(x) can be approximated by x with an accuracy of < .1. (That is, solve |sin(x) − x| < .1 for x)

Students used the following three graphical methods to solve the problem. The fourth method listed was suggested in a follow-up class discussion.

1. Transform the absolute value, inequality problem into one of finding the zeros of the function
$$f(x) = |\sin(x) - x| - .1.$$
The zeros can be approximated from a plot of f by digitizing the x intercepts. An alternate method is to use a root finding algorithm to obtain the zeros of f. These zeros are the endpoints of the desired interval.

2. Plot the function, f, defined by f (x) = |sin(x) − x| and the line y = .1 and then digitize the points of intersection of the two curves. The x coordinates of these two points are the endpoints of the desired interval.

3. Draw a multiplot of f (x) = sin(x) +.1, g(x) = sin(x)−.1, and h(x) = x and then digitize the points where the line y = x intersects the other two curves. The x coordinates of these points are the endpoints of the desired interval.

4. (Numerical method suggested during a follow-up class discussion.) Numerically approximate the desired interval by evaluating f(x) = |sin(x) − x| over the increasing sequence of numbers, {.01 n} until a value exceeds .1. Let .01n* be the largest term of the sequence for which f(x) is less than .1. Since sine and x are odd functions, we know that [−.01n*,.01n*] approximates the desired interval with accuracy < .1.

Like problem-solving, conceptual understanding is enhanced by considering the concept in different modes, e.g., the graphic, numeric, and symbolic modes of a CAS. Determination of extremal values is an example. Consider using two labs to geometrically develop the first and second derivative tests for extremal values. The purpose of the first lab would be to help students develop a graphical interpretation of the meaning of the numerical signs of the first and second derivative. This might be done by asking students to make up several examples of functions satisfying given sign conditions on the first and second derivative. For example, make up and plot three functions whose first derivative is positive and whose second derivative is negative. Describe the general shape of the curves (i.e., discover a pattern). Repeat the example several times choosing different combinations of numerical signs for the first and second derivatives.

The purpose of the second lab would be to lead students into discovering the idea of critical numbers followed by discovering the geometrical version of the first and second derivative tests. This might be done by asking students to construct a multiplot of the graphs of $f(x) = x^3$ and its first derivative and then to look for (shape) relationships between the curves. Repeat the exercise for the function $g(x) = \sin(x)$. Ask the students to continue repeating the exercise with other functions of their own choosing, until they can recognize one or more patterns. (It might be necessary to direct their attention to focusing in on extremal values or the zeros of their functions.) After the students have "discovered" critical numbers and recognized that some of them correspond to extremal values and some do not, have them repeat the exercises looking at multiplots of a function and its first two derivatives. The objective is to connect the "curve shape" intuition developed in the first lab with the location of extreme points. The students are now prepared to tackle a discovery type homework assignment such as:

Determine the behavior of a function in a neighborhood of a critical point by analyzing how the graph of the derivative function meets the x-axis at the critical point. Note that a graph can meet the x-axis in four essentially different ways:

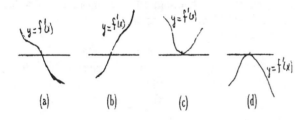

For each of these four plots of $y = f'(x)$, determine:
   (i) the numerical sign of the derivative on either
     side of the critical point;
   (ii) whether the graph of f is concave up or down
     or neither;
   (iii) if the critical point corresponds to a relative
     maximum, relative minimum, or neither.
Describe your thinking and the experimenting that you did in arriving at your answers. Describe how you tested your answers.

The interactive nature of a CAS lab changes the role of the instructor to one of coach and resource person. This, in turn, changes the instructor's focus from teaching to student learning and, in particular, to helping students learn how to learn. Observing and working with students in a lab gives an instructor a better insight into the level of student understanding than is obtained through a conventional class or test. In general, a laboratory setting forces both the instructor and student to deal more openly and honestly with student learning. In contrast, tests directed toward recall of facts and techniques provide little information on how a student tackles a new problem.

Implications

I am excited about and look forward to the changes that I believe will result from incorporating CAS labs into undergraduate mathematics courses. Here are a few examples of expected changes:

1. Students working together.

Most mathematicians work in a team setting, not in isolation. Experiencing the synergistic results of group work are instructive, rewarding and personally empowering. It is exciting and gratifying to see students taking responsibility for their partner's understanding as well as their own. Two students working together seems to be the optimal arrangement. They can trade off on a keyboard whereas three students becomes a crowd and often results in one doing the keyboarding and two discussing last nights game. Laboratories should be designed to encourage collaboration, e.g., room for two students to work at a terminal, chairs with coasters to facilitate students switching places at a keyboard, wide enough isles to allow instructors to easily move throughout the lab, etc. We need to encourage group activity and defy the myth that mathematics is best learned in isolation. Experiencing incorporating another person's thinking into our own is an important ingredient in learning.

2. Development of Problem-Solving Skills.

Too often problem solving means finding a similar worked example to mimic or plugging into a known algorithm. This is a natural consequence of our computational-based instruction with an emphasis on facts and techniques. To be an effective problem solver one needs to estimate, consider alternatives, experiment, conjecture, test, prove, and analyze results. Attitude is a major ingredient in problem solving. CAS's, by freeing students from the constraints of computation and by offering a multimodal environment, provide a setting conducive to developing an inquisitive, experimental approach. Recall the standard textbook problem of forming an open box of maximum volume by folding up the sides of a piece of cardboard? A slight modification turns this problem into an interesting problem solving exercise.

Step 1. Ask each student to form an open box by tearing out a square from each corner of an 8.5 x 11 in piece of paper and then folding up the sides. Having formed the boxes, ask pairs of students to estimate the dimension of the corner squares whose removal produces a box of maximal volume? minimal volume?

Step 2. (students working in pairs) Assign different solution approaches (analytical, numerical, graphical) to different pairs of students to use in determining the dimensions of the corner squares whose removal produces a box of maximal volume.

Step 3. Hold a group discussion on contrasting the three solution methods, on how people formulated their estimates, and on what they might do differently when estimating in the future (based on this experience).

The existence of a laboratory setting and the availability of a CAS "tool" are not sufficient for developing problem-solving skills. A coach (instructor) is needed. (There are many useful analogies that can be drawn between teaching mathematics and coaching an athletic team.) Students working together are likely to consider more alternatives than they would working separately. The same is true of experimenting, especially when they are occasionally prodded by a roving instructor's comment or question. Drawing connections is an important aspect in making conjectures. The ability to draw connections is dependent, in part, on looking at lots of special cases and this is where a CAS becomes a major contributing factor in problem solving. "What ifing" a problem is easy to do with a CAS, although it usually takes encouragement and illustrating on the part of the instructor to initiate students into this activity. The instructor's role in developing problem-solving skills is to comment, question, suggest, but not to tell, to encourage collaboration and "sticking with it." These can all be done much more effectively on a small group basis in a lab setting than in a lecture class.

3. Approximation and Error Analysis.

Finding an approximation that satisfies a given error bound or determining an error bound for a given approximation are basic operations in analysis. These operations usually involve extensive computing and thus are not emphasized in today's calculus courses, in spite of the fact that approximation is the "backbone" of analysis. The growing use of CAS's will change this emphasis and make approximation and error analysis a major theme in calculus. Integration provides an example. Integration is conventionally taught in terms of "closed form" techniques even though very few functions can be integrated in closed form (e.g., how many arc length integrals can be evaluated in closed form?).

While numerical integration (Trapezoidal Rule, Simpson's Rule) is widely applicable, it is only briefly mentioned in today's courses because of the amount of computation involved. Note that the steps involved in approximating an integral to within a prescribed accuracy (differentiating, bounding a derivative, solving for the number of subintervals, and computing the numerical sum) are easily implemented using the symbolic, graphic, and numeric environments of a CAS. The availability of CAS's, I expect, will reverse the roles of closed form and numerical integration.

4. Exploratory Work.

Lynn Steen, Past President of the Mathematical Association of America, wrote:

> The public perception of mathematics is shifting from that of a fixed body of arbitrary rules to a vigorous active science of patterns. Mathematics is a living subject which seeks to understand patterns that permeate both the world around us and the mind within us.

Recognizing patterns often depends on being able to analyze numerous special cases. To the person computing with pencil-and-paper, numerous usually means two. However when working with a CAS, numerous can mean as many as you want because the computations are done by the computer. Developing a formula for $\int x^n e^x \, dx$ is an example of a pattern recognition exercise. It involves looking at several special cases (i.e., particular values of n) making a conjecture, testing the conjecture, and then proving the conjecture. Most of my students computed the above integral for n=0 through 6 before they were able to make a conjecture. Their procedure was to define f as a function of x and n by $f(x,n) = \int x^n e^x \, dx$ and then use a CAS to evaluate $f(x,0), f(x,1),...,f(x,6)$. Once they had made a conjecture, it was then easy to "test" it by comparing their conjecture for n=10 with the CAS display of $f(x,10)$. Most students gave an intuitive proof of the conjecture that involved analyzing n applications of integration by parts. A more advanced class might have given a mathematical induction argument.

5. Graphical Analysis.

The order of graphing and analysis are interchanging. Graphing has traditionally been an application of analysis. However with a good graphics package, graphing will "lead" and sometimes replace the analysis. For example, graphics packages will greatly change the treatment of extremal values. Numerical integration is another example. For instance, determining a bound on the fourth derivative (as required in computing the error bound in Simpson's rule) is easy to do with a graphics package (i.e., read off a bound from a plot of the fourth derivative). Solving equalities and inequalities will be easily accomplished by graphically finding the zeros of the corresponding functions. The (in)famous exercises of finding a δ given an ε will be done by reading the zeros off a plot of the appropriate function. In general, a graphics package that provides for digitizing points or has a built in root finding algorithm frees the user from the severe algebraic limitations on factoring. As a result, I expect that transforming questions into ones of finding zeros of a function (which will be done graphically) will become one of the most used techniques in calculus. The saying that "a picture is worth a thousand words," can be translated in mathematics to mean that a graph is worth a thousand computations. Future generations will point to the 1990's as the time when we learned how to teach and understand mathematics graphically. Surely, the high school calculator of the 90's is a graphing one.

6. Improvement in Student Writing.

Laboratory reports provide a natural vehicle for incorporating writing into a mathematics course. Some CAS's presently provide for word processing programs and more will in the future. The importance of writing has long been recognized as a means of communication. We are now beginning to realize the important role that writing plays in learning and understanding mathematics. Even "rough drafts" aid a student in organizing thoughts and formulating problem approaches. A "polished" report requires the student to describe a logical reasoning process, to make connections, to interpret results, to make conjectures, etc. An underlined 10 does not qualify as a lab report, even though we often accept an underlined 10 as an answer on a homework assignment. Mathematicians pride themselves in using mathematical notation correctly, but students generally do not. A CAS can be used to help address the issue of the importance of using notation correctly. Since a computer does not accept (as many instructors do) missing or ambiguous notation, students are held to a higher level of notational rigor when using a computer. Furthermore the non-judgmental nature of computer error messages, as compared to an instructor's, lessens the emotional barrier that students often raise over notational issues.

7. Student Perception of What is Important in Mathematics.

Students normally measure the importance of an activity

by the amount of time spent on it and the percentage of examination points allotted to it. Since most of a student's "mathematical time" on both homework and tests is devoted to algorithmic computation, it is not surprising that students view mathematics as a collection of formulas (to be memorized), and "to do" mathematics is to compute. Moreover, the restriction to "hand computable" exercises misleads students to think of mathematics in terms of "nice" exercises that fit into well defined categories and yield simple looking answers. The use of CAS's can change this by enabling computation to be viewed as a means rather than an end and by freeing up time for concentrating on concepts, motivation, development, structure, and application. An exploratory approach using CAS's as tools can change the emphasis away from "recipes" toward obtaining a global conceptual understanding through the analysis of special cases.

Students will find that a mathematics course which includes a CAS lab is a very different "ball game" than what they have played previously. The "rules" and expectations will be different, the level of involvement will be greatly increased as will be the uncertainty factor. Their mathematics will become more challenging, less comfortable, and much more "alive" than is presently the case. A student wrote on an evaluation:

> The aspect of the labs that I have been most impressed with is that many of the more difficult problems draw the student, through his or her own reasoning, into areas that have not yet been covered. Although this extra often makes the labs difficult to complete, the benefits justify the struggle because this preparation makes the text much easier to understand.

Because students will find a CAS lab based course different from the "standard" lecture type course, it is extremely important that the instructor clearly communicate his or her expectations. I believe that the major source of student frustration associated with a CAS lab based calculus course is the attempt by students to apply the "old rules" to the new situation. Time spent on a problem is one example, open-ended, multistep problems involve a very different time scale than do most of the exercises found in a standard calculus text. Approximation and error bound analysis is another example. Students are accustomed to (what they believe are) exact answers. The fact is that although students are comfortable using a calculator, many do not realize that most of the printouts are approximations. The uncertainty factor of an approximation is often unsettling to students whose experience in mathematics has been very deterministic in the sense of "plugging into" known algorithms.

Assessment in a CAS lab based course needs to be different than in a lecture course. The objectives, goals, procedures, exercises, etc. in a CAS lab course need to be carefully spelled out (and communicated to colleagues as well as students). These will be different than those in a lecture course and thus should be tested in a different way.

What should be tested about graphing for students who have access to a graphics package on a calculator or computer?

What level of hand skill computation is expected? For example, calculus texts all present the product rule for differentiation accompanied by several exercises on differentiating a product of *two* factors. Do we expect students to be able to extend the procedure to products of three factors? (four? five?)

What level of recall of differentiation is expected one year after the course is finished? Is a student expected to be able to differentiate

$$f(x) = \frac{\sqrt{3x^4 - \cos(x)}}{x^2 + 4}$$

by hand (from memory) or by hand after looking up differentiation in a text or by computer (from memory) or by computer after calling up a Help program on differentiation?

Evaluating $\displaystyle\int_0^1 \sqrt{1 - x^2}\, dx$ might be an appropriate exam question in a lecture course, but for a CAS based course the question would be better changed to:

Approximate: $\displaystyle\int_0^1 \sqrt{1 - x^3}\, dx$ with an accuracy of 0.1.

One thing is certain, student perception of mathematics as being primarily hand based calculation, will not change until the use of calculators and computers are allowed in testing.

Using a CAS as a teaching tool involves making fundamental changes in course goals, content, and pedagogy. (Electronic "plug and chug" is no better than hand "plug and chug," in fact it is probably much worse.) These changes require meaningful support from colleagues as well as college and university administrators. The mathematical associations provide forums for debating the issues and organizational structures for development of curriculum reform. The National Science Foundation and private foundations are sources of funds for equipment and development of experimental programs. However, it is within individual departments and administrations where support must be developed in order to produce meaningful change. There are "costs" to individual schools. For example, maintaining a computer lab involves lab assistants and technical maintenance. This means new positions for most schools. Space for a computer lab is a major consideration. "Start up" time is necessary for faculty members to acquaint themselves with current technology and to adapt their curriculum. Academic recognition and support for curriculum development is crucial (time spent on curriculum development is time not spent on research).

Laboratories and CAS's will aid the committed instructor in realizing the objectives of the calculus reform movement, but laboratories and CAS's will not by themselves bring about the desired changes. The major challenges are pedagogical. How can mathematics laboratories and technology be used to improve student learning? The CAS question is not what CAS's can or cannot do, but how can we use them to do what we want done? This of course, begs the questions: What do we want done? What are our goals? What is our level of commitment to these goals? What level of support do we require? Are we going to be an active player or are we going to observe from the sidelines? We have often told our students that "mathematics is not a spectator sport." The rapidly developing technology and the emergence of the calculus reform movement now challenges instructors with a similar statement, "teaching is not a presentation sport."

Recent Developments in the Use of Computers to Visualize and Symbolize Calculus Concepts

David Tall
University of Warwick

In recent years the growing availability of personal computers with high resolution computer graphics has offered the possibility of enhancing the visual mode of thinking about mathematics in general and the calculus in particular. Interactive computer software can be used to give insight to students, teachers and professional mathematicians in a manner that could not have been imagined a decade ago. Nevertheless, the technology brings with it the challenge to re-assess what is important in the curriculum and this is proving to be the more difficult task for professional mathematicians with a rich experience of pre-computer technology. This paper therefore begins by considering different ways in which we process information and focuses on the need to complement sequential/deductive thinking with a global grasp of interrelationships. It then concentrates on different aspects of the calculus from numerical, symbolic and graphical viewpoints and places recent developments in visualization using the computer within this broad perspective.

Human information processing and the computer

Individuals develop markedly different ways of mathematically thinking. Whilst some professional mathematicians will accept only those things which they can deduce logically step by step from carefully specified axioms, others demand an overall framework in which they can see a network of interrelationships between the concepts. The former viewpoint is a necessary pre-requisite for the *formalization* of mathematical concepts, the latter is invaluable for their *development*, both in mathematical research and in mathematical education. Yet, despite the need for both modes of thought, traditional mathematics teaching – especially at the higher levels – is usually more concerned with the sequential, deductive processes of the former rather than the holistic, predictive ones of the latter.

Mankind's success as a species is enhanced through the invention and use of tools to extend human capabilities. Many of these tools compensate for limitations in evolutionary design, be it spectacles to improve vision, the telephone to extend hearing and communication, or the supersonic jet to give the power of flight. In *The Psychology of Learning Mathematics* (1971), Skemp made the perceptive comment that humans have built-in loud-speakers (voices) but not built-in picture projectors, causing the fundamental mode of human communication to be verbal rather than pictorial. He demonstrated how a geometric proof could be presented in pictures which convey the same information as a verbal-algebraic proof, in a form that may be more insightful for many individuals. His purpose was not to show a preference of one form of communication over another, but to question the dominance achieved by algebraic symbolism and to examine the contribution made by visual symbols.

Not long ago the computer was a purely sequential, symbolic device, with the operator punching in a sequence of symbols and receiving the response in a similar form. But recently developed graphical interfaces now allow the user to communicate with pictures, providing a tool that compensates for the human lack of a built-in picture projector. Modern computer operating systems are replacing symbolic typing of commands by visual pointing at icons to simplify the interface with the computer. Furthermore the computer is able to accept input in a variety of ways, and translate it flexibly into other modes of representation, including verbal, symbolic, iconic, graphic, numeric, procedural. It therefore gives mathematical education the opportunity to adjust the balance between various modes of communication and thought that have previously been biased toward the symbolic and the sequential.

A style of learning that uses the complementary powers of sequential/linear thought processes on the one hand and global/holistic processes on the other is said to be *versatile* (Brumby 1982). The computer with suitable software is a powerful tool to encourage versatile learning (Tall & Thomas 1988b).

The implications of the computer in curriculum sequencing

Traditional mathematics usually introduces learners to sequential techniques and develops each one in a comprehensive way before introducing the learner to higher order concepts. Research has shown that the absence of a broad enough range of experience can lead to the abstraction of a false principle that later proves extremely difficult to eradicate, for instance "subtraction makes smaller" (an implicit property of counting numbers that causes conflict when negative numbers are encountered) or "multiplication makes bigger" (implicitly true for whole numbers but false for fractions). In these examples the hierarchy of concepts implies that one cannot avoid working in the limited context before broadening it (counting numbers before negatives, integers before fractions). But in the case of more advanced concepts it may be possible to reorganise the sequencing of concepts, using the computer to provide a rich environment in which *formally* complex concepts can be met *informally* at a much earlier stage.

A good example is the introduction of the derivative in the calculus. Formally this requires the notion of the limit of $\frac{f(x+h)-f(x)}{h}$ as h tends to zero, so logically the introduction of the derivative must be preceded by a discussion of the meaning of a limit. To make the notion of limit simpler, the limiting process is first carried out with x fixed and only later is x allowed to vary to give the derivative as a function, giving the following sequence of development:

(1) Notion of a limit, graphically, numerically and/ or symbolically,

(2) For fixed x, consider the limit of $\frac{f(x+h)-f(x)}{h}$ as h tends to zero,

(3) Call this limit $f'(x)$ and allow x to vary to give the derivative as a function.

If this sequence is followed, however "intuitively" it is done, then there are cognitive obstacles at every stage, which are well-known to every perceptive classroom teacher and have deeper connotations which have been detailed elsewhere (e.g., Cornu 1981, Tall & Vinner 1981, Orton 1983a,b).

One may conjecture that the interposition of a long chain of sub-tasks in building up a concept may *impede* conceptualization, because properties that arise in the restricted contexts en route lead to serious cognitive obstacles. There are two ways to attempt to solve this dilemma: one is to research the cognitive obstacles so that they may be addressed appropriately at a later stage,

the other is to use a "deep-end" approach (Dienes 1960) in which the whole concept is met early on in a rich, but more informal, context designed to offer a cognitive foundation for a more coherent concept image.

In the case of the derivative the "deep-end" approach can be done by first considering the informal idea of the gradient of a curved graph through magnification. It is based on the idea that a differentiable function is precisely one which looks "locally straight" when a tiny portion of the graph is highly magnified (Tall 1982). The limiting process is then an *implicit* idea used as a tool, rather than the *explicit* focus of study. "Local straightness" is the generative idea of this graphic approach to the calculus, which proves to be both a *cognitive* and a *mathematical* foundation for the theory.

It is instructive to note that in 1981-2, when the Second International Mathematics Study considered the way that calculus was taught, in the report emanating from Ontario (McLean *et al* 1984), not one respondent mentions an idea equivalent to a "locally straight" approach, the favoured methods being the "intuitive" geometrical idea of a "chord approaching a tangent" or numerical or algebraic limiting processes. The "locally straight" idea has some precedents, for example in curricula in Holland and Israel, but in each case the absence of appropriate computer software in the early eighties prevented it from being fully realized in practice. It also exists in a more visionary way in non-standard analysis (e.g., Keisler 1976) in which an "infinite magnification" is used to reveal a differentiable function as a infinitesimal straight line segment. The computer brings a practical (though inexact) model of this abstract theory.

Numerical, symbolic & graphical representations of calculus concepts

The first computer applications in the calculus were numerical - using numerical algorithms to solve equations, calculate rates of change (differentiation), cumulative growth (integration and summation of series) and the solution of differential equations. All of these can be performed in a straightforward, but sometimes inaccurate, manner using simple algorithms, and then improved dramatically by using higher order methods. However, unless interpreted imaginatively, tables of numerical data may give little insight and the concentration on the calculations may tend to obscure the underlying pure mathematical theory.

One method to improve matters is to engage the student in appropriate programming activities so that the act of programming requires the student to think through the processes involved. This may be done in any one of a number of computer languages, but it is preferable in a language that encourages the use of higher level of mathematical thought. Thus unstructured BASIC may allow numerical data to be calculated, but a structured language which allows the development of functional concepts is likely to be more useful. It is one thing to be able to calculate the numerical area under a graph $y=f(x)$ from $x=a$ to $x=b$ using the mid-ordinate rule for strips of width h. But if the language allows the specification of the area as a function area(f, a, b, h) of the function f, the endpoints a,b and the strip-width h, then many further constructions are possible. For instance, the area $s(n)$ under n equal width strips under $f(x)=x^2$ from $a=0$ to $b=1$ is

$$s(n) = \text{area } (x\text{\textasciicircum}2, 0, 1, 1/n)$$

and its value may be studied as n increases. Or the area-so-far function asf(x) under $f(x)=1/x$ from $x=1$, taking strips width 0.1 is

$$\text{asf}(x) = \text{area}(1/x, 1, x, 0.1)$$

which may be studied as a function of x.

This may be performed in a structured form of BASIC (such as BBC BASIC), which allows such functions to be defined (in an appropriate syntax, such as FNarea(f$,$a$,$b$,$h$) for the area function under the graph represented by the string expression f$ from a to b with step h). However, this language has frustrating technical limitations, such as the need to specify expressions as strings of characters (which may include the name of a function which itself can be given by a multi-line procedure) and the lack of mathematical data types other than numbers and character strings. TRUE BASIC has many marvellous facilities, but lacks even the EVALuation operator of BBC BASIC which allows a string such as "x\textasciicircum2-2" to be evaluated for a given value of x. On the other hand, the Interactive SET Language, ISETL (Schwarz *et al* 1986) is particularly conducive to such programming activities, being explicitly designed to allow manipulation of such mathematical constructs as sets, ordered sets, functions, relations, quantifiers, and so on.

Apart from programming, two quite distinct strands of development have taken place with computers. One concentrates on symbolic manipulation, using software to manipulate algebraic expressions and carry out symbolic differentiation and integration. This initially required the power of main-frames to cope with the recursive routines that were necessary and led to such symbolic manipulation systems such as MACSYMA, Maple, Reduce and SMP (Van Hulzen & Calmet, 1983).

Implementations have subsequently appeared on micros, allowing various levels of sophistication, including Maple, Reduce, MuMath (Stoutemyer *et al* 1983), Derive (Stoutemyer & Rich, 1989), *Mathematica* (Wolfram, 1988). More specialized programs have been designed to trace through various calculus techniques (for example Maths Workshop's "Symbolic Calculus" for the BBC computer implements recursive techniques for differentiation of combinations of standard functions, as well as offering *ad hoc* calculation of integrals). Other symbolic systems are also beginning to incorporate such step-by-step tracing of algorithms such as the differentiation of a composite function. However, whilst this gives the user a greater chance of understanding how to carry out the algorithms of formal differentiation, it does not help give insight into what a derivative *is*.

A second approach uses high-resolution graphics on micro-computers to translate the numerical methods into graphical representations. During the nineteen eighties, the number of such software packages has grown extensively (e.g., Kemeny 1986, Tall 1986a, Bach 1988, Tall *et at* 1990). Some early packages simply programmed calculus ideas numerically on the computer, but the later ones take the level of sophistication of the learner into account and attempt to present the ideas in a meaningful way.

These two strands accentuate the two different ways of mathematical thinking described earlier. The symbolic manipulator offers sequential manipulation to lead to a symbolic result, the graphical programs represent a vast amount of numerical data pictorially, giving the user the opportunity to gain an overall grasp of the information. Some symbolic manipulators offer the facility to trace the symbolic methods step by step, to see how the computer software progresses through the solution. The graphical representation, on the other hand, allows the user to see the numerical solution build up in real time.

In practice these strands are *complementary*, and each has different strengths. For example, not all differential equations have symbolic solutions, so numerical methods are essential. And even where symbolic methods are available, they often need geometric interpretation. In Tall (1986c) I gave the following example of a differential equation which was set on a national mathematics examination paper in the U.K.:

$$y\frac{dy}{dx} \sec 2x = 1 - y^2.$$

It is easily "solved" by separating the variables to get

$$\frac{y}{1-y^2} \, dy = \cos 2x \ dx \qquad (*)$$

and integrated to give the "general solution"

$$-\tfrac{1}{2}\ln|1-y^2| = \tfrac{1}{2}\sin 2x + c,$$

but what does this *mean*? By regarding the differential equation (*) as specifying the direction of the tangent vector (dx,dy) to the solution curve through any point (x,y) enables a "direction field" of short line segments to be drawn in the appropriate directions through an array of points in the plane (figure 1).

$$dy/dx=\cos 2x(1-y^2)/y$$

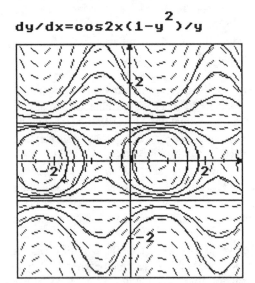

figure 1

It can be seen that some solutions are closed loops whilst others may be conceived as functions in the form $y=f(x)$. The symbolic solution in this case is of little value without a graphical interpretation of its meaning whilst the graphical interpretation alone lacks the precision of the symbolism.

The need for these complementary strands is consonant with empirical research into the success of students in coping with a first year university course in real analysis. Robert & Boschet (1984) hypothesized that there is a better prospect of successful mathematical learning for the student with knowledge (however imperfect) in many contexts than for a student with knowledge in one context. They found that the weakest performances over the year were by students having initial knowledge in very few contexts (usually numerical) whereas the more successful students also had initial knowledge in graphical and symbolic contexts. They found their hypothesis verified repeatedly in empirical experiments, and suggested that the crucial difference appeared to lie not in the mere existence of the prior knowledge but in the difference between two very different ways of thinking: the reductive effect of functioning in a single context as against the liberating effect of bringing several different ways of seeing the same problem from different viewpoints.

Research into the effectiveness of computer approaches

Much of the research and development of recent years has gone into the actual design and implementation of the software and there are only isolated reports of the testing of the materials in practice. Simons (1986) describes the use of hand-held computers programmed in BASIC to enhance the teaching of calculus. He reports that:

> ... the introduction of a personal computer into a course of this nature, whilst enhancing teaching and presentation in many areas, raises profound problems.   (Simons 1986, page 552)

There were evident gains in the "immediate usefulness" of the work, but a substantial number of staff, "with long experience of teaching mathematics", but "little practical experience with a computer or numerical analysis" did not like the course. Simons suggests that "the aversion displayed by some members of staff" lies in the "feeling of uncertainty" in applying a numerical method.

> The traditional mathematician ... is clearly aware that for every numerical method a function exists for which the method produces a wrong answer. ... The statement that nothing is believed until it is proves is the starting point for teaching mathematics and introducing the computer forces the teacher away from this starting point.   (*ibid.* p.552)

However, we should not infer from this that programming is of little value. On the contrary, though early use of programming did not always show enhancement of conceptual ideas in mathematics, more recent uses of programming environments to concentrate on specific constructs have shown very deep insights (Tall & Thomas 1988a). Programming in a language such as ISETL (which is designed to enhance the user's understanding of concepts through having to specify how processes are carried out) shows positive gains in conceptual understanding (Dubinsky, to appear).

The symbolic approach has powerful advocacy from several quarters. Lane *et al* (1986) suggests ways in which symbolic systems can be used to discover mathematical principles and Small *et al* (1986) reports the effects of using a computer algebra system in college mathematics.

In the latter case the activities often consist of encouraging students to apply a technique, already understood in simple cases, to more complicated cases where the symbolic manipulator can cope with the difficult symbolic computations.

However, Hodgson (1987) observed:

> In spite of the fact that symbolic manipulation systems are now widely available, they seem to have had little effect on the actual teaching of mathematics in the classroom.
>
> (Hodgson 1987 p.59)

He quoted a report (Char *et al* 1986) of experiences using the symbolic system Maple in an undergraduate course where students were given free access to the symbolic manipulator to experiment on their own or to do voluntary symbolic problems which they could elect to count for credit. He noted a "somewhat limited acceptance of Maple by the students":

> While many explanations can be put forward for such a reaction (little free time, no immediate payoff, weaknesses of the symbolic calculator for certain types of problems, absence of numerical or graphical interface, lack of user-friendliness), it is clear that the crux of the problem concerns the full integration of the symbolic system to the course in such a way that it does not remain just an extra activity. This calls for a revision of the curriculum, identifying which topics should be emphasized, de-emphasized or even eliminated, and for the development of appropriate instruction materials.
>
> (*ibid.*)

Since this was written, the interface of Maple has been considerably improved (particularly on the Macintosh computer), and efforts have been made to enhance its user-friendliness, particularly for educational purposes. Just as the introduction of programming into mathematics courses received an initial mixed reaction only to show its greater value when used for explicit conceptual purposes, so symbolic manipulators may overcome their initial drawbacks as more imaginative *conceptual* uses are invented in teaching the calculus.

Heid (1984) reports her own research into an experimental calculus course which used the symbolic manipulator MuMath and appropriate graphical programs to introduce the concepts for twelve weeks, with practice of routine symbolic techniques only being studied in the final three weeks. She concluded that:

> Students showed deep and broad understanding of course concepts and performed almost as well on a final exam of routine skills as a group who had studied the skills for the entire fifteen weeks.
>
> (Heid 1984 p.2)

Based on the data from her experiment she formulated a number of conjectures, including the following:

> When concepts form the major emphasis in an introductory calculus course (assignments, class discussions, tests), and the computer is used to execute routine procedures:
> ... student understanding of course concepts will be broader and deeper ...
> ... student thinking will re-focus on the decision-making aspects of problem-solving ...
> ... students will remember concepts better, and be better able to apply them at a later time ...
> ... students will process information related to the concepts in larger "chunks"...
> A computer graphics approach to concept development in mathematics classes will result in better student performance on tests of "far transfer" of course concepts...
> Students will do more internal consistency checks when they work on conceptual problems if they can use the computer to process routine algorithms...
> Exposure to calculus skills through the use of symbol manipulation programs will not automatically improve the ability to perform these skills by hand or the ability to perform related algebraic manipulations...
> Use of symbol manipulation programs as tools will not eradicate algebra-based syntax errors...
>
> (*ibid* p.57 *et seq.*)

These conjectures, based on practical experience, are consonant with other related research.

Tall (1986b) reports the building and testing of a graphical approach to the calculus, using software designed to allow the user to play with examples of a concept, to enable the abstraction of the underlying principle embodied by the software. Such an environment is termed a "generic organizer". Generic organizers were designed for magnifying graphs (to see examples and non-examples of those that are "locally straight"), moving a chord along a (locally straight) graph to build up the gradient function, solving the reverse process of knowing the gradient and seeking the original function, numerical calculations of areas under graphs that emphasize the conceptual ideas, graphical solutions of first, second, and simultaneous first order differential equations (Tall 1986a, Tall *et al* 1990). In this research only the differentiation part was formally tested with pupils aged 16/17.

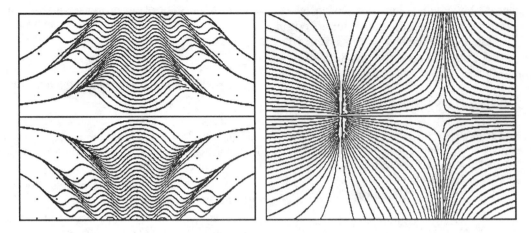

figure 2

Using matched pairs selected on the pre-test, experimental students scored at a statistically significant higher level than the controls on global/holistic skills, including sketching derivatives, recognizing graphs of derivatives, relating the derivative of a function to its gradient function, and in explaining the notions of gradient, tangent and differentiation from first principles from a geometrical viewpoint. At the same time, more traditional logical/sequential tasks, such as explaining symbolic differentiation from first principles or routine differentiation of polynomials and powers, showed no significant difference. These results are very much in line with Heid's work.

Artigue (1983, 1987) and her colleagues developed a teaching programme for differential equations which first introduced qualitative ideas of families of solutions through studying pre-prepared computer drawn pictures. They then used these ideas to match pictures of the solution curves to the corresponding differential equations. For example, students were given different differential equations:

$$y' = \frac{y}{x^2 - 1}, \ y' = y^2 - 1, \ y' = 2x + y, \ y' = \sin(xy),$$

$$y' = \frac{\sin(3x)}{1 - x^2}, \ y' = \sin x \ \sin y, \ y' = y + 1$$

and the same number of corresponding pictures to match, of which two are shown in figure 2.

They then related different methods available for solving specific differential equations using both qualitative and algebraic approaches and concluded by considering the qualitative theory of differential equations, with proofs of theorems based upon pictures of solution curves, dealing with barriers, trapping regions, funnels, attractors

and so on. The students used both pre-prepared pictures and interactive computer programs, for instance to explore the phase portraits of differential equations depending on a parameter. At the end of the course the students showed themselves capable of giving meaning to the qualitative approach, to describe and draw solutions without symbolic differentiation and to coordinate algebraic and graphical representations.

Blackett (1987) introduced the relationship between the algebraic representation of a linear relation and its straight line graph to younger pupils (aged 14/15) using graph-plotting software. He taught three experimental classes of low, average and above average ability, who were matched with corresponding control classes. Those using the computer were able to tackle activities that might be too demanding using paper and pencil, for example, to instruct the computer to draw $y=x+20$ and $y=2x+10$ on the same axes to include their point of intersection (requiring some investigation to determine appropriate scales). The post-test showed a significant overall improvement in performance in all the groups, except for one control class taught by a teacher who used the pre-test to teach the pupils specific tasks likely to arise in the post-test. Detailed analysis of the responses revealed that these control students scored higher than the corresponding experimental students in questions that were almost identical to the pre-test, but they scored considerably lower on tasks with even tiny conceptual differences. Blackett reports that:

> Pupils who had been taught to answer specific questions rather than the underlying concepts experienced difficulty whenever new questions varied, even slightly, from those they had met previously. These results appear to highlight the effects of encouraging instrumental as opposed to relational understanding.   (Blackett 1987, p.93)

He also noted substantial discrepancies between the children's performance on the post-test and traditional school tests. In particular there were a number of children who performed badly on traditional serialist/analytic questions, yet performed well on global/holistic tasks.

Blackett took these exceptional children from the lowest ability class and added them to the highest ability class for an introduction to the idea of the gradient of "locally linear" graph, using the "Graphic Calculus" software. Blackett found them well able to cope and concluded that:

> Students who had achieved a clear understanding of the straight line and its equation, particularly the significance of the gradient, can, with the aid of suitable computer graphics, develop an equally clear understanding of locally linear graphs and the curves associated with polynomial equations. ... There were pupils unable to handle number work successfully but were nevertheless able to demonstrate an understanding of advanced concepts presented in a visual form requiring either a visual interpretation or a drawing, rather than a calculation, as an answer. (Blackett 1987, pp. 127, 128)

His experiment indicated that these pupils, aged 14/15, were able to perform the task of sketching a global derivative at a level comparable with the 16/17 year old experimental students in Tall (1986b).

A theme running through all these pieces of research is the power of graphic computer software to improve the versatility of students thought processes, but with little significant change in their ability to cope with symbolic manipulation. On the other hand there is as yet little evidence that symbolic manipulators improve students manipulative ability, although there are hopeful indications that students may use them in suitably designed tasks.

A recurring observation is the difficulties experienced by teachers, both at university and in school, to come to terms with the new technology. Great experience of student problems in a pre-computer culture can sometimes be a hindrance in trying to predict what difficulties students may have when using the new technology. We are at present in the throes of a paradigmatic upheaval and cultural forces operate to preserve what is known and comfortable, and to resist new ideas until they are proven better beyond doubt.

In several countries the ability to implement a graphical approach on current microcomputers is leading to the production of such programs and the development of new curricula. For example, the School Mathematics Project in the U.K. is now designing a curriculum in which the first introduction to gradient is via "locally straight curves" and limiting processes are postponed to the second year of the course.

More recent developments in graphic approaches to calculus

Technology moves on apace. The new RISC (reduced instruction set chip) processors are becoming generally available in micro-computers at a lower price than primitive eight bit chips four or five years ago. The greater speed of these processors enable far more complex activities to be carried out on the computer in real time, including the plotting of models of nowhere differentiable curves. For example, it is easy to calculate the *blancmange function*, $y=\mathrm{bl}(x)$ (figure 3) to any desired accuracy and to draw it extremely quickly, provided the software is sensibly programmed. (Some symbol manipulators such as *Maple* and *Derive* are self-defeating –

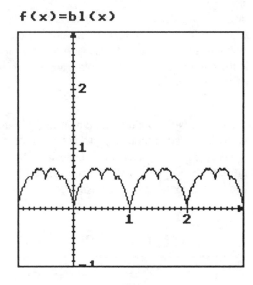

figure 3

as first published – being designed to allow any desired accuracy, and therefore failing to provide optimum, even adequate, speed.)

This is calculated using the saw-tooth function $y=\mathrm{s}(x)$ where y is calculated for given x as follows:

let $t=x-[x]$ be the fractional part of x
if $t<1/2$ then let $y=t$ else $y=1-t$.

The nth approximation to the blancmange is then found

---

by adding together a sequence of sawteeth each a half the previous one:

$$\mathrm{bl}(x,n) = s(x) + \frac{s(2x)}{2} + \ldots + \frac{s(2^{n-1}x)}{2^{n-1}}.$$

As n gets larger, the sawtooth $\frac{s(2^n x)}{2^n}$ gets very tiny and contributes little to the sum, so that $\mathrm{bl}(x,n)$ stabilizes to look like the wrinkled blancmange $\mathrm{bl}(x)$. This is easily seen graphically and the limiting process can be translated into a formal argument (Tall 1982).

More generally, the function which I call the Van der Waerden function, $\mathrm{van}(x,m)$, can be calculated by adding together teeth to get

$$\mathrm{van}(x,m,n) = s(x) + \frac{s(mx)}{m} + \ldots + \frac{s(m^{n-1}x)}{m^{n-1}}$$

and taking a sufficiently large value of n to see the picture stabilize to give (an approximation to) the non-differentiable function $\mathrm{van}(x,m)$. These functions have the property that they *nowhere* magnify to look straight, so they are not differentiable anywhere.

In this way it is easy to build up a number of different functions which are continuous everywhere yet differentiable nowhere. The ready availability of such functions has interesting consequences for student's understanding of concepts in mathematical analysis.

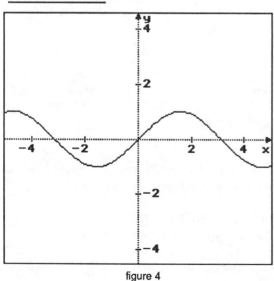

figure 4

Figure 4 shows two graphs superimposed: $y=\sin x$ and $y=\sin x + n(x)$ where the function $n(x)$ is a tiny Van der Waerden function. (It is calculated from $v(x)=\mathrm{van}(x,3)$ as

$$n(x) = \frac{v(1000x)}{1000}.)$$

Figure 5 shows the two graphs magnified to reveal the graph of $y=\sin x$ as locally straight, but $\sin x + n(x)$ as being nastily wrinkled. This informal idea represents the generative idea of the essential difference between a differentiable and a non-differentiable curve which are indistinguishable to a normal scale.

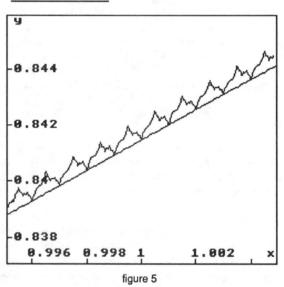

figure 5

Pictures drawn with today's technology have relatively large pixels, so the graphs look rough. This phenomenon can be turned to advantage, to underline that what is being drawn is a *model* of the graphs of the functions rather than the graphs themselves. What matters most is the quality of the picture in the student's mind, not on the screen!

Figure 6 shows the area function under the blancmange function being drawn in real time. A static picture in an article is completely inadequate to represent this potent dynamic idea. Here the line of dots represents the graph of the area function from 0 to the current point x. A straight line is drawn through the last two points, representing the gradient of the area function. As the area function grows, the gradient of the area function changes; and it visibly changes in a smooth way. The computer is giving an approximate model of an area function that is everywhere differentiable once, but not twice.

f(x)=bl(x)

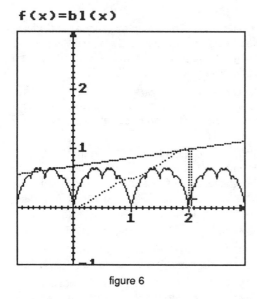

figure 6

y=[x]+blx

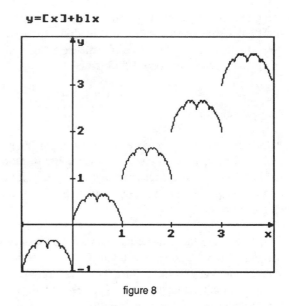

figure 8

Figure 7 shows the area function being calculated under the discontinuous function $y=x-[x]$ (where $[x]$ denotes the integer part of x). The dots are the "area so-far" under the curve from $x=0$. Clearly the area function has 'corners' at those points where the original function is discontinuous.

tinuous, nowhere differentiable blancmange function of figure 3. Because bl(x) has values between 0 and 1, a solution curve to the differential equation moves up and down (smoothly!) with gradient between 0 and 1.

f(x)=x-[x]

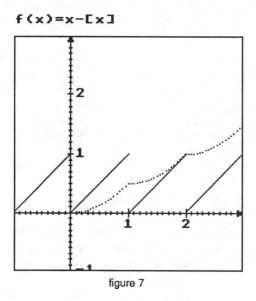

figure 7

dy/dx=bl(x)

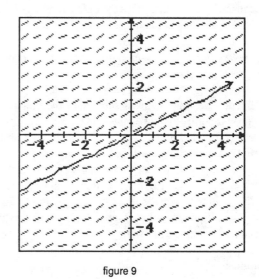

figure 9

Figure 8 shows the curve $f(x)=[x]+bl(x)$ where bl(x) is the blancmange. As $f(x)$ is discontinuous at every integer and continuous, but not differentiable, everywhere else, what will the area function look like? Where will the area function be continuous and where is it differentiable?

Figure 9 shows the drawing of a solution to the differential equation $\dfrac{dy}{dx} = bl(x)$ where bl(x) is the everywhere con-

Hubbard and West (1985) found that students had difficulty in understanding that a differential equation can have a solution when it is not possible to express the solution in "closed form" (made up of a combination of standard functions). To help students they developed interactive programs for the Macintosh computer that used the mouse interface to point at areas of interest to home in on singularities and to investigate the qualitative behaviour of solutions. The "existence" of a solution now depended on the ability to draw its graph, which could be done numerically without knowing a formula for it. Such experiences with a computer proved to give powerful insights into theorems of existence, uniqueness and behaviour of solutions of differential equations.

The way ahead

In the next few years, software of this nature can but get more and more powerful, with increasing use of more flexible input devices, such as using a mouse to point at a part of the picture that looks interesting, to zoom in and take a closer look.

Of greater importance in the software will be the development of flexible environments that unite numerical, symbolic and graphical facilities. Good software is not designed by encrusting the existing materials with more and more options. What is absolutely essential is to take into account all possible methods of information processing that can make the ideas simpler to handle. Such software is more likely to be successful if it uses the capacity of holistic/global thinking to complement the mathematician's traditional methods of deduction.

Back in 1971 Skemp made the pertinent observation that the teaching of mathematics by logical methods is good in that it shows that mathematics is a structured and not arbitrary science, but is weak in that it shows the *product of mathematical thought* rather than the *process of mathematical thinking*. Skemp subsequently emphasized this point by distinguishing between *building* and *testing* concepts. Testing involves subjecting the concepts to rigorous enquiry, to make sure that their construction and proof is founded on a firm logical base, but before they can be tested they must be built, and building is a cognitive act which is not served solely by demonstrating formal proofs. Using a logical deductive approach to the calculus is but one side of a two-sided coin. The other may be illuminated by graphical insight from well-designed computer software.

References

Artigue M. 1987: 'Ingénierie didactique à propos d'équations differentielles', *Cahier de Didactique No 15*, Edition IREM, Paris 7.

Artigue M. & Szwed T. 1983: *Représentations Graphiques*, Edition IREM, Paris Sud.

Bach J.O. 1988: GrafMat-programmeme, Matematiklærerforeningen, Copenhagen.

Blackett N. 1987: *Computer Graphics and Children's Understanding of Linear and Locally Linear Graphs*, unpublished M.Sc. Thesis, The University of Warwick.

Brumby M.N. 1982: 'Consistent differences in Cognitive Styles Shownfor Qualitative Biological Problem Solving', *British Journal of Educational Psychology*, 52, 244-257.

Cornu B. 1981: 'Apprentissage de la notion de limite: modeles spontanés et modeles propres', *Actes du Cinquiéme Colloque du Groupe Internationale P.M.E.*, Grenoble.

Char B.W., Fee G.J., Geddes K.O., Gonnet G.H., Marshman B.J. & Ponzo P. 1986: 'Computer Algebra in the Mathematics Classroom' in *proc. 1986 Symposium on Symbolic and Algebraic Computation*, Assoc. Comput Mac. 135-140.

Dienes Z.P. 1960: *Building up Mathematics*, Hutchinson.

Dubinsky E. (to appear) 'Reflective Abstraction', in *Advanced Mathematical Thinking* (ed. Tall, D.O.)

Heid K. 1984: *Resequencing skills and concepts in applied calculus through the use of computer as a tool*, Ph.D. thesis, Pennsylvania State University.

Hodgson B.R. 1987: 'Symbolic and Numerical Computation: the computer as a tool in school mathematics' in *Informatics and the Teaching of Mathematics* (ed. Johnson D.C. & Lovis F.) North-Holland, 55-60.

Hubbard J.H. & West B.H. 1985: 'Computer graphics revolutionize the teaching of differential equations', in *Supporting Papers for the ICMI Symposium*, IREM, Université Louis Pasteur, Strasbourg.

Keisler H.J. 1976: *Elementary Calculus*, Prindle Weber & Schmidt, Boston.

Kemeny J.G.1986: *Calculus*, True BASIC Inc.

Lane K.D., Ollongren A. & Stoutemyer D. 1986: 'Computer based Symbolic Mathematics for Discovery', in *The Influence of Computers and Informatics on Mathematics and its Teaching* (ed. Howson A.G. & Kahane J.-P.). Cambridge Univ. Press, 133-146.

Maths Workshop 1987: *Symbolic Calculus*, 45 Carson Rd, London E16 4BD.

McLean L.D., Raphael D. & Wahlstrohm M.W. 1984: 'The Teaching and Learning of Secondary School Calculus: Results from the Second International Mathematics Study in Ontario'.

Orton A. 1983a: 'Student's understanding of integration', *Ed. Studies in Math.*, 14, 1-18.

Orton A. 1983b: 'Student's understanding of differentiation', *Ed. Studies in Math.*, 14, 235-250.

Robert A. & Boschet F. 1984: 'L'acquisition des débuts de l'analyse sur R à la fin des études scientifiques secondaire françaises', *Cahier de didactique des mathématiques* 7, IREM, Paris VII.

Schwartz J.T., Dewar R.B.K., Dubinsky E. & Schonberg E. 1986: *Programming with Sets, An introduction to SETL*, Springer-Verlag.

Simons F.H. 1986: 'A course in calculus using a personal computer', *Int. J. Math. Ed. Sci., Techn.*, 17 5, 549-552.

Skemp R. R. 1971: *The Psychology of Learning Mathematics*, Penguin.

Small D., Hosack J., Lane K.D. 1986: 'Computer Algebra Systems in Undergraduate Instruction', *The College Mathematics Journal*, 17 5, 423-433.

Stoutemyer D. *et al* 1983: *MuMath*, The Soft Warehouse Honolulu, Hawaii.

Stoutemyer D. & Rich A. 1989: *Derive*, The Soft Warehouse, Honolulu, Hawaii.

Tall D.O. 1982: 'The blancmange function, continuous everywhere but differentiable nowhere', *Mathematical Gazette*, 66 11-22.

Tall D. O. 1986a: *Graphic Calculus*, (software for BBC compatible computers), Glentop Press, London.

Tall D.O. 1986b: *Building and Testing a Cognitive Approach to the Calculus Using Interactive Computer Graphics*, Ph.D. Thesis, The University of Warwick.

Tall D.O. 1986c: 'Lies, damn lies and differential equations', *Mathematics Teaching* 114 54-57

Tall D.O., Blokland P. and Kok D. 1990: *A Graphic Approach to the Calculus*, (software for IBM compatible computers), Sunburst Communications Inc.

Tall D.O. & Thomas M.O.J. 1988a: 'Longer Term Effects of the Use of the Computer in the Teaching of Algebra', *Proceedings of P.M.E.12. Hungary,* 601-608.

Tall D.O. & Thomas M.O.J. 1988b: 'Versatile learning and the computer', *Focus*, 11, 2 117-125.

Tall D.O. & Vinner 1981: 'Concept Image and Concept Definition in Mathematics, with particular reference to limits and continuity', *Ed.Studies in Mathematics*, 12 151-169.

Tall D.O. & West B.H. 1986: 'Graphic Insight into Calculus and Differential Equations', in *The Influence of Computers and Informatics on Mathematics and its Teaching* (ed. Howson A.G. & Kahane J-P.), Cambridge University Press, 107-119.

Van Hulzen J.A. & Calmet J. 1983: 'Computer Algebra Systems', in *Computer Algebra. Symbolic and Algebraic Computation* (2nd edition, ed. Buchberger B., Collins G. & Loos R.) Springer-Verlag.

Wolfram S. 1988:*Mathematica*, Wolfram Research Inc.

Supercalculators as Laboratory Instruments

Thomas Dick*
Oregon State University

Introduction

This paper suggests that the traditional recitation section can be readily converted into a calculus laboratory with the use of the new generation of calculators. First, we make some philosophical comments regarding the objectives a calculus laboratory should have. Second, we address some common concerns raised regarding the use of these instruments in the classroom. Finally, we offer a few examples of the type of specific laboratory activities we hope capture the spirit that we believe calculus laboratories should have. All our remarks are based on actual experiences and observations of the use of HP-28S calculators in the calculus sequence at Oregon State University.

Recitation vs. Laboratory

The "recitation" model of calculus instruction has been firmly in place at many large institutions for several years. Under this model, students typically attend lectures 3 - 4 times per week, often in large halls with literally hundreds of students in attendance. This instruction is supported by smaller recitation sections (perhaps 20-50 students) meeting 1-2 times per week where graduate assistants or instructors go over additional examples and answer questions about assigned homework problems.

Even the word "recitation" suggests what goes on all too often in these sections: teaching assistants dutifully working step-by-step on rote example after example with students dutifully transcribing every chalk mark to their own notebooks. The questions asked, if any, are offered only by those students who are caught up with the homework (a small minority). To be sure, there are many recitation instructors who are able to bring some gems of insight to, or clarify some subtle concepts from, the main lectures. But the notion of intellectual discovery on the part of the students is a foreign concept to the general conduct of recitation sections. While many of the sciences may also be taught with large lectures supported by recitation sections, a laboratory experience is often considered an essential component of the science student's instruction. The laboratory activities used in many of the biological and physical sciences are generally intended to provide students with two types of learning experiences:

1) the opportunity to gather first-hand observations of phenomena and then to analyze and reflect on these observations in light of the theoretical concepts,

2) the opportunity to engage in guided inquiry and exploration and form conjectures based on experimental evidence.

The intent of these types of activities are clear-- for students to appreciate the spirit of the scientific method, they must partake of it directly. If we believe that a laboratory model should be utilized in calculus, then it is not so much the technical particulars of the science lab that we should emulate, but rather it is the spirit of the scientific method that we should strive to capture. Analysis and reflection, exploration and conjecture-- are not these the cornerstones of mathematical inquiry?

Technology has created a calculus laboratory instrument which can be used to allow students firsthand observations of calculus phenomena. Numeric, symbolic, and graphic capabilities are now available on hand-held super calculators. Because of their portability and ease-of-use, these machines open up opportunities for experimentation and guided inquiry for *every* student in calculus. Super calculators provide the means to effectively convert recitations to inquiry-based calculus laboratories.

Concerns about using supercalculators in the classroom

Here are some responses to the concerns we have heard voiced most often in opposition to the use of calculators in calculus.

l) The time needed to train students to be proficient on a calculator will be prohibitively long, and will rob the course of time better spent on instruction.

Nonsense. Yes, because of its very power, it will take more than 30 minutes to get comfortable with a super calculator. But students neither need to learn to be expert programmers of the calculator nor do they need to even be aware of

*Some of the activities discussed in this paper were developed in a project supported, in part, by the National Science Foundation (award #USE-8813785) and the Lasells Stewart Foundation

all the "direct" functionality of the calculator to be able to make good use of it. At one time short sessions on how to use a slide rule were routinely given near the beginning of many introductory science courses. If instant proficiency was not expected of students with those simple instruments, then it does not seem unreasonable to commit some initial start-up time to bringing students up to speed on a few uses of a super calculator. It does not take much time at all to get students comfortable with the key sequences necessary for basic graphing, numerical solving routines, and symbolic differentiation. And the dividends to be reaped from this small initial investment of time can be handsome indeed.

2) It is too difficult to manage the classroom because some students will press the wrong key and get lost.

There's no doubt about it-- a calculator laboratory will demand more preparation than a run-of-the-mill recitation. But there are some simple things which can be done to make such sessions go smoothly. When doing a guided example with a class, frequent "signposts" should appear which inform the students exactly where they should be. One way to accomplish this is to use an overhead display of your calculator screen, so that students can monitor what they should be seeing on their own screens and correct themselves quickly. If such a device is not available, some overhead transparencies can be prepared in advance of class showing what the current screen should look like after every 3 or 4 keystrokes of the example. As for helping a student who has accidentally stepped off the common track our experience has been that neighboring students tend to provide assistance quickly, spontaneously and free of charge.

3) Students will use the calculator mindlessly as a black box.

In fact some students hope that this will be the case. However, any dreams of optical scanners which can read word problems and then print out worked solutions quickly vanish. These machines don't think. Students who don't think either are quite helpless with or without these machines. "Aren't there some strictly computational calculus exercises which can simply be plugged right into these calculators to get the answer?" Yes. But doesn't that suggest we should be rethinking the importance of asking students to perform strictly computational tasks rather than forbidding the calculators in the classroom.

4) The super calculators are too expensive.

The history of calculator prices would suggest that this is a temporal concern. But even at current prices, students using them throughout a calculus sequence (not to mention other science or engineering courses) will find the "per course" cost quite small. In fact, the quality slide-rule of the student days of yesteryear was more expensive than today's most powerful hand calculator (if the price is adjusted for inflation to current dollars). And if the price of a symbolic/graphical calculator still seems too expensive, then much can still be done with simply a scientific graphics calculator.

5) The super calculators are not as powerful as the best available symbolic/graphical software.

They never will be. Surely, when a hand-held device more powerful than software like Mathematica or Maple is produced, there will be some more impressive software package available. Just as with computers, if we put off buying them because they're still improving, we'll simply never buy them. Super calculators already have features found only on software running on mainframe computers a few years ago. As for their advantages over microcomputer software, calculators come with their own hardware and "palmtop" beats laptop for portability any day. I believe the issue of portability is one that is undersold by many. Departments with responsibilities for teaching calculus to literally thousands of students can find the logistical problems of using microcomputers in labs, homework, and especially *testing* very difficult. The handheld super calculator solves many of these problems quite nicely.

**Suggestions and Examples
of Supercalculator Laboratories**

Once one has made the decision to employ a laboratory approach in teaching calculus, there comes the hard work of designing or finding good laboratory activities. The choice to make substantial use of technology in the labs, either in the form of computers or super calculators, adds another dimension to this task. In this section we offers some suggestions about how to think about calculus labs and the use of technology, and some specific ideas for labs that we have tried and found successful.

In Oregon State's Calculus Project, we meet four days a week and set aside an average of one day a week for a laboratory. (Sometimes it has made more sense to have two labs in the same week, or none at all. In a large lecture model where the labs take the place of a recitation, one would not have this same flexibility.) Lab days which immediately precede an exam date are used instead for review. This leaves approximately 12 "true" labs per semester, or 8 per quarter. The true lab days fall into three categories: skill labs, applications labs, and exploratory/discovery labs.

Skill labs center on the skills needed to use a super calculator *intelligently*. It is extremely rare to find a college calculus student who does not have experience with a scientific calculator, but very few students (and very few calculus instructors, for that matter) have really wrestled with and analyzed some of the effects the numerical limitations of a machine can cause. A good working knowledge of the symbolic algebra syntax and the navigational skills for graphing needed for using the super calculator on a daily basis require some explicit instructional time. A common theme throughout all of the labs is having students *actively* making connections between graphical, numerical, and symbolic information. Most importantly, students learn how to use the graphical information the calculator can provide to monitor the reasonableness of their computations, a skill vastly more valuable than memorizing any number of symbolic tricks for special cases. Moreover, it provided a new foundation for motivating topics in the lectures.

In the applications labs, the super calculator is not on center stage, but plays an important supporting role with its graphic and computational capabilities. Exploratory/discovery labs are designed to either introduce a particular topic through a sequence of examples by a guided discovery approach, or to allow students an opportunity to gain some "hands on" feeling for a concept through guided exploration and discussion. It is extremely important to include examples which confront the limitations of the machine. Here is a description of the labs we have used through a first-year calculus course. The first three labs below are introductory skill labs used during the first two weeks of classes to bring students "up to speed" on the use of the super calculator.

1. Numerical Analysis

Investigate the effects of underflow, overflow, round-off, and cancellation errors. All of these appear naturally in the numerical investigation of limits. In particular, cancellation errors occur when investigating the limits of difference quotients. Practice identifying when and what kind of error is likely to occur in computational situations, and some strategies for dealing with them.

2. Graphical Analysis

Graphical behavior can be hidden by window position, scaling, and the numerical limitations of the machine. This lab provides instruction and practice in navigating in the graphics environment to get the best views of a graph. The notions of local and global behavior are introduced. Discussing *how* the graphics image is determined goes a long way in helping students understand how the numerical limitations covered in the first lab will affect graphing.

3. Solving equations and inequalities

This lab is devoted to discussing how the super calculator can be used as an aid in problem solving. Instruction in using the algebra syntax and equation solvers is covered. The uses of graphing and numerical evaluation as both heuristics and monitors of symbolic computations seem to be quite new revelations to many students, and this lab sets the tone for the super calculator applications labs.

4. Limits and continuity

The intuitive notion of limit lends itself well to a graphic presentation. In our first introductory calculus lab, we start by having students graph the function $y = (x-1)/(x-1)$ on their calculators. Why was there a hole in the graph? We were off and running with a discussion of domains and the need for the language of limits to effectively describe the behavior we saw. (In the process, many students learned for the *very first time* the importance of the proviso "$x \neq 1$" on the "identity" $(x-1)/(x-1) = 1$.)

The graphs of $|x|/x$, $1/x$, $\sin(1/x)$, and $(\sin x)/x$ as well as many other examples quickly followed so that students could see a wide range of functional behaviors. It's important to note that these examples are not new or exotic, and that many calculus teachers draw attention to them or other very similar examples. What was different in this lab was that the students were producing the graphs themselves and were actively engaged in making their own connections between the concept of limit and the graphical behavior that they could see. Students seem much more convinced that $(\sin x)/x$ has a limit of 1 as x approaches 0 by simply looking at the calculator-generated graph of $y = (\sin x)/x$ than they are by any of the "proofs" given in calculus books.

5. Local slope and differentiability

In this laboratory students zoomed in on the graphs of functions at particular points until they appeared linear, and then approximated the local slope by using the usual two-point formula for lines. This set the stage for the notion of derivative as a locally linear approximation. A copy of this laboratory handout is included in section 3 of this volume.

6. Extrema and the first derivative test

In this laboratory, students gain skill in sketching the graph of the second derivative directly from the graph of the original function. A sequence of examples is explored leading to the development of the first derivative test for extrema.

7. Second derivatives - concavity, inflection points, and the second derivative test

In this laboratory, students gain skill in sketching the graph of a derivative directly from the graph of the original function. A sequence of examples is explored leading to the development of the second derivative test for extrema. Inflection points are emphasized as points of greatest (in absolute value) rate of change.

8,9. Optimization applications

Two of our labs are devoted to the applications of calculus to solving problems of optimization. The emphasis here is on mathematical modeling and the role of the supercalculator as a problem solving tool.

10. Numerical Integration

Following a discussion of Riemann sums and left, right and midpoint methods of numerical integration in a previous class meeting, this lab applies those techniques using a program entered prior to the start of the lab. The program takes a function f, the endpoints of a closed interval $[a,b]$, a number n of equal subdivisions for this interval, and a real number r between 0 and 1 to choose a point from each subinterval (0 for left endpoint, 1 for right endpoint, and 0.5 for midpoint) as arguments, and returns the corresponding approximation of the definite integral. The students use these programs on a variety of examples, noting the relative sizes of errors. This can be used to motivate the trapezoidal and Simpson's Rule approximations.

11. Antiderivatives and the Fundamental Theorem of Calculus

Area functions are plotted with the numerical integration provided by the super calculator. These functions are plotted by hand and then their derivatives are sketched. The purpose of this lab is for students to discover the

fundamental theorem of calculus for themselves.

12,13,14. Applications of definite integration

Three of our labs are devoted to the applications of definite integration to finding arc length, area, volume, work, force, etc. The emphasis is on modeling by Riemann sums (i.e., setting up the correct definite integral), since the numerical integration capabilities take care of the computations.

15. What is π?

Sequences are introduced in the context of approximations. Trigonometry and the method of exhaustion are applied to approximating π as the area of regular polygons inscribed and circumscribed about a unit circle.

16. Sequential convergence

Numerical and graphical explorations of sequences.

17. Newton's Method

After deriving the recursive formula in a previous class meeting, the students implement Newton's Method for several functions by entering the formula into the super calculator's SOLVER. After evaluating the formula for an initial guess, the SOLVER allows the students to immediately feed the output back in as the next guess. Students see first-hand the nature of recursion, the speed of convergence of Newton's Method, and its sensitivity to the initial guess. When the method fails, the graph of the function can make clear the reason for failure.

18. L'Hospital's Rule

The notion of local slope is used again to motivate L'Hospital's Rule. First, here's a simple problem from algebra: given two non-vertical lines which cross at the point a on the x-axis (see picture), compare the ratios of the y - coordinates at x = b to the ratios of the respective slopes of the lines.

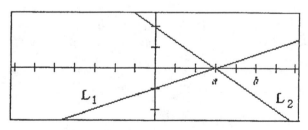

L_1 is the line $y = m_1(x - a)$ and L_2 is the line $y = m_2(x - a)$.

A few simple calculations yield that the ratios are the same. Now, suppose that we have two functions f and g which have a common zero and are differentiable at x = a . Since f and g are approximately "locally linear" at x = a , zooming in on their graphs can bring us to a picture exactly like that above. This suggests that the limit of f(x)/g(x) as x approaches a is the same as the limit of f'(x)/g'(x) in this case. L'Hospital's Rule is thus motivated in a very natural way.

19. Series convergence.

Numerical and graphical explorations of series.

20. Improper integrals

The super calculator's numerical integration routine can give good results on improper integrals with a vertical asymptote at an endpoint of the interval of integration. By making use of change of variables *(dy* instead of *dx)* the calculator can be used to evaluate some improper integrals with an infinite limit of integration. Analyzing *a priori* convergence vs. divergence of an improper integral thus becomes a skill with direct and relevant importance to these computations.

21. Taylor polynomials - Graphical convergence

Building the coefficients of a Taylor polynomial is greatly aided by the use of a super calculator. Comparing the graphs of a function and a sequence of Taylor polynomials provides a powerful visual picture of the notion of convergence of a sequence of functions.

22. Power series and intervals of convergence

Intervals of convergence for Taylor series and power series are explored both graphically and numerically in this lab.

23. Polar coordinates and complex numbers

Most students come to freshman calculus with surprisingly little exposure to complex numbers, and even less to polar coordinates. A super calculator can express points in the plane in either polar or rectangular form, and convert between the two forms. In this lab we explore the graphical interpretations of complex number addition, multiplication, and exponentiation, including DeMoivre's Theorem.

24. Polar graphs and parametric equations

In this lab, students dynamically explore the distinction between common locus and "collision point" of two parametrically defined curves.

Concluding remarks

Super calculators are instruments which can be used to transform the usually passive calculus recitation experience into an active hands-on laboratory. Making that transition takes some work, but the benefits can be substantial. If the medium is the message, then a laboratory approach to calculus can teach students that mathematics is something we *do,* and not just a spectator sport. If you choose to try it, then you might be surprised at a change in yourself. The laboratory instructor is truly more of a "guide on the side" than a "sage on the stage," and that's an uncommon experience for many of us. In this case, the "revitalization" of calculus can apply to both student and teacher.

Microlabs and Cooperative Learning: New Ways to Teach Calculus

Donna Beers*
Simmons College

Introduction

Recently there has been a flood of articles in our professional journals on reforming and renewing the calculus. Of especial interest to us has been the issue of teaching calculus more effectively, and we have been experimenting with some new approaches. Although our decision to experiment was arrived at independently, it is supported by two reports, *Mathematics for a New Century* and *Everybody Counts*:

> ...Few students can learn from lectures and homework alone. ...Effective teaching for today's student requires a more diverse repertoire of approaches...[8].

> Lecture less; try other methods [5].

Since Fall, 1985, the Simmons College Mathematics Department has required that students in a sophomore-level statistics course attend a weekly computer lab, in addition to three 50-minute lectures. Our positive experiences with these labs led us in Fall, 1988, to introduce a mandatory weekly lab into all sections of Calculus I and Calculus II. Some of these would be spent using personal computers, others would be paper-and-pencil. These labs are the subject of this paper.

Microlabs: Our Objectives

In introducing computer labs into our first-year calculus course, our objectives were these:

·To reinforce and make vivid precalculus concepts by exploiting the beauty and power of computer graphics to shore up students' backgrounds in the theory of functions (e.g. domains, ranges, and properties of the logarithmic, exponential, and trigonometric functions) and analytic geometry (e.g. addition of functions, translations of graphs, and conic sections). Students in Calculus I frequently complain that the review of precalculus given in calculus texts is too cursory to be of much help. Furthermore, their backgrounds are not homogeneous: Some have had little experience with trigonometry, others are unsure of logs and exponentials. Therefore, in our first computer labs we try to fill in gaps and provide a common background.

·To complement and enhance the lecture portion of the course by increasing the number and range of examples through computer generation. For example, it is time-consuming and difficult to get decent plots, by hand, for functions like (sinx)/x and x sin(l/x). We want students to use the computer to save time, perform tedious computations (like generating tables of values for functions), and double-check work done by hand.

·To strengthen student intuition and ability to think spatially. Demana and Waits discuss using computer graphics to examine the "end behavior[3]" of functions (i.e., their behavior to the far right and far left [3]), and note that by viewing the graph of the rational function
$y = (a_n x^n + a_{n-1} x^{n-1} + \cdots + a_0)/(b_m x^m\ b_{m-1} x^{m-1} + \cdots + b_0)$
in a huge viewing window, students can see that it behaves like $y = (a_n/b_m)x^{n-m}$.

Regarding spatial thinking, we want students to find out whether a function is odd or even and to utilize symmetry in sketching graphs.

·To foster inquiry, and, along with this, use of the computer as a tool for investigation. We regard it of paramount importance to get students more actively/thoughtfully engaged in their learning. One of the keys here surely must be to ask questions which require more than simple yes/no answers. As examples: What is the connection between the degree of a polynomial and the number of turns in its graph? When a function like (sin x)/x defies algebraic manipulation, how do you compute limits for it? What is $\lim_{x \to 0} (\sin x)/x$? What does the graph of (sin x)/x look like near x = 0? Here the stage is set for students to use the computer to graph the function and, as they do so, to think about how computing a limit relates to selecting an appropriate domain and using graphics features like "window" and "zoom-in."

---

*An earlier version of this paper was originally delivered at the <u>Spring Conference on the First Two Years: Teaching the Mathematical Core</u> held at the University of Hartford, March 31-April 1, 1989

·To give a richer calculus course than was before possible, by covering nonstandard topics such as elementary numerical methods of integration and root-finding [9].

·Above all, to highlight the geometry of the calculus ideas under study, wherever appropriate. Students often get the impression that the secret to cracking a limit problem always resides in algebraic manipulation. Unfortunately, they get the same idea when it comes to definite integration. According to Paul Zorn,

> The definite integral...is for many students not a number, defined as a limit, but rather an algebraic process: antidifferentiate and plug in the endpoints [9].

The advantage of computer graphics is clear:

> Numerical and graphical approaches to the definite integral are not only better at "getting answers"; more important, they also lead to deeper and more useful understanding of the integral as an analytic object [9].

To sum up, for our first time around, we were not setting out to teach programming or to find exciting, uncontrived real-world applications of the calculus, although we are aware that this is a serious deficiency of most texts. Instead, we set our sights on doing a better job of teaching the calculus, as measured by student understanding and achievement.

Microlabs: Their Structure

At Simmons College, all computer labs for mathematics courses are conducted in a library classroom equipped with a dozen Compaqs, a dozen Macs, and projection capability. The calculus software we purchased is MicroCalc, which only runs on IBM or IBM-compatibles, so only the Compaqs are used. The lab can comfortably handle twenty-four students, working in pairs. As noted earlier, labs run for an hour and twenty minutes, this being the minimum amount of time in which students can comfortably move around in the MicroCalc environment and also do some significant mathematics. For each session we prepare a handout that states its purpose and the MicroCalc modules that will be featured. Handouts are designed so that students are led, through experimental investigation, to draw conclusions or make conjectures. For example, in the review-of-precalculus lab, students use the MicroCalc program, GRAPH OF Y = F(X), to graph $y = x^2$ and then superimpose on it the graphs for $y = (x-2)^2$, $y = (x+2)^2$, etc.

Eventually they are asked to generalize and conjecture how the graphs for $y = (x-c)^2$ and $y = (x+c)^2$ compare with $y = x^2$, for any constant c. Similar investigation-type questions lead students ultimately to discuss how the graph for $y = a(x-c)^2 + b$ may be derived from the graph for $y = x^2$ through appropriate translations and contractions or dilations. This same approach is taken with trig functions, for example in reviewing how to graph $y = A \sin(Bx + C)$. In Calculus II, on the other hand, one of the first labs for the course asks students to use the program RIEMANN SUMS to compute upper and lower sums for $y = x^2$ over [0,1] when [0,1] is partitioned into 4 subintervals, then 8, etc. Next the students are asked to compute upper and lower sums for $y = 2x^2$ and to compare them with the corresponding ones for $y = x^2$. Of course, the point here is to lead students to the generalization that

$$\int_a^b c \cdot f(x)\, dx = c \cdot \int_a^b f(x)\, dx$$

In a similar experimental way, they are led to two other calculus generalizations:

$$\int_a^b (f(x) + g(x))\, dx = \int_a^b f(x)\, dx + \int_a^b g(x)\, dx$$

and

$$\int_a^b f(x)\, dx = \int_a^c f(x)\, dx + \int_c^b f(x)\, dx$$

We think that the experimental style we have taken in our labs is in line with current thinking that students learn mathematics by doing.

MicroCalc version 5.1

MicroCalc, developed by Harley Flanders at the University of Michigan at Ann Arbor, is a menu-driven software with two-and three-dimensional graphics. It has online help and documentation that can be accessed by typing the command MANUAL. Flanders calls MicroCalc "interactive" since the user must choose options and enter functions and numbers. MicroCalc is sufficiently comprehensive that it may be used to varying degrees in Calculus I through IV. It has about 50 programs (e.g. Limits, Secants and Tangents, Approximate Integrals, Taylor Polynomials, Parametric Plane Curves, Parametric Surfaces, Direction Fields) and 5 utilities -- Scratch Pad, Series/Product Generator, Constant Editor, Function Editor and File Manager. The Series/Product Generator, Function Editor, and Constant Editor are resident, i.e., accessible from every one of the 50 programs. The Editor allows users to save up to 10 functions and to alter these; in particular, a function together with its first seven derivatives may be saved using the Editor utility.

MicroCalc takes a minimum core memory of 512K, assuming that no software is resident when MicroCalc is started except DOS, which must be version 2.0 or higher.

MicroCalc is not an example of computer-aided-instruction. Rather, its objectives are:

-To provide a tool for calculus experiments, thereby providing a laboratory environment for the teaching of calculus...
-To remove the drudgery of calculations...
-To provide a tool for checking the results of hand calculations (cf. MicroCalc Manual).

On the last point, MicroCalc will not teach you to differentiate , but you can use it to double-check answers worked out by hand.

MicroCalc has beautiful graphics. Students in Calculus II particularly enjoy its solids of revolution. Other advantages include its user-friendliness, the portability of up to ten functions from one program to another, good documentation (each program is preceded by a brief description of its purpose, its required input, and its output), and the fact it seldom crashes. Students have to specify both the domain and the (approximate) range for any function they want to graph, which reinforces those ideas while forcing them to think about scaling, round-off error (especially as this arises when using the "zoom-out" option), and the importance of getting more than one view of a function (i.e., trying several domains for the function), an idea discussed at some length by Demana and Waits in [4]. Incidentally, the "zoom-in" feature convinces students that every curve -- looked at close up -- looks almost like a line, and it is an easy jump from this to the notion of linear approximation, something also noted in [3].

All software has disadvantages which, of course, can be turned to advantage by precipitating a discussion of what went wrong. In earlier versions of MicroCalc tan x on $[-2\pi,2\pi]$ was graphed as though it were continuous. There was no option for graphing piecewise defined functions or for computing limits at infinity. Also, every base raised to a power was converted into the product of a logarithm times an exponential, resulting in incomplete graphs (e.g., $f(x) = x ^(-1/3)$ was only graphed for $x > 0$), a problem noted by L. Carl Leinbach in his review of MicroCalc 3.01 in [7]. These difficulties sparked some interesting discus-

sions. It should be noted, however, that since these difficulties have been noted, the developer of MicroCalc, Harley Flanders, has corrected them and they no longer occur in MicroCalc 5.1. For example, rationals are now treated as a separate data type, so functions like $x^(-1/3)$ are graphed for all real x.

Other difficulties have occurred when students are interpreting the output of a program. For example, consider a simple differentiation problem like this: If

$$y = F(x) = sqrt(2)*x^(3/2) - sqrt(3)*x + sqrt(6)$$

then the differentiation module gives this output:

F(x) = 1.41421 3562(x^1.5) - 1.72305 0808x + 2.44948 9743
dF/dx = 2.12132 0344(x^0.5) - 1.73205 0808

In addition, MicroCalc has no symbolic integration (O.K., you can get around this by having MicroCalc differentiate your answer). Finally, we mention one other inconvenience. With the TANGENT module, the user automatically gets the graph of the original function and the equation for the tangent line at the point which the user specified. However, in theSOLIDS OF REVOLUTION module only the graph of the solid is given. To get its volume, it is necessary to enter the APPROXIMATE INTEGRATION module.

Does MicroCalc 5.1 fulfill the objectives of its author? We think it does. It complements the lecture portion of a course by providing a laboratory tool for experimentation. Is it "interactive?" Jon Barwise in a recent article in the Notices states:

A successful piece of courseware commands the student's active and sustained attention. It is not enough to show. The student has to get involved by doing things. The program should provide a setting for a constant, dynamic, two-way interaction with the material [1].

Measured against this standard, MicroCalc is not truly interactive. As noted earlier, user involvement is limited to choosing options and entering functions and numbers.

Cooperative Group Learning: Paper-and Pencil Labs

On Sunday, March 12, 1989, the Boston Sunday Globe ran an article entitled: "Educators say those who learn together learn better [2]." On Wednesday evening, March 15, 1989, ABC Evening News carried a story about cooperative group learning, asserting: "This is the way the world runs." Both media publicized a philosophy of education that is only just beginning to catch on in the U.S. (particularly at the elementary school level), although research on it is at least fifteen years old (cf.[6]). The assumption underlying this model is that children learn better when they cooperate in their learning rather than compete for grades, as in the traditional classroom setting. Here, the old motto: "I swim, you sink; you swim, I sink" gets replaced by "We sink or swim together." Advocates claim there is evidence which shows that children learn more and retain longer when they learn cooperatively rather than competitively, and they are more positive about school and each other, regardless of differences in ability, ethnic background, or handicaps (2]). The obvious question, then: Would cooperative learning be appropriate at the college level? The Globe article points to the research of Richard J. Light, professor at the Harvard School of Education. He reports that when questionnaires circulated by the Harvard Assessment Seminar among college students and college graduates asked what the most serious omission in their education was,

> Many named the failure to learn to work cooperatively in groups. Ninety-one percent of students and alumni called group learning a very important skill, but only sixteen percent said that their college had helped them to do it [2].

Some time ago, we became aware of the cooperative group learning model through our colleagues in the Simmons College Department of Education and Human Services. During 1987-88, the year before the calculus lab got started, we began experimenting in a limited way with this method. We first tried it in Calculus I with maximum/minimum applications. Since word problems are the bane of Calculus I students, we thought that by having students work together in small brain-storming sessions they might be less anxious and do better than if they were left to work alone. The results were mixed, partly due to our inexperience with this approach, but it seemed worthwhile to try it on a larger scale.

The organizational details are these. A problem set is given to each student in the class. Students are assigned to heterogeneous (as measured by test grades) groups of four to six students. Each group chooses a leader whose job is to make sure that every member of the group shares her ideas on each problem with every other member. After the group has reached a consensus on a solution, she also must check and make sure that every group member understands how to solve each problem. At the end of a session, each group turns in one paper, typically the one belonging to the

group leader, and every member of the group receives the grade earned by the team paper. As an added feature, we also have each group designate a recorder, i.e., one who keeps track of the first attempts at the problems and why these got shot down. The point here, of course, is diagnostic -- we want students to analyze their mistakes in order not to repeat them. The role of the teacher is to monitor, intervene and provide necessary skills when they are needed. When studying one-sided limits, for example, a natural function to use is the greatest integer function, $F(x) = [x]$. Although the handout carefully defined this function and evaluated it for several values of x, every member of one team graphed the function incorrectly. In their table of values, the only nonintegers that students considered were $\pm 1/2, \pm 3/2, \pm 5/2$, etc. The resulting graph had no discontinuities!

We call the cooperative group learning labs "paper-and-pencil" labs because the computer is not used during these sessions. In fact, we often use these labs to study functions that MicroCalc cannot graph, like the greatest integer function and the Dirichlet function (i.e., the function that assigns all rationals the value 1 and all irrationals the value 0), or to give further practice with concepts discussed in lecture. For example, in the third lab of Calculus I, the subject was limits. One of the problems given was to find

$$\lim_{x \to 1} \frac{x^5 - 1}{x - 1}.$$ It became clear that some students remem-

bered synthetic division, others remembered polynomial division, and still others thought the answer was 1 (since "substituting 1 for x yields 0/0 = 1" (sic)).

Whether they are used for practice or to cover nonstandard examples, these sessions can enrich a calculus course. One immediate benefit from cooperative group learning is that students have to "talk calculus," and we are convinced that when a student can articulate an idea, it's hers. Perhaps the real value of this approach is intangible: When students are responsible for each other's learning, this increases their self-confidence, boosts morale, and produces an *esprit de corps* among team members. Also, the moral suasion of one's peers (e.g., "I know synthetic division. Do you? Should you?") often carries more weight than the admonitions of a teacher.

Conclusions and Questions

We are all beginners in exploring ways to teach that effectively use the new technologies, including hand-held calculators and personal computers. It is likely that the key to success will be to balance the traditional lecture/exams approach to teaching with newer approaches. Microlabs and cooperative group learning techniques are just two ways to complement a calculus course. Each has its strengths: Computer graphics make the geometry of calculus ideas vivid and strengthen intuition as well as deepen insight (e.g., they can reinforce the definite integral as a limit of sums). Computers also relieve students of time-consuming numerical computations and double-check work done by hand. Paper-and-pencil labs provide a supportive context in which to practice new ideas as well as bone up on old ones. They also give students the opportunity to verbalize mathematics, which perhaps helps them learn more and retain what they have learned longer.

Our limited experience with these new approaches leads us to believe they have improved student performance, as measured on exams. But we end with two questions:

·Is the improvement we see due to calculus labs or to increased contact hours with our students?

·Do computer graphics really deepen understanding or do students feel they do not know what is going on unless they can do it for themselves?

References

1. Jon Barwise and John Etchemendy, "Creating Courseware," **Notices of the American Mathematical Society** 36, no.1 (1989) 32-40.

2. **The Boston Sunday Globe,** March 12, 1989, p.B19 & p.B23.

3. Franklin Demana and Bert Waits, "Problem Solving Using Microcomputers," **The College Mathematics Journal,** vol. 18 (1987) 236-241.

4. Franklin Demana and Bert Waits, "Pitfalls in Graphical Computation, or Why a Single Graph Isn't Enough," **The College Mathematics Journal,** vol. 19 (1988) 177-183.

5. *Everybody Counts: A Report to the Nation on the Future of Mathematics Education (Summary),* **Notices of the American Mathematical Society,** vol. 36(1989)227-236.

6. David W. Johnson and Roger T. Johnson, **Learning Together and Alone: Cooperation, Competition and Individualization,** Prentice Hall, Englewood Cliffs, N.J. 1975.

7. L. Carl Leinbach, "MicroCalc 3.01," **The College Mathematics Journal,** vol.19 (1988) 367-368.

8. Lynn Steen, "Mathematics for a New Century," **Notices of the American Mathematical Society,** vol. 36 (1989), 133-138.

9. Paul Zorn, "*Mathematica* in Undergraduate Mathematics," **Notices of the American Mathematical Society,** vol.35 (1988) 1347-1349.

Mathematics Computer Laboratories
for Education and Research

Ed Dubinsky
Purdue University

I have been involved in setting up and running a number of different kinds of computer labs at a number of universities. In 1983 at Clarkson University, I set up a lab with IBM terminals connected to a mainframe computer. This was for a summer program to retrain college,e mathematics teachers to teach computer science. The following year, this facility was converted to a stand-alone lab with 30 Dec Pro 350 microcomputers. This became a permanent facility at Clarkson and was used, under my direction, for the summer program and for teaching and research during the academic year. In 1985 I set up another terminal lab (this time it was Unix based) in the mathematics department at University of California, Berkeley. It was a temporary arrangement for the purpose of teaching a single course. In 1988 I set up a microcomputer lab consisting of 20 Macintosh SE Computers at Purdue University. In fall 1989, the lab was doubled in size. The Macintosh lab at Purdue is intended to be a permanent facility, mainly for educational use. It is open for about 100 hours per week and is used extensively by a number of departments here. The mathematics department has exclusive use of the facility for about 60% of the time that it is open and I am in charge of that operation. Most of its use by the mathematics department is for our experimental calculus course which we describe in Section 2 of this volume.

In addition to setting up and running computer labs, I have advised other universities in the establishment of such facilities.

In my experiences with setting up computer labs for mathematics, I have found that there are a number of conditions which must be met as part of constructing a useful and used facility. In particular there are two requirements that I think are absolutely essential for success. First, adequate hardware must be available or else there must be funds to purchase it. Second, there should be a dedicated individual who is very interested in constructing the facility and is willing to be intensely involved in all aspects of planning, implementation and continuing operations.

Once these two prerequisites have been satisfied, the other conditions are generally within the realm of possibility to find or obtain at most academic institutions. This is one of the jobs of the dedicated individual whom I will refer to as the *organizer* of the facility.

After hardware and an organizer, the necessary conditions include: physical space for the laboratory, technical staff, operating personnel, software, courseware, funds (for start-up and continuing operations), documentation, teachers who are ready to use the facility for their courses, and researchers who have projects that involve computers.

In the remainder of this report, I will describe these conditions separately. There are many decisions to make and the appropriate choices are not always clear cut. A major factor should be the use that is intended for the lab. Therefore, there needs to be, in the early stages, a great deal of thinking about what sort of activities are expected, the number of people who will use the lab and the number of hours per week that it will be open.

The organizer must be experienced with, or willing to learn about, all aspects of computer labs. He or she must have at least some familiarity with a number of computer systems and some of the software that can be run on them. It is also important for this individual to be knowledgeable in at least some of the areas of education and/or research in which the lab is to be used. The best situation is when the organizer is a faculty member who intends to be a major user of the facility.

The organizer

This person should have a strong personal desire to establish a computer laboratory. He or she must be ready to launch a major effort of persuasion and other forms of influencing people in order to meet the conditions necessary to establish a useful and used facility. Other members of the institution will not only have to be convinced of many things, but educational efforts regarding the potential use of a computer lab will be necessary.

Often, things will go wrong. The computers will not arrive on time (perhaps the order was lost), the tables are bigger than necessary and they will not all fit in the room, no one knows how to load the network software, the room does not have a pencil sharpener, or the telephone rings too loudly. All of these "glitches", major and minor, are a threat to the success of the facility. It is the job of the organizer to be prepared to deal with any unanticipated circumstance. Therefore, he or she must be flexible, resourceful, and most of all, able to function in a crisis atmosphere.

The hardware

Select the hardware as early as possible. There is no formula for picking the particular type of computer. This and other decisions must be based on a balance between needs and intended use on the one hand, and availability of funds on the other. One must decide on the type and number of computers, type and number of printers, additional hardware for networking, and a certain amount of peripheral equipment such as wiring, connectors, tables, chairs, etc.

When choosing the type of computer, one must consider, in addition to cost, also reliability (and possibility of repair or replacement), capabilities, available software, and personal preferences of the people who will be using the facility. The main choices for computers are: Macintosh, IBM PC, PC compatible, and terminals connected to a mainframe. There are a number of advantages and disadvantages for each alternative. The people who will be using the facility should have as much involvement as possible early in the decision-making process.

When thinking about the number of computers, the form of usage is an important consideration. For educational purposes, many people believe that two students to a computer is optimal. This turns out not to work for everyone and so a ratio of 1.5 people per computer is a reasonable estimate. For research, however, one person per computer is usually necessary.

The present facility at Purdue contains 40 Macintosh SE computers plus peripheral equipment in a room that can hold about 60 users plus support personnel. An important feature of computer equipment is that costs tend to decrease with time. This is because almost all of the expense for the manufacturer is in development. The production cost is quite low for most items. On the other hand, newer and more expensive versions of products are constantly appearing. A balance must be struck between buying the latest models and waiting for the prices to go down.

Acquiring the equipment is one problem. Installing it properly, maintaining it and repairing or replacing broken items is another. A very skilled and experienced technical staff is a major ingredient in establishing and continuing a smooth operation. Maintenance contracts for computers can be very expensive. An alternative is to purchase more equipment than is needed so that replacement can be delayed. Another possibility is for the institution to become an official maintenance agency for the manufacturers of the main equipment that is used. This is what is done at Purdue but it is only possible for a very large organization. An important consideration in choosing between alternatives is that (unlike the situation a few decades ago with vacuum tube machines) computer breakdowns are relatively rare. If care is taken not to abuse the machines in the operation of the lab, it is reasonable to assume that one will not have to pay for replacements very often.

Funds

Funds are, of course, essential. In addition to hardware and software necessary to start the lab, there is a certain amount of expense involved in the continuing operation of the facility. I have already dealt with hardware costs and the cost of software really depends entirely on the use of the facility. Specific operating costs will also depend on local conditions, but it may be useful to list some items that should appear on a budget.

- •Supplies (paper, disks, printer cartridges, telephones, etc.)
- • Maintenance of computer equipment
- • Replacement of obsolete computer equipment
- • Operating staff

The organizer should prepare a complete estimated budget for all costs of establishing the facility and for at least the first year or two of operations. Obviously, determination of availability of funds is one of the earliest concerns in the entire operation. Most likely there will have to be some fund raising activity and this will be another responsibility of the organizer. It is generally quite difficult to get external sources to support continuing operation of the facility, but for establishing it there are many possibilities for various kinds of financial support. The terminals used in Clarkson's 1983 lab were loaned by IBM and the Pro 350 computers installed the following year were donated by the Digital Equipment Corporation (DEC). The Berkeley lab in 1985 used existing facilities. The equipment for the Macintosh lab at Purdue was obtained through an NSF grant with matching funds from Purdue. Within the University the cost was distributed through several departments including the computation center. In all of my experiences, the ongoing costs of operation were borne by the institution.

Space

Space can be a serious problem in an institution which is crowded. The organizer should try to obtain an early allocation of a specific area for the lab, although my experience is that the imminent arrival of equipment is one of the more effective aids in getting an administrative decision. It is essential that there be a separate room dedicated to the computer lab. In addition to computer equipment it should have a desk (for the person who is supervising the room), file cabinet, blackboards and storage closets.

The room should be large enough not only for the machines that it will hold, but also for the maximum number of people that are expected to be in the room at one time. The Purdue facility houses 40 computers, three laser printers and two file servers. It is approximately 27 *ft*. x 41 *ft*. and has two large structural columns (totalling about 30 *sq. ft.)* so a little more than 1000 *sq. ft.* is available. The maximum number of people in the room at one time is not expected to exceed 65.

The configuration of computers in the room is important both from the point of view of having the maximum number of machines (with sufficient room to move around) and to provide a reasonable working atmosphere. My preference is to arrange the machines is such a way as to suggest that people work together in small groups, rather than as isolated individuals. This makes sense pedagogically and also is close to what the student is likely to experience after leaving the University. We have found that a cluster arrangement with clusters of varying size satisfies both considerations. A sketch of the configuration used in the Purdue facility is given on page 58 of this volume.

Usage

The use of the facility is an area in which the organizer again must begin to plan early. The best situation is to have at least one person (possibly the organizer) or group ready to use the lab for a significant portion of the time available. Of course every effort should be made to inform all faculty members of the lab's existence and capabilities. It can be useful to organize demonstrations and even tutorial sessions for the faculty on how to use the facility.

Provision should be made for scheduling use of the lab. There are some situations in which the planning has been so poor, that the facility is underused or not even used at all. In many cases, however, after an initial period, the demand

exceeds the availability of time and space. Priority systems must be established and an overall equitable scheme worked out and made public.

Faculty should be given an opportunity to reserve the lab long in advance. It must be remembered, however, that not everyone is efficient at long term planning. Some people will reserve time and not use it. Others will realize only at the last minute that they need the facility. Therefore, it is important to have some flexibility built into the scheduling scheme.

One useful feature of the Purdue facility is its size. Most uses of this lab involve either individuals or classes ranging from 15 to 40 students. This means that there is always plenty of room for more than one activity at the same time. On occasion we will schedule two classes in the lab simultaneously. Otherwise, whenever a single class is in the lab, the unused machines are available for open use.

In general, a mix of scheduled use and free availability of machines seems best.

Software

Software largely depends on the use that will be made of the computers. The simplest procedure is to have each individual be responsible for providing the software that is to be used for her or his class or research as the case may be.

There is, however, a certain amount of software that should be provided by the facility. Networking software is extremely important. This will permit the lab to hold large files (including individual software systems) centrally in a very high capacity storage facility (e.g., a disk). These files can then be quickly loaded on individual machines with much smaller storage facilities when needed. Another use of networking software is to permit a user to send files to the printer from an individual machine, or even to communicate with other users. A faculty member can distribute instructions to students and collect assignments electronically with such software.

There are also a number of general software systems that could be provided for everyone. These include editors, word processors, data base managers, and mail facilities. From a global point of view, it is usually much less expensive to purchase a site license that will permit a piece of software to be used on every machine in the facility, than to have each individual user purchase it separately.

There is a fair amount of public domain software that can be very useful and is free. The computer lab should make information about such software available to all users. On the other hand, there is the problem of "viruses" which can enter the system on software that is not carefully checked. The lab staff needs to become familiar with "viruses", how to prevent them, how to detect them and how to remove them.

There is a wide variety of software that can be and is being used for Calculus. These are discussed in some detail throughout this book. Our particular choices for our experimental calculus course are described in our article beginning on page 47 of this volume.

Documentation

The documentation of software provided by the facility will be necessary for most users. Many people will not be familiar with the basic operations of the computers in the lab and there should be some written material to help them. In addition, there will be various protocols of operation for the lab (signing on and off, removing temporary files, should machines be turned off or left running when finished, how to print a file, etc.) that should be readily available for all users.

Great care should be taken with preparation of documentation because this can have a strong effect on the usefulness of the facility. Many people need just a little of the right kind of information presented in the right way to get started before they can become very productive with computers. On the other hand, it can be difficult for someone who is very familiar with a computer operation to understand the needs of the novice. There are specialists who are experienced in preparing the kind of documentation that is needed and their services should be used.

Operations

It is a long, arduous process to plan for a computer laboratory, obtain funding, acquire equipment, set everything up and establish a functional operation. Once this is accomplished, there is a great tendency to relax and assume that the lab will more or less run itself from now on. This is a mistake. Unless careful planning has taken place and continues, there is a great danger that the lab will not be used or that it will not function, or function inefficiently with breakdowns, delays, confusion and frustration.

Here are some specific concerns for the continued operation of the facility.

1. Just as establishing the facility requires an overall organizer, its continued operation should be directed by a single, capable individual. This could be the same person as the original organizer or, if desirable, this is a reasonable point at which to turn the work over to someone else. Obviously, the director must have administrative, organizing and leadership skills. He or she must know something about computers and their use in education and research. If the overall operation is efficiently organized and there is a competent staff, this does not need to be a full time job. It can be something like a major committee assignment for a faculty member.

2. However one decides to handle the maintenance problem (see **Hardware** above), there needs to be access to technical people. Many of the breakdowns that occur are trivial (e.g. a wire is loose) and can be fixed instantly by a knowledgeable person.

Another point is that enhancements of equipment are always becoming available and will need to be installed. The size of the technical staff required will vary with the size of the facility and the institution, but it probably does not need to be very large. Often a single individual who is readily available, but not assigned to the lab full time will suffice.

3. One needs to decide the days and hours of operation of the facility. Of course this varies with many factors and will require initial guesses followed by revision. It is also necessary to provide for different kinds of operation and establish a number of policies. Who will have the right to use the lab? How will priorities be established? When a class is meeting in the lab, will individual users be allowed to use available machines? Assuming that the facility is large enough, will more than one class be held simultaneously in the lab? Will it be permitted for faculty to hold examinations in the lab? A schedule for the operation of the lab must be worked out in advance. It is essential that open lab hours be scheduled. For example, we have found that for calculus, students need to have quite a large amount of lab time, both scheduled and during open hours.

4. It is absolutely essential to have a staff person

in the lab at all hours of its operation to supervise the activity. These individuals will answer questions, load software, keep records, note breakdowns and provide for repairs, advise on policy changes and enforce the existing policy. Depending on size, it may be necessary to have more than one person supervising the lab operation. In the Purdue facility, when it was inaugurated, one person was enough. Now that its size has doubled and faculty are using it more and more, we are finding it necessary to have two staff people present, especially at certain peak hours.

It may also be necessary to have staff working a small amount of hours during times when the facility is closed. This can be used for software development, eliminating bugs in the operation, and general preparations.

My experience has been that undergraduate students are excellent for the operating staff. They will know or quickly learn about the operation of the computers on a level that goes beyond the knowledge of faculty or even technical staff. They enjoy the work, are enthusiastic and reliable. And they learn a great deal. Moreover, they are relatively inexpensive—especially if one can argue that this is part of their education.

5. Finally, an outreach activity should be part of the operation of the facility. It will not work to establish the lab and wait for people to come and use it. Information must be disseminated about what resources are available and how to use them. Presentations can be made and there should even be tutorials on how to use the hardware and software, not only at the very beginning, but as a regular ongoing activity of the facility.

Timing

Again I emphasize the need for long range planning and working to have appropriate things in place at appropriate times. Work on the present Purdue lab began in the Fall of 1987 with the writing of proposals. The first version of the lab was in operation for the beginning of classes in Fall 1988 and the present version was operational a year later. Throughout those two years we obtained funds, purchased equipment, organized space, recruited and trained staff, prepared courses, and continuing in the present, maintained and operated the facility. It has been and still is a complex, long term activity and has required the efforts of a number of very competent individuals working hard to achieve what appears to be a successful operation.

Computer Selection for Iowa's New Computer Labs
K. D. Stroyan
University of Iowa

Introduction

The University of Iowa has successfully run computer lab courses with calculus and linear algebra for over 15 years. These labs are described in another article in this volume. For about 7 years we have been working on "calculus reform" in the form of a special section of calculus we call Accelerated Calculus. We wanted to revise our Accelerated Calculus curriculum in more fundamental ways and include up-to-date scientific computing as part of that revision. We felt that scientific programming languages like MATLAB, MACSYMA, Maple, or Mathematica could replace BASIC in our computing lab with the added benefits of greater computing potential and a serious start on our students' scientific computing education. This goal led our decision making process - we wanted up-to-date computing power and as much ease of use as that would permit. Since we were going to devote a major amount of labor to developing materials, we felt that unit price was a secondary consideration. Low cost old technology would be gone before our development could be completed.

If your computing plans are less ambitious or you feel you need low unit cost to serve more students, there a several other software products running on less expensive machines that you need to consider, such as Calculus T/L, MATHCAD, Derive, ...

Software Selection

I purchased a Mac II computer, Maple, MATLAB and Mathematica last year. We have MACSYMA on a University owned computer. During the summer I made an evaluation of software, especially Maple, MATLAB and Mathematica. I attended the St. Olaf conference on Computer Algebra Systems in Calculus during this period of evaluation where I discussed choices with many other people. My choice was Mathematica with the "Notebook" front end. This combined choice required looking at both software and hardware. Different versions of the software operate differently and I felt some were much less friendly than others. Mathematica was not my immediate first choice, but worked its way up there. I believe Mathematica Notebooks offer the best available combination of well integrated high-power scientific computation and ease of use. Our aim is to make serious use of computing, but to focus on calculus, so ease of use is important. We want graphics, numerics and symbolics, probably in that order of importance. These three aspects of computing are well integrated in Mathematica.

Maple and MATLAB are excellent programs, but are difficult to use compared with Mathematica Notebooks. However, both hardware and software change very fast and need to be re-evaluated right before you purchase equipment.

I don't really like learning programming languages and reading computing manuals for their own sake. I like using the computer to get interesting mathematical results or using it to illustrate mathematical ideas for my classes. I did not work long enough to become an expert in each software package, but only to form an impression of how each package could serve my needs. After 5 months of intensive use, I am pleased with the choice, but am still not an expert in Mathematica and have encountered a few mathematical problems that I could solve with Maple, but not Mathematica.

In the final analysis, only Macintosh and NeXT received consideration because of the Notebooks, but this fact only emerged as part of the hardware selection.

Hardware Selection

We let hardware bids in July and ordered equipment in September. We received quotes from IBM, Apple, Hewlett Packard, Silicon Graphics, Apollo, etc. and finally decided on a NeXT network of 10 eight megabyte stations running two 660 Mb magnetic hard drives from two file servers (with one laser printer.) The system cost $75,000. Things have changed already, even for NeXT who now offers an 'HD 40 only' machine for $5,000. We could have purchased 3 more workstations... The price of memory chips has dropped... Total price is the important figure and different products have to be configured differently to accomplish roughly the same task.

You often have to discuss the bid with a vendor so that you both understand what is desired. Several vendors were very helpful to us. You probably need to discuss your needs and priorities with them in detail and run the versions of your preferred software on their machines as we did.

Equipment was shipped 3 Oct. '89 (version 1.0, only 3 days late) and up and running within a week. NeXT offers Macintosh-like ease of use and Mathematica Notebooks, but runs a genuine UNIX network and has a host of bundled software besides Mathematica (including developers tools). We decided that a UNIX network offered us more 'future.' One of the things that was very hard to estimate in selecting hardware was the amount of system programmer labor that each vendor's equipment would require. Fortunately, as UNIX systems go, NeXT is quite easy to set up and use. We hired a computer science graduate student UNIX expert to help set up our network and try to optimize the two hard drives. He had a fair amount of labor for the first few weeks, but barely looks in on us now. I can manage the user accounts with NeXT menus without knowing any UNIX. I would not want to manage separate disk drives on the individual machines compared to loading courseware on the master account and having student access through the network. Even with the relatively low-tech current use, we are pleased with the choice of a NeXT network. In the future we will make use of E-mail and other network functions. We do not know whether or not we will link our network to the University computer center.

We could have afforded color monitors if we had chosen Macintosh computers instead of NeXT. (We wouldn't have gotten a UNIX network nor as much additional software.) I was not sure if sacrificing color would matter or not. The first day I thought it was a loss, but when I went back to my color Mac II monitor a few days later felt that the higher resolution and bigger size of the NeXT screen are a better choice than the smaller color screen.

Satisfaction

We are very satisfied with the choice of both NeXT hardware and network and with Mathematica. NeXT/Mathematica still has some bugs, but they are relatively minor. It is very easy to start using Mathematica with Wolfram's *Mathematica* book, but difficult to use that documentation to write more complicated procedures. Perhaps the new book, *Programming in Mathematica,* by Maeder will help. NeXT has a relatively friendly front end on their version of UNIX which makes a very functional easily maintained student computing network. One of our magnetic hard drives failed during the semester, but once we diagnosed the problem, NeXT shipped us a replacement overnight. Their service and support has been excellent.

Funding

Forty percent of our NeXT network was purchased with NSF "ILI" funds. That is a good program, you should apply if you have serious plans to use computing in calculus and your school is willing to share at least 50% of the cost of student computing equipment. Contact the National Science Foundation, Program for Instrumentation and Laboratory Improvement, Washington, D.C. 20550.

Part II
Examples of Established
Calculus Courses with Laboratories

Constructing Calculus Concepts:
Cooperation in a Computer Laboratory

Ed Dubinsky and Keith Schwingendorf
Purdue University

Although an important part of any mathematical endeavour is individual thinking, the construction of mathematical ideas does not really flourish in solitude. From the superstar research mathematicians who invariably take off from the results of others ("stand on the shoulders of giants" in the words of Isaac Newton) and feel a deep seated need to communicate (i.e., publish) their results, to the struggling students who take comfort in the realization that mathematics is just as challenging for everyone else as it is for them, there is every indication that mathematics is a social activity.

We believe that mathematical knowledge, in education as well as research, grows as a result of people constructing their own mathematical concepts and synthesizing their individual thoughts with the ideas of others. We will try to show in this paper how a computer laboratory can be a very good teaching environment in which to get students constructing mathematical ideas by working together as a team.

Our work is based on a developing theory of learning so we begin with a sketch of this theory. We consider its relation to other points of view and give an indication of the role of computer labs in fostering the construction and cooperation that this theory calls for. In Section 2 we go into some detail on the particular programming language we use in our laboratory and give some examples of how it can help get students to make mental constructions appropriate for learning mathematics. In Section 3 we describe our particular laboratory environment and all of the nuts and bolts that go into this particular implementation of our approach to helping students learn calculus. In Section 4 we consider one major mathematical topic — the Fundamental Theorem of Calculus — and give a full explanation of our treatment, including some indication of its effectiveness. Finally we offer some overall evaluations of our approach from various sources.

1. Theoretical basis for construction and cooperation in computer labs

Construction

We start with the observation that how you teach is determined by what you believe about how people learn and about the nature of the subject you would like to help your students learn. We are convinced that the ineffectiveness of mathematics education at all levels is largely due to the teaching methods chosen by mathematics faculty based on traditional beliefs about learning and about mathematics.

We reject these beliefs and the choices they imply, and we propose below alternatives based on a constructivist[1] theory of knowledge and its development in an individual. It turns out that it is possible to design instruction in mathematics that is consistent with our theories and the beliefs to which they lead us. Moreover, our experience has been that a computer lab can be used to implement such instructional treatments, and the results in terms of student learning can be refreshingly positive.

Our description, in the following pages, of our theoretical analysis and the corresponding instructional treatments is necessarily brief. A more complete discussion, together with references to the literature that reports on our results, is given in [6,7,8].

Beliefs and choices — what we reject

We reject the idea that people learn mathematics spontaneously by listening or watching while it is being presented. We do not feel that mathematics is learned effectively by working with many examples and trying to extract their essential features. Nor do we believe that having students solve problems in other fields is the only way to motivate them to study mathematics.

[1] We refer here to the epistemological constructivism of J. Piaget and not the mathematical constructivism of L.E.J. Brouwer. See [13] for an introduction to the former and its distinction from the latter.

Regarding the nature of mathematics in general or Calculus in particular, we do not believe that it is a body of knowledge existing outside ourselves and waiting to be discovered. Moreover, we insist that mathematics is more than a collection of techniques for solving standard problems or for solving "real-world problems."

Because of these beliefs, we do not choose a teaching methodology that is restricted to giving lectures, showing applications, having students work on set problems, and testing their performance on examinations.

Admittedly, these rejections are sharply at variance with "conventional wisdom" but, after all, we are in a situation in which conventional methods (of teaching mathematics) do not appear to work. Perhaps it is time to reconsider our basic beliefs about the nature of mathematics and how it is learned.

Our beliefs and choices — what is mathematical knowledge

We believe that mathematics is a set of ideas created by individual and collective thought over many hundreds of years. But its nature is dynamic, not static. Mathematical knowledge is not something you *have* but rather something that you (might) *do*. Let us offer a definition as a basis for discussion.

> *A person's mathematical knowledge is her or his tendency to respond to certain kinds of perceived problem situations by constructing, reconstructing and organizing mental processes and objects to use in dealing with the situations.*

This is a very general statement that might apply to many disciplines, but when we consider some of the details of the responses we are talking about, it will become clear how our definition applies to mathematics.

All of us who teach are familiar with the distinction between the problem which the teacher sets before the student and the student's perception of that problem. We have every opportunity to make the situation as clear as we can, but eventually the time comes when the student decides what he or she thinks the problem is and it is that perception that determines the student's response. We might think, for example, that the "problem" is to understand that the derivative of a particular function is a useful linear approximation to it near a point. But our students will often conclude that the "problem" is to find a formula for the derivative of that function. It is not very easy to get

them to revise their interpretations so it is very important to be concerned with how our students initially perceive the problems we set before them.

Another familiar experience is the inconsistency of mathematical understanding within a single person. Like the athlete, the mathematics student is sometimes able to function at a very high level of sophistication, but on other days, with similar or even identical problems, he or she is much less successful. Thus mathematical understanding is not about what a person is surely able to accomplish, but what he or she has a tendency to do.

In connection with this, we can think of a person faced with a problem situation, not as bringing forth immutable pieces of mathematical knowledge, but rather as reconstructing what he or she has previously constructed in order to deal with the present situation. It is this reconstruction which is at the root of this inconsistency. Sometimes it produces tools that are less powerful than those which the person has previously used. On other occasions, stimulated by the special difficulty of the actual problem, it can produce something that is more powerful, more sophisticated and more effective. In the latter case, we can say that growth of the person's mathematical knowledge has taken place.

If what we are saying has any relation at all to what really goes on with students, then it has some very disturbing implications for testing, indeed for our overall system of evaluation. If inconsistency is a normal aspect of a person's intellectual performance, if it is actually an important part of intellectual growth, then it is not clear what is being measured by an examination which is a one-shot, do or die situation. If we decide that learning mathematics is both constructive and cooperative, then in addition to rethinking how we teach, we must also reconsider the means by which we evaluate the results of that teaching.

Our beliefs and choices — what is learning mathematics and how computer labs can help

The part of our definition that has to do specifically with learning mathematics is the construction and reconstruction of mental processes and objects. It is in this area that we can become specific to mathematics, and it is here that we encounter the mental activities that are so amenable to group work in a computer laboratory.

Roughly speaking, processes are built up out of actions on objects and ultimately converted into new objects which are used for new actions that are converted into new processes and so on, as a person's mathematical knowledge spirals up to higher and higher levels of sophistication. Let's look at some details.

Numbers are *objects*. *Actions* such as arithmetic calculations can be performed on them. When an action such as adding three to a number is repeated with different numbers, there is a tendency to become aware of and *interiorize* this action into a *process*, x + 3. This leads to algebra. Processes can also be constructed by composing two processes, say adding three and squaring, which gives $(x + 3)^2$, or by reversing a given process, say adding three, to obtain the process of subtracting three or x - 3. A single process such as adding 3 can be *encapsulated* to become an object, in this case, the expressiom x + 3. Now the standard algebraic manipulations with expressions can be seen as actions on these new objects. Below is a schematic display of these different kinds of constructions.

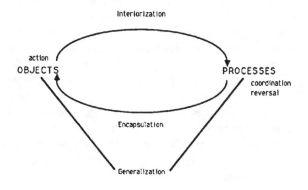

Construction of objects and processes

Here we can introduce a major connection between our use of the computer and this theoretical analysis. It turns out that each of these mental steps: actions on objects, interiorization of actions into processes, composing processes, inverting processes and encapsulating processes into objects can be represented in terms of computer tasks. Moreover, when students are set to performing these tasks, we believe that they tend to perform the corresponding mental activites and begin making constructions. This is our basic justification for using a lab in connection with calculus instruction.

Of course the computer tasks that are reasonable to perform depend on the particular computer system and how it is used. For example, the mathematical notation $(x+3)^2$ can represent two kinds of mental constructions. One is the process obtained by composing the "add 3" process with the "square" process. The other construction is the object obtained by squaring the object x + 3. Both representations

are important, although the former carries greater mathematical content, whereas the latter is more formal and can lead to performing manipulations quickly and efficiently, but perhaps without much understanding of their content.

In any case, there are computer systems appropriate to either interpetation. Computer Algebra Systems emphasize the object view of expressions like $(x + 3)^2$. On the other hand, having students write procedures that implement the functions given by such expressions emphasizes the nature of a function as a process.

There are, of course, much more mathematically sophisticated situations. In Calculus, most of the examples in which mathematical concepts are interpreted in terms of objects and processes have to do with functions. The input/output point of view treats a function as a process, as does the interpretation that points on a graph come from evaluating the function at the *x*-coordinate and taking the answer for the y-coordinate. Lab work that has students implementing functions by writing computer procedures tends to get the students interiorizing the processes of those functions.

On the other hand, considering differentiation as an operation that transforms one function into another, thinking about iterating the composition of a function with itself, or seeing that the solution of a differential equation must be a whole function (rather than a number or even a structured collection of numbers) requires the interpretation of a function as an object.

In order to model such an interpretation, a computer language should treat functions as what computer scientists call *first-class objects*. This means that functions are data like numbers and can be operated on, collected together in sets or sequences, accepted as inputs to other functions and can be the resulting object output by the execution of a procedure. Research [14] suggests that one way of helping students construct mathematical objects in their minds is to let them perform operations on what is to become an object.

Sometimes it is necessary to conceptualize a function simultaneously as a process and as an object. To understand the notation f'(3), for example, requires the idea of transforming a function to its derivative (object) and evaluating at 3 (process). We suggest that this can only happen if the object conception of the function was constructed by encapsulating the process conception corresponding to evaluation. In this case, the subject is able to go back and forth between the two interpretations. Thus the computer language should not only allow functions as processes (procedures) and functions as objects (first-class objects), but also its syntax should allow easy passage between the two interpretations.

Functions are not the only context in which this theory operates. Mathematical induction is an important tool in Calculus which is often (wrongly) assumed to be well understood by the students. Predicate Calculus is critical for understanding the ε - δ formulation of the limit, and it seems likely that the difficulty students have with limits and related concepts is at least partly due to their poor comprehension of quantification. For details about these and other relations between the theory described here and various mathematical topics, we refer to [4, 5, 6, 7, 8, 9, 11].

Other ways computer labs help mental construction

In addition to the specific construction of mental objects and processes, several other mental constructions can be specifically benefitted by a calculus laboratory. One is the issue of *generalization*. It is important to distinguish between generalization and reconstruction. Both involve using previous mental schemas to deal with new situations. In the case of generalization the schema that is used in the new situation is completely unchanged. This works sometimes, but it often leads to trouble.

Mathematics education abounds with examples of false generalization. Looking a little bit at computational understanding that precedes Calculus, consider the "classic" error of writing

$$\sqrt{x+y} = \sqrt{x} + \sqrt{y}$$

This is not a random mistake. A student who thinks this relation holds may have a strong schema for the distributive rule. So strong that it is applied whether or not it is appropriate. A similar example is the error of thinking that the derivative of $\sin^2 x$ is $2 \sin x$. This could come from a misguided application of the power rule.

Reconstruction on the other hand, involves taking a rule, or schema apart to see what makes it tick and putting it together in the new situation — not just to get an answer, but to make some kind of sense. When the schema is related to something which the student has constructed on a computer, such as a procedure, it is more concrete and there a greater possibility that the action, being explicit for the student, can change.

Returning to the general definition of mathematical knowledge, we note our assertion that a person constructs mathe-

matics in order to deal with perceived problem situations. This means that the student must know what the problems are. One important use of a computer laboratory is to give students experience with a number of problem situations. Even when students have not yet developed the mathematics necessary to solve these problems, computer tools can help them to see clearly what the problem is. Often, for example in the case of approximating derivatives at a point or areas under a curve, the computer will allow students to work with a problem and even get a "solution" that might be a satisfactory approximation. Then the mathematics becomes a methodology for organizing and formalizing activities which are already in the students' mental repertoires. This can provide a motivating factor that we suggest is at least as important as the fascination we are all supposed to have with "real world" problems.

Finally, we can mention the issue of timing. The mental constructions that a person makes in learning mathematics amount to building new knowledge based on knowledge that already exists for her or him. There must be objects for processes to transform and processes to encapsulate into objects. Whatever takes place in the classroom, it will not be successful if it is based on incorrect assumptions about what mathematical entities are present in a students' mind. Another purpose of a computer laboratory is to give all students a chance to get in their minds the phenomena that subsequent mathematical discussion will help them explain and understand.

Cooperation

Cooperation as a basis for introducing a laboratory aspect into a Calculus course is, at least for us, a relatively new idea. Indeed our original design did not emphasize group work. It has grown with us as we have become experienced with having our students work in small groups, and as we have come to understand the benefits of this kind of approach. We consider three aspects of cooperation in a laboratory environment: reflection, support systems, and "preparation for life."

Reflection

In the previous section we considered mental activities such as interiorization, encapsulation, coordination, and reversal and indicated in a general way how the use of computer activities can get students performing these mental activities so as to make appropriate constructions. There is another ingredient that is essential in order to interiorize an action to a process or encapsulate a process to an object. That is reflection. According to constructivist theory, both interiorization and encapsulation arise, at least in part, as a result of conscious reflection on the problem situation and methods of dealing with the problem — both successful and unsuccessful.

One way to heighten your awareness of what you are doing, for example when making calculations, is to explain it or teach it. Some of the computer tasks we have students perform are by way of "teaching their actions to a computer," by writing programs.

Another way of getting students to be aware of mathematical relationships is to have them explain them to each other in a group. In addition to establishing formal teams and allowing group submission of assignments, we have tried to set up our labs so that students are encouraged to work together and talk to each other.

A lab can have an atmosphere in which as soon as one student figures something out it spreads throughout the group and then throughout the entire class. It is of course true that sometimes what spreads is more of a "how to" than an understanding. Nevertheless, whenever a student is forced to make an explanation to someone else the act of explanation carries with it a modicum of intellectual growth.

A computer lab allows us to give large assignments with many parts and encourage students to divide the work. Often, the parts are similar enough so that no one loses any value by not doing all of them. Moreover, the collection of several parts can lead to a group result that is not feasible for a single student to achieve. The only thing that is really lost is the repetition.

On another plane, we use the computer lab to enhance the learning of students after they have taken our course (or even other courses). Some of our students become lab assistants who play an essential role in helping new students with minor technical details. With our encouragement, however, the role of the undergraduate lab assistants is expanded to include the role of explaining mathematical ideas to students who may be no more than half a year younger than themselves. This form of cooperation leads to mathematical awareness and reflection, and hence the construction of mathematical concepts, on the part of the lab assistant.

Support systems

Recent research [15] and development projects [3] suggest that students will do much better in mathematics if they have the right kind of human support system. This is not only important for developing self confidence and helping a student deal with the many kinds of frustration that are endemic to doing mathematics. It is also the case that working with a group can help one form and reinforce effective study habits. Many young people need little more in the way of support than acknowledgement from their peers that working hard, studying and doing well in school are reasonable ways to live one's life. Others need more specific assistance such as models for allotment of time and organization of work. In general, most students will do better in mathematics (and other intellectual activities) if they find themselves in an environment where a positive attitude towards studying mathematics is not only accepted but is an admired mode of behavior.

What does all that have to with a lab? Not much more than that a lab can become a place where an atmosphere such as we have been describing is developed and nurtured.

Preparation for life

Finally we offer a simple observation about how mathematics is done and used in "real life": that is, in the workplace. Although we remain convinced that not even research in mathematics at the highest levels is a solitary activity, this point is arguable. But for the kind of relationship to mathematics that most of our students will have after they leave school and work in industry or education or government, there is little question that doing mathematics is a team sport in the workplace.

The first thing anyone should think of doing when stuck is to ask a co-worker. This is not always easy to do nor is it easy to respond to. The questioner needs to learn how to formulate questions. The respondent needs to know how to share information, to answer the question that was asked and not the question one thinks should be asked, or to point out that the wrong question has been asked. All of these are skills that can be learned through teamwork in a computer laboratory. And then, perhaps, our students will not feel so strongly that what they did in school has little to do with what they are supposed to do in work.

2. Fostering mental constructions via programming

We use three different software systems in our Macintosh computer labs — **ISETL**, a mathematical programming language, **MAPLE**, a symbolic computer system, and graphics software. Actually, there are only two distinct systems because the graphics system is already contained in one of the others (**MAPLE** at the moment, but we are switching to the new graphics capabilities of **ISETL** in

Fall, 1990). We keep them separate because there are conceptual differences.

The three software systems correspond neatly to three different ways of thinking about functions. For **MAPLE**, as for all symbolic computer systems, a function is basically an algebraic/trigonometric expression and is an object as such. For **ISETL** a function is a process in that it is implemented by a computer procedure (called func in **ISETL**) that students write, and it is an object in that funcs are first class objects and are treated as data. A graph is also an object with the main feature of being visual as oppposed to the analytic nature of the other two interpretations.

Incidentally, as a technical point, we should mention that, strictly speaking, the statements in the above paragraph are not completely true. It is possible to write procedures in **MAPLE** (and other symbolic computer systems). These procedures have some, but not all, of the properties of first class objects (in particular, the result of applying a function in **MAPLE** cannot be a function). In **ISETL**, procedures have full status as first-class objects and can be returned by other procedures. On the other hand, although it would be possible to implement symbolic manipulation in **ISETL** this would not be very convenient and there are no plans to do so. Finally, both **MAPLE** and **ISETL** have (or shortly will have) graphics capabilities sufficient for calculus.

It may well be that the development of software will be in the direction of constructing a single software system that will combine all features of all three of these systems. We are not sure that would be a good idea (for Calculus). In our course, we have used the differences to help our students develop an awareness of the different interpretations of functions and to reflect on the (very important) activity of using several representations and switching back and forth between these representations within a single problem situation. Now that graphics are coming to **ISETL** and it will be feasible to reduce work with **MAPLE** to a minimum, it is not clear whether this is the best thing to do. We have inaugurated a research project to help us answer that question.

In this section we give some details about how we use each of the three software systems to help students construct mathematical concepts in their minds. First, we consider **MAPLE** which is, in our opinion, most valuable in helping students develop the necessary manipulative skills for doing mathematics. Our use of **MAPLE** is not very different from what most projects are doing and so our discussion is very brief. Next, we explain how we coordinate two systems in our approach to graphics. Finally we have a

long discussion of **ISETL**, its general features and how we use it to help students interiorize actions to construct processes and encapsulate processes to construct objects.

Learning manipulative skills with MAPLE

The laboratory work that we do with the symbolic computer system **MAPLE** alone involves students using the system to simplify complex expressions, find derivatives and antiderivatives, evaluate limits and definite integrals, and solve the standard problems of calculus as well as some new problems that are accessible to them because they have such powerful tools.

This work has considerable merit and students learn the manipulative skills of calculus about as well (and sometimes better) than they do in traditional courses. Since this is about *all* that most students learn in the traditional courses today, we are satisfied that our laboratory course achieves at least as much as the traditional courses. We will also be arguing in this paper that since manipulation is only a small part of what happens in our laboratory course, our students learn much more than do those in traditional courses.

Having said this, we hasten to add that our use of a symbolic computer system and the results we are obtaining do not differ very much from what other people have achieved. For a more detailed description of the effect of symbolic computer systems on students learning manipulative skills in Calculus, see [12].

Doing graphics in a different way

As we indicated above, at the beginning of our project, the only powerful graphics system we had was the one embedded in **MAPLE**. This is a reasonable system, provided that one is content to produce graphs of functions that come from expressions. There are, however, other kinds of functions that are important in Calculus. These include functions defined in parts — that is, given by different exressions in different parts of the domain, functions created by a procedure (such as an approximation to the derivative of a given function), and functions which are sequences.

The mathematical programming language **ISETL** can implement all of these kinds of functions (and more), so we had to devise a means of using **ISETL** for defining a function and **MAPLE** for its graph. This is not very difficult. Suppose, for example, it is desired to sketch a graph of the following function (which results from fitting a curve to data giving the compressibility of a substance as a function of its volume — the different branches correspond to the state of the substance, solid, liquid, gas)

$$Z(V) = \begin{cases} 1 + \dfrac{(V-2)^3(V-6)}{54} & \text{if } 0 \leq V < 6 \\ \dfrac{V^3}{3 \times 10^3} - 0.048V^2 + 1.73V - 7.7 & \text{if } 6 \leq V < 90 \\ \dfrac{4V^{\frac{3}{2}}}{\sqrt[5]{V^3} \, 8 \times 10^5} - \dfrac{1}{V-100} & \text{if } 90 \leq V \end{cases}$$

The first thing that must be done is to implement this expression with an **ISETL** procedure and assign it to a variable name, say Z. This is easy to do and we will see how it looks when we discuss **ISETL** syntax in some detail below.

Next, the student decides on the region of the independent variable over which to sketch the graph, for instance the interval [0,300], and the number of points to sample, say 100 (here the student learns about the trade-off between using many points for greater accuracy and waiting a long time for the graph). The following command is then given in **ISETL**

graph(Z,0,300,100, "data");

The result is to create a file named "data" outside of **ISETL** containing a set of 100 ordered pairs of data points for this function. The student than leaves **ISETL** (this is unnecessary if Multifinder is available), enters **MAPLE** and gives the following two commands,

read data;
sketch(");

The result of the first command is to display the data on the screen and the second causes the graph to be drawn.

This complex use of (what is conceptually) three systems has its advantages and disadvantages. It is cumbersome and does not lend itself to quickly displaying different parts of the same graph. On the other hand, it forces the student to act and to wait, and there is hope that this will enhance awareness of and reflection on how the computer is making this graph. In particular, our approach could improve the student's understanding of the connection between a function and its graph. Hopefully the research project, in which we will compare our approach with approaches using a single software system, will help us decide whether to keep this method or move to something that will be much more convenient when the **ISETL** graphics are ready.

A final point about graphs. As far as we know, every graphics system samples points and uses splines to connect them. This can lead to difficulties when a function has singularities and we have seen examples in which the system will connect dots smoothly across a discontinuity. To avoid dealing with a problem that is, as far as a computer lab for calculus is concerned, purely technical, our sketch tool simply puts the dots on the screen. This is admittedly primitive because the same density of dots is used no matter what the function is doing. It will have to be improved. It does give us some interesting tasks to set before the students, however. We have them put the graphs of several functions on the screen and make a hard copy. Then it is their task to connect the dots. They have to use their understanding of the processes of the functions to distinguish between the several collections of dots.

Fostering mental constructions with ISETL

The syntax of **ISETL** consists of very simple and standard control statements together with a number of expressions that correspond very closely in meaning and notation to fundamental mathematical concepts. Our experience is that students have very little difficulty learning to use the language. Their main efforts are devoted, as they should be, to understanding the ideas behind various mathematical constructs so as to be able to use them in writing **ISETL** programs.

Because of **ISETL**'s clear mathematical structure, we are able to introduce the reader to its basic operations with a minimum of detailed explanation. We do this in the next paragraph. In the following paragraphs we give a number of examples of specific **ISETL** constructs that we use to get students to make several important mental constructions. For interiorization we consider using **ISETL** funcs to express functions with split domains, the **ISETL** tuple for sequences and series, and the func again for constructing a process corresponding to the composition of two functions. For encapsulation, we explain how implementing composition in **ISETL** as a binary operation on two functions that produces a function helps students construct an object conception of function. We also consider the operation of approximate differentiation (for which there is a simple **ISETL** implementation) as having a similar effect.

Generalities about ISETL

ISETL, which stands for *interactive set language* is an interactive, interpreted, high-level programming language that runs on the Macintosh or PC (under MS-DOS). It contains the usual collection of statements common to procedural languages but a richer set of expressions than is usually available. The objects of **ISETL** include integers, floating point numbers, the two boolean values (true and false), character strings, finite and heterogeneous sets and tuples (sequences), funcs (procedures) and maps (sets of ordered pairs).

One uses **ISETL** by entering an expression to which the system responds by evaluating and returning a result. An expression can involve arithmetic operations on numbers (integers or floating point), boolean operations, or operations on character strings. Assignments can be made to variables, and expressions can combine variables and constants. The domain of a variable is determined in context dynamically by the system and there is no need to declare data types, sizes, etc. Many important mathematical operations on these data types are implemented directly in **ISETL** and are used with a single command. In addition to the usual arithmetic, they include mod, max/min, even/odd, signum, absolute value, random, greatest integer less than, concatenation (of strings) and the standard trigonometric, exponential and logarithmic functions.

The power of **ISETL** begins to appear with the complex data types of set, tuple, func, and map. Syntax such as

$$\{7..23\};$$
$$\{-4,-1..40\};$$
$$\{9,7 . . 0\};$$

can be used to construct sets of finite arithmetic progressions of integers. It is also possible to construct a set containing any data types whatsoever (including other sets) simply by listing them. For example, the following set has cardinality 5.

$$\{8-1, \text{"t"} + \text{"he"}, 1.2, \{1,3,4,2\}, \{1..4\},3<2, \text{"the"},7, \text{false}\};$$

Students don't often see this right away because they count elements of this set which are sets (such as $\{1,3,4,2\}$) according to their cardinality and will count repeated elements twice. Working with **ISETL** helps them learn about elementary properties of mathematical sets.

Once such sets have been constructed, one can then construct complicated subsets by using a set former notation that can generally be understood by anyone who knows the mathematics. Consider the following **ISETL** code.

$$\{k**3 : k \text{ in } \{-N,-N+2..N\} \mid k**2 \text{ mod } 4 = 2\};$$

This is the set of cubes of even integers (of absolute value less than or equal to N) whose squares are congruent to 2 mod 4.

Standard set operations are implemented with single command syntax. They include union, intersection, difference, adjunction, tests for membership or subset, power set, cardinality, and selection of an arbitrary element. It is possible to iterate over a set to make loops conveniently but the operations of existential and universal quantification over a set are implemented and they often render loops superfluous. For example, in the following code, the first and second lines construct respectively the sets of all positive even integers and all primes less than or equal to N and the third checks the Goldbach conjecture up to N. The last line returns the value true.

```
E := {2,4..N};
P := {p : p in {2..N}|(forall q in {2..p-1}|p mod q /= 0) };
forall n in E |(exists p,q in |n = p+q);
```

All of this is very interesting and has been used in important ways to help students learn various topics in mathematics [1,4,5,9,10], but it has little to do with calculus directly. We will consider funcs and tuples in the next few paragraphs and see how they are used to get students to make mental constructions that are essential for understanding calculus. Our introduction to **ISETL** is far from complete (for example we made no mention of smaps). For more information see [1].

Interiorization — the func

Consider the compressibility function that was described (mathematically) on page 53. The mathematical expression defining Z as a function of V has a great deal of meaning for the mathematician, but usually it is totally incomprehensible to the calculus student.

Before explaining how we deal with this difficulty, let's review what we would like to be going on in the student's mind relative to this function. We would like her or him to construct an action and interiorize it as mental process. What is this process? One can think of it in terms of a transformation of the volume V. One has a particular value for V. It is necessary first to check that this value is nonnegative and, if so to determine whether it is less than 6. If that is the case, then a particular calculation is made with V. If not, then one must see if it is less than 90. If that is so, then a different calculation is made. If not, then still a different calculation is required. Thus, in one way or another, a value for V is transformed into a value for Z.

This is the sort of dynamic activity that we would like to be going in the student's mind. We would like her or him to be able to work with it, to realize that the choices correspond to something meaningful in the situation (the state of the substance) and to be able to imagine various mathematical operations performed on it. The way we get students to perform such an interiorization is to ask them to write a procedure that implements this function and to perform various operations on the procedure in the computer laboratory. Here is how that procedure looks as an **ISETL** func (note that almost all of the complications are due to the mathematical formulas — and not the programming)

```
Z := func(V);
    if V < 0 or V = 100 return "Out of domain"; end;
    if V <=6 then return
        1 + ((V-2)**3 * (V-6))/54;
    elseif V <= 90 then return
        V**3/3.0e3 - 0.048*V**2 + 1.73*V -7.7;
    else return
        4*V**(3/2)/(5*sqrt(V**3 - 8.0e5)) - 1/(V-100);
    end;
end;
```

Our claim is that writing this func requires the student to go through precisely the same mental manipulations as we described above. In this way, students construct the process in their minds as they construct the func on the computer.

Interiorization — the tuple

The concepts of sequence and sum can also be seen as processes which the student must construct. One can imagine the process for a sequence as running through the positive integers, and at each step, one grabs a number. That is an infinite sequence.

The process for sum can come from the sequence of partial sums. Here the student must encapsulate the sequence process to get an object. Then he or she again imagines running through the positive integers. This time at each step say n, the sequence is grabbed (it is, after all, an object) and de-encapsulated to its process, at least up to n and all the numbers are added up. This gives the nth partial sum. There are a number of ways of dealing with this in **ISETL**. Let us suppose for simplicity that "infinity" is not an issue. (There are **ISETL** approaches that take infinity into account, but we are only beginning to learn how to use them.) We "approximate" infinity with large positive numbers N. With this point of view, one can express the alternating harmonic sequence in **ISETL** as follows.

$$a := [(-1)**(i+1)/i : i \text{ in } [1..N]];$$

Then, the sequence of partial sums can be written (the symbol %+, is what **ISETL** uses for Σ)

$$Sa := \%+[a(1..i) : i \text{ in } [1..N]];$$

We do not suggest that these expressions are any easier to understand than standard mathematical notation — they are almost identical. What we claim is: the fact that they are computer objects and students can calculate with them leads to the students constructing useful mental models for these mathematical entities. For example, students can investigate convergence of this series just by writing

$$Sa(i);$$

for large values of i (making sure to increase N when necessary) and see that it appears to converge.

An interesting alternative investigation comes from replacing the original sequenc a by the following,

$$b := \{x : x \text{ in } a\};$$

b is the *set* of all elements of the *sequence* a. Then computing Sb analogously to Sa will give interesting results because each time b is evaluated, its elements will appear in a different order. Riemann's theorem is not far away at this point.

Interiorization — composing two processes

Understanding the composition of two specific functions requires that the student has constructed processes for the individual functions and then that he or she coordinate these two processes in series to form a new one. Once the individual processes are constructed, the coordination or composition is simple. This process is similar in mathematics, and, psychologically, students who have written funcs to construct the two functions can immediately form a new func corresponding to their composition.

Suppose for example, one is investigating the function Z for compressibility in terms of volume V. It is possible to imagine an apparatus that varies the volume with time, but keeps the temperature constant so that the expression for Z remains valid. Suppose for example, the volume is varied with time according to the following function:

$$V(t) = \begin{cases} 26.7\,t^2 & \text{if } 0 \le t \le 50 \\ \frac{4}{3}\pi\,t^3 & \text{if } t > 50 \end{cases}$$

At this point, students do not have much difficulty in implementing this with an **ISETL** func:

```
V := func(t);
   if t < 0  return "Out of domain"; end;
   if t <=50 then return
      26.7t**2;
   else return
      (4/3) * Pi * (t**3);
   end;
end;
```

The student can now easily write a func representing the function Z_t that gives compressibilty in terms of time.

```
Zt := func(t);
         return Z(V(t));
      end;
```

This helps students coordinate two processes. They have little difficulty figuring out the meaning of expressions like

$$Z_t(10) = Z(V(10))$$

and they can overcome the difficulty presented by the branches. Working with this kind of composition of functions sets students up for a more meaningful experience with the chain rule.

Encapsulation — the operation of composition

Once students have a good grasp of the composition of two specific functions, they are ready to think about an operation that takes two functions and produces a new function. This requires several encapsulations. The functions to be composed (general, not specific) must be thought of as objects and then — a major psychological difficulty for students — it is necessary to imagine an operation whose result is not a number, *but a whole function*. The code is simple, but students labor over it. It is the right struggle, for it consists of their developing the ability to convert processes to objects and then go back and forth between the two. Here is the **ISETL** code.

```
co := func(f,g);
   return  func(x);
      return f(g(x));
   end;
end;
```

It is a great triumph for students to write such code and then come to understand the meaning of an **ISETL** expression like

$$(Z \text{ co } V)(10);$$

Once students have it well in hand, issues such as the compatibility of the range of V with the domain of Z or various properties (such as limit or continuity) of the composition of two functions become accessible to them.

Encapsulation — approximate differentiation

There is an even more important example that is very similar to what we have done with having students compose two functions and then construct an operator that will compose any two functions.

Return again to the compressibilty function Z. We will discuss below some of the investigations we ask students to make with a func like the following,

```
Zs := func(V);
         return (Z(V + 0.0001) - Z(V))/0.0001;
      end;
```

Once again we note that the value of 0.0001 is not the issue here (although we do have students make investigations in which it is). We can have students make tables of this function, draw its graph and compare it with the original function Z. Obviously there is great interest in what happens in the different branches and at the cut points.

Psychologically, the above func corresponds to using the process of the function Z to produce another process, the difference quotient function, or what we call the approximate derivative of Z. In analogy with composition, the next step is clear. We have students write the following func.

```
D := func(f);
      return func(x);
         return (f(x + 0.0001) - f(x))/0.0001;
      end;
   end;
```

After the experience with composition, this is not so hard. The student must imagine the incoming function as an object to be acted on, take it apart to get its process, use the process to construct the new process for the difference quotient, and wrap that up as a function (object) which is sent back as the answer.

Mathematically, the student who has written this func has a tool that will allow her or him to use things like D(Z) as a function whose properties can be investigated. Issues like different expressions for the derivative in different branches and computation of the derivative at the seam by using the definition tend to become accessible to students who have made these constructions. Finally, as we will see below, it forms a critical step in the student's construction of the ideas surrounding the fundamental theorem of calculus.

3. Environment and implementation

In this section we describe our computer laboratory and its operation. We begin with some principles that established our goals and helped us with the main design decisions. The question of hardware and configuration was dealt with in detail in the article in Part 1 of this volume. The major considerations for software selection were given in the previous section of this article. We believe that our principles were reflected in each of these aspects of a computer laboratory.

Principles

We were guided in our design of the laboratory by the following principles or goals.

> The lab should be a place where students engage in guided discovery relative to the mathematical phenomena of calculus.

> The lab activities should stimulate the students to make some of the specific mental constructions of objects and processes discussed in the previous section.

> The atmosphere of the laboratory itself and the activities that are assigned should encourage the students to engage in group work.

> Students should think of the computer lab as a comfortable place to go and to work in the company of their colleagues and with the support and encouragement of faculty, staff and assistants.

> Every effort should be made to minimize distractions and frustrations that students might experience in trying to get the hardware and the software to function as it should.

The two variables that we have a chance to manipulate in trying to live up to these principles and achieve these goals are first, the lab itself and the way in which it operates and second, the design of instruction that is at least partly implemented in the lab. In this section we concentrate on the former.

Hardware

In this day and age the selection of computer hardware is not critical once one has decided to have a networked microcomputer laboratory. Networking is essential for the practical requirements of distributing material to the students, printing and submitting completed assignments. It also has the pedagogical value of fostering communication between students (if it has a mail facility) and encouraging group work.

The decision to use microcomputers (rather than terminals linked to a mainframe or workstation) is a temporary choice dictated by the fact that, today, the most developed, convenient to use and inexpensive computing systems appropriate for educational purposes exist on microcomputers. There is no educational reason why this has to be the case forever. If the situation should change, then alternative choices would be reasonable.

Given the choice of a networked microcomputer laboratory, one must decide between Macintosh and PC compatible computers and one must determine various parameters such as size, speed, numbers, etc. Of course, the equipment must be sufficient to run the software that will be used and accomodate the number of students one expects to serve. Beyond that, the selection can be determined by personal preference and the resources available. At the present time we are using 20 Macintosh SE computers and a laser printer linked with Appleshare networking.

Software

The most important things that we have to say about software are contained in the previous section. This is because the only thing that really matters about software is how it is used and the extent to which it is effective.

There is one major point that we should make about software that is more connected with the computer laboratory than with the content of the course. The software should be as easy as possible for the students to use. This means not only that it should be "user-friendly" but there should be documentation that is readable, accurate and helpful. We had to spend a fair amount of time trying to produce documentation for both **ISETL** and **MAPLE** that had these properties.

Configuration

Purdue University is experiencing a serious space problem. It was easier to get financial support for the equipment than it was to get a room to house it. The room we got was small and poorly ventilated. It had the advantage of being readily accessible for both students and faculty. The figure at the head of the next column shows the configuration of the laboratory.

The Macintosh next to the LaserWriter (LW) was used for a file server and was not otherwise available. There were 20 computers available to the students with a total of 40 chairs. The machines were very reliable so that we never had less than 18 functioning computers. In general, there were about 25 students in the laboratory during the course.

The configuration of the lab encouraged group work in two ways. The fact that there were more students than machines meant that they had to work in teams. Also, the arrangement of machines on tables encouraged groups of varying size. The five tables accomodate 6, 6, 8, 8, and 12 students, respectively. The nature of the groups tended to vary throughout a lab sesssion. A table with 8 students would sometimes have 4 groups of 2 members each, and other times it would be only 2 groups with 4 members each. On occasion the students at the table would form a single group of 6-8, all talking, sometimes heatedly, and often about the work at hand.

This arrangement of tables and the flexibility supported individual differences as well as the fact that there were times when a student might want to work alone and other times when he or she benefited from being part of a team. We could observe the students as they flowed back and forth between groups and set their work situation according to the task at hand and/or their needs at the moment. This relaxed, friendly atmosphere contrasted with the intensity of the work that was set before them and the former made the latter considerably more palatable. We made every effort to contribute to both the friendliness and the intensity.

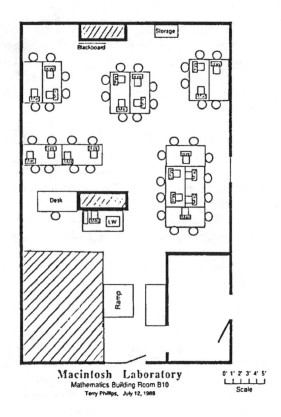

Macintosh Laboratory
Mathematics Building Room B10
Terry Phillips, July 12, 1988

The laboratory is one place where we tried to contribute towards our goal of minimizing distraction and frustration that can be a part of working with computers. The software itself is relatively free of programming distractions. Although the students had a great deal of difficulty, most of it was mathematical, and this is at it should be. After all, if the students were *not* having difficulty because of the mathematics involved in working their computer tasks, then it is likely that they were not learning very much.

Relation to the other sections of this calculus course

This same course was taken by about 2150 students. The first two semesters each carry 5 semester hours of credit. It meets three times a week in large lecture sections (about 450 students) and twice a week in recitation sections of about 40 students. For the third semester, the course is reduced to four credit hours and it meets four times per week in classes of about 30 students.

By agreement with the department, our course was constrained to follow, as closely as possible, the content and syllabus of the regular course. Our students did not take the same hourly exams, but they did take the same common final exam in the first two semesters.

Operation of the course

Our course met three times per week (Monday, Wednesday and Friday) in a classroom and twice per week (Tuesday, Thursday) in the computer laboratory. Each meeting was scheduled for a period of 50 minutes. In addition, the students had a free period scheduled after each of the lab sessions during which time they had the option of continuing to work in the lab.

Each week the students were given a lab assignment with a list of (mainly) computer tasks to perform in the lab and a homework assignment with (mainly) problems from the text assigned according to a schedule similar to that used by the regular sections. The homework assignments we gave were not as extensive as those given to the regular sections but, together with the lab assignments, the students had considerably more work to do than was required in the regular sections.

The students were expected to do a significant amount of work on the lab assignments during the scheduled lab hours but were not expected to complete them without returning to the lab during its open hours. The discussions in the class meetings were based on the work that students were expected to have done on the computer tasks.

The lab assignments were due each Monday and the homework assignments were due each Tuesday. They were corrected by a student grader based on an answer sheet prepared by us and were returned to the students. The examinations during the course were given in the evenings. They consisted of questions designed to elicit the understandings that students had constructed of various concepts in calculus. There was no time restriction for the exams.

4. A case study - The Fundamental Theorem of Calculus

The example we give here is not a single lab assignment, but a coherent sequence of assignments that were given over a period of time. In this sense, our example is not typical (almost all of the other tasks were given in a single assignment). It is, however, typical of our assignments in several aspects. First, this assignment involves a coordinated use of both **ISETL** and **MAPLE**. Second, it is typical in that it involves not only an opportunity for students to discover important mathematical ideas, but also it offers a stimulus for students to make various important mental constructions. Finally, this assignment, as did many others, makes use of certain mathematical and programming tools that we had to provide.

The Fundamental Theorem of Calculus is not mentioned in the course until the 13th week of the semester. There is a series of computer assignments, however, that are spread throughout the previous 12 weeks and are designed to help the students make a number of mental constructions that will prepare them for a discussion of this theorem.

About half of these assignments are designated as Lab Assignments. The others are called Homework Assignments. This distinction only refers to whether the students are expected to do most of the assignment during scheduled lab hours or entirely on their own time. All of the work is to be done on the computer in the laboratory either during scheduled lab times (Lab Assignments) or during open hours in the lab that are scheduled at various times (Homework Assignments).

Appendix 1 contains copies of the 11 assignments that were given on this topic. We discuss each of them briefly to indicate our expectations of what the student should gain from working on the task. These tasks represent only one problem in the entire Lab or Homework Assignment indicated. Generally there were 10-12 such tasks in each assignment.

Lab Assignment 2

A partition is a process consisting of marking points on a line. It is also an object in that it is a sequence of demarcation points. Because the process is so simple, we consider that doing it once is sufficient to get the students to interiorize it. Collecting the points in a tuple is expected to help the students encapsulate the process into an object.

Homework Assignment 2

The purpose of this assignment is to move the student from thinking about a single partition to being able to consider the formation of partitions in general. We hope that they will coordinate the tuple, the func and mathematical notation into a single scheme for partitions.

Homework Assignment 3

The purpose of this assignment is to get the students to interiorize a process of concatenating two partitions to form a new one. This is done by having them write **ISETL** code to implement such an action. Our goal is to get them ready to think about the fact that if $a < b < c$ then the integral over $[a,c]$ is the sum of the integral over $[a,b]$ and the integral over [*b,c*]. (See Homework Assignment 12.)

Homework Assignment 5

The process to be interiorized this time consists of forming the usual Riemann sum for a partition and a function, with the evaluation point chosen to be the left endpoint of each interval. This is applied to a reasonably complicated function. It is one, however, that the students have worked with so it is not completely unfamiliar.

Actually, we consider the Riemann sum to be two processes. One is the analytic calculation of the sum of values of the function multiplied by lengths of subintervals and the other is the graphical intepretation as an approximation to the area under a curve. The purpose of this assignment is to get the students to coordinate these two process into a single process. Our intention is that coordinating the **ISETL** and **MAPLE** implementations will help achieve this synthesis.

Homework Assignment 6

As with partitions, we now would like the students to deepen their process conception of Riemann sum from something that is done with a particular function to a process that can be applied to any function. We try to achieve this by having them write an **ISETL** func that implements the process, and having them apply that process to several functions. Naming the process (Riem-Left) may also get them to encapsulate it into an object, although this is not an important goal for us at this point.

The fact that the three functions to which they apply their func are not very specific (a mystery function, a function obtained as the the result of a previous operation) is intentional. The vagueness is expected to contribute to their sense of generality for this process.

Note also that we have them apply their func for Riemann sums to the result, Df, of applying the approximate derivative operator to a function. This is, of course, not an accident but is designed to start them on the road to the Fundamental Theorem.

Lab Assignment 8

If the process of adding up all of the small rectangles from left to right to get an area is interiorized, it should be possible to reverse it in the sense of going from right to left. The students make the simple discovery that this has the effect of negating the result obtained by going from left to

right. Our intention is that they should also discover the reason for this in the details of the calculation.

Lab Assignment 9

Another variation of the RiemLeft process consists of taking the evaluation point at the right of each interval. The students are expected to discover this just from the name RiemRight.

Lab Assignment 10

This is in the same spirit as the previous assignment, this time for midpoints. Another point of this task is to discover something about the relation between the mesh size and the accuracy of the approximation. The examples given are intended to point out that this relation can depend on the particular function.

Lab Assignment 11

The purpose of this assignment is to further deepen the students' process conception of Riemann sums by coordinating it with another activity, of comparable difficulty, on which they worked in previous assignments. This is the process of estimating the maximum value of a function on an interval by partitioning the interval and picking the demarcation point at which the function takes on its largest value.

The coordination of the two processes leads to a fourth version of Riemann sums, RiemMax. Again, the method by which we help the students construct these processes is to have them implement the actions in **ISETL**

Lab Assignment 12

This the *piece de résistance*. We can discuss it more easily in terms of the solutions that we hope for. First of all, the func Int could look like this.

```
Int:= func(f,a,b);
            return func(x);
              if a <= x and x <= b then
                return RiemMid(f,a,x,25);
              end;
            end;
        end;
```

We feel that the student really must encapsulate the Riemann sum process into an object in order to be able to use it here with a variable upper endpoint. Related work in class and on assignments has them apply this mental construction to defining functions by integration. Next there are the last two columns which print3 is supposed to produce. The values, for a particular x are given by

$$Int(D(f),a,b)(x)$$

and

$$D(Int(f,a,b))(x)$$

Make no mistake about it, these two pieces of code are extremely difficult for the students to come up with. They struggle with it for long periods of time, but most of them do eventually produce this code and apply it to get results that are reasonable enough for them to try to interpret.

The two pieces of code are, of course, **ISETL** implementations of the two parts of the Fundamental Theorem of Calculus. We see in the code all of the mental constructions that the student must make. In the first, it is necessary for the function f to be an object which is transformed by the *process* D to another object D(f). Only then can the process Int transform this into yet another object which is a function. This function, which is an object, must be de-encapsulated to obtain its process (which is known only through a deep knowledge of Int and D) and applied to the value x. This is how the student is expected to interpret the numbers that appear in the third column of print3.

For the second piece of code, the whole thing is reversed. The process Int transforms the object f into another object which is again transformed into a new object by D. This object is again de-encapsulated to its process and applied to the value x. This is how the student is expected to interpret the numbers that appear in the fourth column of print3.

The student cannot miss the fact that the second and fourth columns of this table agree. It is interesting, amusing, and above all gratifying to watch their reactions to the third column. They seem to feel intuitively that this column should also agree and many of them think that an error has been made. They puzzle over this for a while and discuss it among themselves. After not too much time in the laboratory, a few students will notice, and the point spreads around the room like wildfire, that this column differs from the second and fourth by a constant. Someone pipes up with a phrase like, "Oh, so that is what the constant of integration is all about!" One can almost see the light bulbs illuminating over the heads of the students.

At this point, they are ready for a proof of the Fundamental Theorem of Calculus.

Homework Assignment 12

Here we coordinate the earlier work (Homework assignment 3) on concatenating two intervals with what we expect is a serious understanding of the Riemann integral and give the student an opportunity to discover the fact that the integral is additive over unions of adjacent intervals.

Evaluation

It is extremely difficult to evaluate the effects of giving students a sequence of tasks such as these. We have already suggested that a person has only a tendency to display knowledge and this makes traditional "objective" tests somewhat suspect. We did not make observations (as in research such as [2,4,9]) to determine if the students appeared to be making the mental constructions that were intended. We do have various results on the students' performance and these were quite high, but so much was going on that it is nearly impossible to identify this particular collection of tasks as one of the causes.

The only direct data that we have is the number of correct and partially correct submissions on the assignments. This information must be treated with care for a number of reasons. The assignments had deadlines, but we did not enforce them. (This is an overall weakness of our first implementation that will be discussed later). The students worked together in groups and although we believe that almost all of them made sincere efforts to complete these tasks and tried individually to understand the work, there is no way to tell how many, in the end, got their final results from their colleagues without understanding them. After all, their scores helped determine their grades and these students were no more free of the pressures to obtain good grades than are most students in higher education.

	A	B	C	D
Lab Assignment 2	18	6	1	0
Homework Assignment 2	19	3	2	1
Homework Assignment 3	24	0	0	1
Homework Assignment 5	20	1	1	3
Homework Assignment 6	16	4	1	4
Lab Assignment 8	9	12	3	1
Lab Assignment 9	20	2	2	1
Lab Assignment 10	14	9	0	2
Lab Assignment 11	18	3	2	2
Lab Assignment 12	12	6	4	3
Homework Assignment 12	15	1	1	8

Results of Computer Tasks on the Fundamental Theorem

Finally, we point out that the idea behind the Lab Assignments is that the students were supposed to do serious work on each of them *before* the topics were considered in class. They were not necessarily expected to complete the task successfully, only to "wrap their minds around the issues." We can give a subjective evaluation in that our observations of students working in labs indicated that they did indeed give serious thought to these problems.

Given all of this, the reader is free to interpret the Table of their scores on submissions of their results on these tasks. The code "A" indicates that the task was completed successfully or nearly so. The code "B" indicates that they made significant progress but did not fully succeed on the task. The code "C" indicates that they were, essentially, wrong and the code "D" indicates that they did not submit anything on this task.

We can perhaps summarize the results by pointing out that, in the end, most students turned in something reasonable on most of the tasks. The sharp drop in the number of those who turned in the last assignment probably has to do with end of semester difficulties that are fairly typical.

There is one final subjective discussion we would like to include relative to these tasks. At the end of the three semesters, we gave an assignment to write an essay (about three pages) entitled "What I learned in Calculus." Many students referred to Riemann sums, definite integrals and their applications. Here are some excerpts from those essays. First, from one of the brightest students in the class who is an engineering major.

> "The integral is equal to the area under the curve of a function. In order to learn this, I dealt with Riemann sums. This is basically just dividing the independent variable into a number of parts within the limits of integration. These parts can then form rectangles by using the value of the function at one of the points as the height. Summing the areas of these rectangles then gives an approximation to the integral. Increasing the number of rectangles under the curve until there are an infinite number of them would actually give you the exact integral. This can be done simply by taking the limit as the number of subdivisions of the independent variable goes to infinity. While I also learned many different techniques for solving integrals and applying them, I feel this initial basic description of the integral was the most important thing I learned, allowing me to solve many different varied problems later."

And second, from a student who, throughout the three semesters, had more difficulty than anyone, almost dropped out on several occasions, and will probably not major in anything connected with mathematics.

> "When taking the integral of equations in a double integral the first integral would divide the region w/respect to that variable and then the next would do it with respect to the other and the limit of the sums of all these was the area. You had to decide the order and limits of integration by looking at the picture of the region and the equations and then decide."

And finally, a student who also did quite poorly in the course tried to be humorous in his essay and ended it by saying that one of the "ultimate truths" he learned in Calculus was

> "Every math problem can be solved by using small boxes."

Nothing decisive, of course, and there is nothing surprising about good responses from the best students. But if we can have even our weakest students walking away from Calculus with thoughts like those expressed in the last two quotes, then perhaps there is some hope for us.

5. Overall Evaluations

The difficulties that we mentioned in connection with evaluating the effects of our instructional treatment of the fundamental theorem of calculus are endemic to evaluation of any aspect of our project, or, in our opinion, to evaluation of any educational activity. Short of conclusions which amount to little more than "those who do well, do well," we do not really understand very much about deciding whether one instructional treatment is better than another, or even if a particular approach is worth the trouble it takes.

The reason for this is that we continue to use variations or polishings of evaluation instruments that are based on implicit assumptions about learning — assumptions we believe are not justified. For example, as we have indicated earlier, a person may know something, but not display that knowledge with any reliability.

There are other difficulties, and we are convinced that a major effort of research and development on evaluation techniques will be necessary before much progress can be made. Such an effort should start with the question, "What is learning?" and go on to ask what it means to have learned something, how does one display what one has learned, and a host of other questions before beginning to think about what reasonable instruments might look like.

In the meantime, we are reduced to presenting what data we have obtained, and letting the reader draw whatever conclusions he or she deems appropriate. Our information is in several categories: performance of students on finals taken in common with all sections of the corresponding calculus course at Purdue, performance of students on exams we designed, comments of students, and comments of various visitors.

Performance of Students on Common Finals

In the first two of three semesters, the students in our special class took a common final examination together with all Purdue students (over 2000) taking first year calculus. These exams are timed (three hours), multiple choice and are designed to test the lowest common denominator of student skills in manipulations and solving the same problems they have practiced throughout the semester.

In our course, we spent very little time with this kind of problem and almost no time in drilling students on these kinds of calculations. After 13 weeks of the 15 week first semester we changed this abruptly. We told the students that the course had ended and that the remaining two weeks would be spent in preparing for the final examination. We spent two weeks engaging the students in the same kind of practice and memorization that the students in the regular sections had been doing all semester.

In the second semester we did this for only a week and with somewhat less intensity.

In the first semester, our class average on the final was 75% as opposed to 69% for the regular sections. In the second semester, our class average was 59.25% as opposed to 59% for the regular sections.

Thus our students, after spending most of the semester in conceptual development relative to calculus and only a short time practicing with techniques, performed on these exams as well as, or better than, students who spent the entire semester practicing techniques. This result is a replication of the findings of Heid [12] and others. In the second run-through of our course taking place at the time of this writing and with different instructors, the results do not seem to be very different.

In the third semester we made a decision not to make this comparison and our students did not take the common final exam. There were two reasons for this. In the first place, there seems little information to be gained, because every time the "experiment" is performed the results are the same.

The second reason is more serious. As the students moved, at the end of the semester, from trying to understand mathematics to trying to memorize it, we began to suspect that this may be doing more harm than good. All semester we worked hard to convince the students that they should think about what they are doing, that even if they do not see how to solve a problem immediately, continued thought over time could lead to success, problems don't necessarily have unique, short answers, and so on. These are new ideas and attitudes for them and they are inconsistent with other messages they are getting in their educational experiences. The students only begin to come around to this way of thinking gradually over a period of time. Then we suddenly shift gears and organize activities that seem to deny everything we have been telling them. We thought we detected a negative effect on their thinking about mathematics from this reversal and we decided that, when not required to do so by the department, our students would not take common final examinations.

Performance of Students on Our Exams

During each of the three semesters we gave two class exams to the students in our section. These tests were designed to elicit information about the students' understanding of the concepts that were studied in class. More than half of the questions were unfamiliar to the students and the questions that were similar to ones they had worked on before tended to be more difficult than, for example, their homework problems.

These exams were given in the evening with no time limit. Students were not permitted to use books or notes, but hand calculators were allowed.

Semester 1, Exam 1	76.9
Semester 1, Exam 2	80.8
Semester 2, Exam 1	63.6
Semester 2, Exam 2	72.9
Semester 3, Exam 1	73.2
Semester 3, Exam 2	78.4

Class Average Percentage Scores on Exams

The exams were graded by the instructor (Dubinsky) and partial credit was given. For this reason, even the numbers in the preceding Table have a subjective component.

The actual distribution is interesting. In general, about half of the students scored above 85%. Most of the rest were in the 55-75 range and two or three (usually the same individuals) had very low scores.

Comments of Students

The students had many opportunities to make written comments about their experiences, usually anonymously. They were uniformly positive. The full set of comments is available from the authors. Following are some excerpts that are typical of the students' attitudes toward the lab and computer component of the course at the end of the three semesters.

> "... our use of the computer was not for technology's sake, but to help us think."

> "... piecing together of information ... had to be done on computers in the lab where everything had to be reduced to general format. Although the process of breaking down a problem, however complex, into coherent parts took place mostly on a subconscious level, there where times when this process brought an ordinary confusing problem into a much more logical light. Riemann sums are the first example that comes to mind. I've heard many people from Ma161 tell me how they didn't understand Riemann sums, but after plodding through the code on the Macintosh and having to understand everything about them, I felt somewhat unable to help the ordinary Ma161 student who didn't have such appropriate means to learn what was going on."

> "... working with Riemann sums on the computer put a physical significance on the work we did in class. It made the physical aspects of the theory recognizable."

> "Since our class went quite in depth into understanding the theories and physical aspects of calculus, I felt that the computer work and class work went very well together."

> "Programming definitely strengthened what we learned because in order to program the computer

we first had to understand exactly what was going on and break it down into parts. So not only did I learn how to use **ISETL** and **Maple** to calculate things quicker than I would be able to by hand, but I also greatly strengthened my calculus knowledge by doing so."

> "In my opinion, Labs are the most beneficial part of this class. Labs indicate to the student what he/she doesn't understand. It helps students to formulate questions."

> "The Macintosh Lab was a great asset to the class because its usage required an understanding of the processes involved and made us "learn" Calculus not just "plug and chug" like the "normal" calculus classes. The major lesson I learned was how to attack problems."

> "Placing the problems on the computer also further embodied the spirit of the course by presenting us with the difficulty of implementing and understanding the method, versus subjecting us to the repetition of many similar problems and much busywork."

> "The use of computers enabled me to work with the concepts behind the mathematics, while the homework made sure I could function by hand. The exploration of the concepts gave, and still promises, greater understanding than is possible through hand written solutions in the same length of time. Indeed, the increased versatility given by the computer increases experimentation and depth of understanding exponentially."

> "The way a concept becomes clearer when a person writes a function in **ISETL** to do it is amazing, and actually being able to manipulate graphs also helps out a bunch when working with functions."

> "In my opinion the labs helped me better understand the meaning behind the things that we covered."

The future

Naturally, we have our own evaluation and, of course, it is completely subjective. We feel that our approach is a vast improvement over traditional methods of teaching calculus. The results that have been obtained are, no doubt, due in part to the small class size, the enthusiasm of the teachers and the newness of it all (Hawthorne effect). But this is not the first time the authors have taught, nor is it their first experiment in innovative methods. Indeed we have, between us, more than half a century of undergraduate teaching experience. Both of us have always been interested in improving our teaching and have tried many things. Never before have either of us achieved the level of success (relative to the particular students) that we are having with this approach. We are convinced that our use of computer activities and small group problem solving to implement a theory of learning mathematics has shown itself to be an extremely promising direction for improving the learning of mathematics.

There are many things that still need to be done and many ways in which our course must be improved. The most important has to do with the content of the course. At the insistence of the mathematics department, we have made our content and the order in which it is taken up essentially identical to that of the regular sections. For the next stage, beginning Fall 1990, we will have two classes of about 60 students each and the only restriction on content is that at the end of each semester any of our students who desire to transfer to regular sections will have the appropriate background to do so.

There are also areas within the context of our present method that need to be improved. The most important has to do with the timing of Lab Assignments. An important component of our method is that the appropriate lab activity must be completed before the matter is considered in class. Our students did not always achieve this and in a few cases they fell seriously behind. We had deadlines but did not enforce them because we were concerned about the effect this might have on the class atmosphere. We were under considerable pressure from the students (who could leave at any time) to minimize the extent to which the demands of our course exceeded that which students experienced in other sections. The workload in our course was greater and initial difficulties with our overall laboratory system (unavoidable for new projects) made the burden on the students frightening.

In our next phase, we should solve most of the external problems and our software systems (especially **ISETL**) will be considerably more convenient to use. We intend to use this easing of pressure from external effects to increase the intensity of the work. In particular, deadlines will be enforced.

The same could be said with respect to participation in

class. Not everyone answers questions. This can mean that some people just don't do their thinking in ways that allow them to formulate reasonable responses to questions. But there is the danger of degenerating into a situation where only a few students do all the talking. On occasion we approached such a situation. Measures will be taken in the next phase to avoid this issue.

Within the parameters of a first attempt we feel that our approach has been very successful. We hope that in the next phase of our course, we will take advantage of our first step and take a number of subsequent steps towards a calculus course that results in students learning a significant amount of mathematics.

References

1. Baxter, N., Dubinsky, E. & Levin, G., **Learning Discrete Mathematics with ISETL**, Springer, New York, 1989.

2. Breidenbach, D., Dubinsky, E., Hawks, J. & Nichols, D., *Development of the process conception of function*, preprint.

3. Conciatore, J., *From flunking to mastering calculus*, Black Issues in Education, pp 5-6, February (1980).

4. Dubinsky, E., *Teaching mathematical induction I*, The Journal of Mathematical Behavior, 5 (1986), pp 305-317.

5. Dubinsky, E., *Teaching mathematical induction II*, The Journal of Mathematical Behavior, 8 (1989), pp 285-304.

6. Dubinsky, E. (In press) *Constructive aspects of reflective abstraction in advanced mathematical thinking*, in L.P. Steffe (ed.), **Epistemological foundations of mathematical experience**, New York: Springer-Verlag.

7. Dubinsky, E., *Reflective abstraction in advanced mathematical thinking*, in (D. Tall, ed.) **Advanced Mathematical Thinking** (in preparation).

8. Dubinsky, E., *A learning theory approach to calculus*, Proceedings of the St. Olaf Conference, St. Olaf (1989) (preprint).

9. Dubinsky, E., F. Elterman & C. Gong, *The student's construction of quantification*, For **The Learning of Mathematics 8**, 2 pp 44-51 (1988).

10. Dubinsky, E., *On Learning Quantification* (preprint).

11. Dubinsky, E. & P. Lewin, *Reflective abstraction and mathematics education: the genetic decomposition of induction and compactness*, **The Journal of Mathematical Behavior**, 5 (1986), pp. 55-92.

12. Heid, K., *Resequencing skills and concepts in applied calculus using the computer as a tool* , **Journal for Research in Mathematics**, 19 (1988), pp 3-25.

13. Selden, A. & J. Selden, *Constructivism in mathematics education: a view of how people learn*, **UME Trends**, E. Dubinsky (ed.), 2, 1 (1990) p 8.

14. Sfard, A., *Operationals vs. structural methods of teaching mathematics - a case study*, **Proceedings of the 11th Annual Conference of the International Group for the Psychology of Mathematics Eduaction** (A. Borbas, ed.) Montreal (1988) 560-567.

15. Treisman, U., *A study of the mathematical performance of Black students at the University of California, Berkley* , Thesis, University of California, Berkley (1985).

Appendix I

Laboratory Assignments Relative to the Fundamental Theoerm of Calculus

MA161 Lab Assignment 2

Tuples can have more than two components. One use for tuples with many components is to represent partitions of an interval. Here is a tuple that represents a partition of the interval from -1.5 to 0.7 with 5 equal subdivisions.

[$-1.5, -1.06, -0.62, -0.18, 0.26, 0.7$]

SUBMIT

1. A partition of the interval from 0.05 to 12 into 17 equal subdivisions. (Handwritten)

2. A diagram of a line with the endpoints of the interval and subdivisions clearly marked.

MA161 Homework Assignment 2

A tuple such as the one in Lab Assignment 2, #2 is often expressed in mathematical notation more compactly as

$$-1.5 + 0.44(i - 1), \quad i = 1, 2, \ldots 6$$

In general to express the partition of the interval from a to b into n equal subdivisions, in mathematics, we write,

$$a + (b - a)\frac{i-1}{n} \quad , i = 1, 2, \ldots (n+1)$$

A very similar notation is used in **ISETL**. For example the first tuple can be represented in **ISETL** as

> [-1.5 + 0.44*(i - 1) : i in [1,2 . . 6]]

Enter this expression and see the resulting tuple written out explicitly. Such an expression is called a *tuple former*.

SUBMIT

1. A tuple former that produces the partition of the interval from 0.05 to 12 into 17 equal subdivisions.

2. A func that will accept numbers a, b with $a < b$ and a positive integer n and will return the tuple of the partition from a to b into n equal subdivisions. *Don't submit an untested* func - *try it on the examples first.*

MA161 Homework Assignment 3

For this problem, review the work that you did on partitions in Week 2 (Lab Assignment 2, #2 and Homework Assignment 2, #7). Also, notice that you will have to record your **ISETL** session.

If t1 and t2 are tuples in **ISETL,** then t1+t2 is a tuple formed by tacking t2 onto the end of t1. Suppose then that we have a partition P1 of the interval from a to b into n subintervals and a partition P2 of the interval from b to c of k subintervals. Then P1+P2 is *essentially* a partition of the interval from a to c into $n + k$ intervals. The only thing is that the point b appears twice. Actually, in all of our applications of partitions this will not matter, but we can remove it for neatness. This is done by removing the first component of P2 before tacking it onto P1. (This value is stored in **x** but we do not use it.) Here is **ISETL** code that will do it and display the new partition.

```
> Take X from b P2;
> P3 := P1 + P2;
> P3;
```

Use tuple formers (see Homework Assignment 2, #7) to construct a partition from -17.63 to 0.0005 with 14 equal subdivisions and a partition from 0.0005 to $\sqrt{3}/2$ with 9 subdivisions. Then use code such as the above to construct a partition from −17.63 to √3/2 with 23 (definitely non-equal) subdivisions.

SUBMIT
1. A record of your **ISETL** session, once it is correct.
2. The last partition, copied from the screen. One decimal place accuracy will be sufficient.

MA161a Homework Assignment 3

The **ISETL** operation %+ can be used to add up all of the quantities of a tuple, or even quantities depending on the values in a tuple. For example, if P is a partition of the interval from a to b into n equal parts and f is some function whose domain includes this interval, then there are many situations in calculus where one would like to consider each subinterval and multiply the value of f at, say, the left endpoint times the length of the subinterval and then add up all of these products. The mathematical notation for this quantity is

$$x_i = \left(a + \frac{(b-a)}{n}(i-1) \right), \qquad i = 1, \ldots, n+1$$

$$\sum_{i=1}^{n} f(x_i)(x_{i+1} - x_i)$$

and using %+, the **ISETL** code for obtaining it is very similar.

```
x := [a+((b-a)/n)*(i-1) : i in [1..n+1]];
%+[f(x(i))*(x(i+1)-x(i)) : i in [1..n]];
```

1. Use the following function and a partition of the interval form -3 to 2 into 17 equal parts to compute the above quantity .

$$H(y) = \begin{cases} \dfrac{1.3y^2 + 6y - 4}{y^2 + 2y + 1} & \text{if } y \le -2.6 \\ y + 1 & \text{if } -2.6 < y \le 1 \\ \dfrac{1}{3y - 2} & \text{if } 1 < y \end{cases}$$

2. Use **ISETL** and **Maple** to produce a graph of H on the interval from −3 to 2, make a hardcopy of it and indicate on the graph a representation of what is being computed here.

SUBMIT
1. An **ISETL** record of your calculations.
2. The numerical answer.
3. Your graph with the indicated representation.

MA161a Homework Assignment 6

In Homework Assignment 5, Problem #4, you worked with using **ISETL** to compute the quantity

$$\sum_{i=1}^{n} f(x_i)(x_{i+1} - x_i)$$

where f is a function and $\{x_i : i = 1,\ldots,n+1\}$ is a partition of the interval from a to b into n subintervals. In this problem you are to put it all together and write and **ISETL** func which accepts a func, say f which represents a function, the endpoints a, b of and interval contained in the domain of f, and a positive integer n. Your func is to construct the partition of the interval from a to b into n equal subintervals and then return the value of the above expression. Call your func by the name RiemLeft, Apply your RiemLeft in the following situations. Choose your own value of n.

1. The function is mys2, $a = -1$, $b = 4$.
2. The function is the absolute value, $a = -1$, $b = 2$
3. The function is Df where f is the absolute value function and D is the operator you developed in Lab 5, #5 $a = -1$, $b = 2$

SUBMIT
1. A copy of your RiemLeft
2. Your three answers, copied from the screen.

MA161a Lab Assignment 8

Return to Homework Assignment 6, Problem #4 and you func RiemLeft. Run this func on the same three examples you used then except reverse a and b . Compare your answers with the answers you got originally

SUBMIT
1. What is the effect of this reversal?
2. Go through RiemLeft and explain exactly why you get this effect.

MA 161 Lab Assignment 9

Return again to Homework Assignment 6, Problem #4 and consider the first function, mys2 on the interval [−1, 4]. Sketch a graph of this function and, using a value of $n = 10$, shade in the quantity on the graph that is represented by the calculation that RiemLeft would make.

Repeat this process except this time, use the calculation that RiemRight would make. Of course you have to guess what RiemRight would be!

SUBMIT
1. Your sketch using RiemLeft.
2. Your sketch using RiemRight

MA161a Lab Assignment 10

In the spirit of Lab 9, Problem #2, write a func called RiemMid. Apply RiemMid to the functions given by the following expressions on the interval [0,1].

$$t^2$$
$$u^{100}$$

You must experiment to determine the choice of n as follows. The exact values that are being approximated as larger and larger values of n are taken (that is, smaller and smaller "meshes") are 1/3 and 1/101 respectively. You must choose your value of n so as to get 3 significant figures accuracy.

SUBMIT
1. A copy of your func RiemMid.
2. The values of n you used to get the required accuracy.
3. An explanation of why the accuracy improves as n increases.

MA161a Lab Assignment 11

You will write a function RiemMax which will be the same as the others except for the following.

1. RiemMax will accept an additional parameter crit which is a tuple consisting of a list of all points at which the function in question could have a relative maximum.

2. Instead of computing

$$\sum_{i=1}^{n} f(x_i)(x_{i+1} - x_i)$$

The quantity $f(x_i)$ will be replaced by the maximum value of f on the interval $[x_i, x_{i+1}]$. This is given by the **ISETL** code

    %+max[f(t) : t in T]

Where T is a tuple of all possible values in the interval $[x_i, x_{i+1}]$ at which f could attain a maximum.

HINT : *If* S *is a tuple,* p *and* q *numbers, then the tuple of all numbers in* S *that are in the interval* [p,q] *together with these two endpoints is given in* **ISETL** *code,*

    [p] + [s : s in S | p < s and s < q] + [q]

As part of your func, you will have to have a construction of a tuple of tuples, corresponding to your partition that gives the possible points at which the maximum could occur.

Run your func on mys2 with the three following choices of the interval $[a, b]$
1. [−1,4]
2. [−1,0]
3. [0,4]
Use $n = 100$ in all cases.
Of course you must figure out the tuple crit by hand.

Think about the following questions

> What property of the function on the interval will guarantee that the results of RiemMax and RiemLeft are the same? What will guarantee that RiemMax and RiemRight are the same? That the results of RiemRight and RiemLeft are the same?

SUBMIT
1. Your func RiemMax
2. The results of applying it in the three given cases.
3. Your answers to the questions.

MA161a Lab Assignment 12

In this problem, which is perhaps the most important of all, you will put together several things that you have been using this semester. First, the approximate derivative func D from Week 5 (use a value of 0.000001 for e), then RiemMid and finally func print1 from Week 6.

The first step in the problem is to write a func Int which accepts a func representing a function f and two numbers a and b representing the interval $[a, b]$ and returns a func whose value, for a number x in $[a, b]$ is the result of applying RiemMid to f with the interval from a to x (note this is x, not b) and 25 subdivisions.

Next adjust the func print1 to obtain the func print3 which will accept only one func representing a function f, an interval $[a, b]$, a positive integer n and a filename. The result of print3 is to place in the file four columns of values. The first gives the numbers x at $n+1$ evenly spaced points in the interval [a,b]; the second gives f at x; the third gives the value obtained by applying D to f, applying Int to the resulting func and then evaluating the resulting func at x ; and the last column gives the value obtained by applying Int to f, applying D to the resulting func and then evaluating the resulting func at x;

SUBMIT
1. A hard copy of your func Int.
2. A hard copy f your func print3.
3. A hard copy of the result of applying print3 to the function $x \sin x^2$ on the interval from 0 to $\pi/2$. Use $n = 10$.
4. Your interpretation of the resulting data
5. A hard copy of the result of applying print3 to the function $\sqrt{1-x^2}$ on the interval [0,1]. Use $n = 10$.
6. Your interpretation of the resulting data. In particular, is there any way in which this example differs from the previous one? If not, explain why they are the same. If so, explain the cause of the difference.

MA161 Homework Assignment 12

Let $a < b < c$ be three real numbers. Suppose that you applied RiemLeft three times to the same function with the same n but three different intervals, $[a, b]$, $[b, c]$, and $[a, c]$. Find a simple relationship between the three answers that you would get.

You can try this on several examples to guess the answer, but you should also attempt to understand, from the meaning of RiemLeft, why your relation must be true.

SUBMIT
1. The relation you came up with.
2. Your explanation of why it holds.

Appendix II

A Sample Exam Given in M161a

We have been asked what difference does teaching Calculus as a laboratory course make in terms of what our students learn? We feel that we can answer this question best by showing an exam that we have given to our students. The average score on the exam was 80.8. This exam was the second exam given in Math 161a. It was given on November 21, 1988. The first exam was given on October 6, 1988.

Math 161a Test 2

Each Problem is worth 10 points (A = 100 points)

1. Find the equation of the normal line to the curve defined by the equation:

$$3x^4 + 4y - x^2 \sin y = 3$$

at the point [1,0].

2. For each of the following two curves, give a graphical description of how Newton's method, starting at the indicated x_0, would succeed or fail to approximate the indicated zero. Label each situation as success or failure, and indicate what happens by drawing right on the graph.

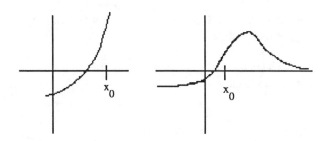

3. Following are the graphs of two equations:

graph1

$$(x^2 + y^2 + 25)^2 - 100x^2 = 10^8$$

graph2

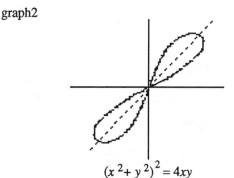

$$(x^2 + y^2)^2 = 4xy$$

(a) On graph1, label the intersections of the graph with the coordinate axes and on graph2, label the points at which the graph is farthest from the origin. Your labels should be the approximate numbers at which these points occur.

(b) For each, draw intervals onthe coordinate axes to restrict the graph so that it defines a function, u for graph1 and v for graph2, but do this in such a way that the composition $v \circ u$ is defined.

4. Find the third Taylor polynomial for the function given by
$$x\sqrt{4-x}$$
about the point 1.

5. Suppose that you applied the func RiemLeft and the func RiemMax to the same function f on the same interval $[a,b]$. Can you be certain of any relation between the two results? Explain your aanswer.

6. Make a careful sketch of the graph of the function given by the expression:
$$x^{\frac{2}{3}}(3x+10)$$
Label with values the important points and show the shape of the curve clearly.

7. Boyles Law for gases states that $pv = c$ where p denotes the pressure, v the volume, and c is a constant. At a certain instant the volume is 75 in^3 and the pressure is decreasing at a rate of 2 (lbs/in^2)/sec. At what rate is the volume changing at this instant?

9. A package can be sent by parcel post only if the sum of its height and girth (the girth is the perimetter of the base) is not more than 96 in. What is the maximum volume that can be sent if the base of the box is a square?

10. The Campbell Soup Company has come to its senses and designed its can so as to minimize the amount of material used. If the radius of the can is increased by 1 percent and the basic design is maintained, what is the corresponding increase in the amount of material?

11. Use the Mean Value Theorem to show that if f is a function whose domain is the interval $[\alpha, \beta]$ and whose derivative is constant value k on $[\alpha, \beta]$, then f is a linear function.

12. Let f be the function given by
$$f(x) = \begin{cases} 2x & \text{if } x \le 1 \\ 3 + x^2 & \text{if } x > 1 \end{cases}$$
find f' and explain the method you used for determining $f'(x)$ for each x.

A Computer Laboratory for Calculus

James F. Hurley
University of Connecticut

Background and Philosophy.

After three years of experimentation with programmable calculators in a special section of freshman calculus, the University of Connecticut Mathematics Department in 1983 applied for and received a grant from the office of the Dean of the College of Liberal Arts and Sciences to set up a computer laboratory, which now contains 25 IBM-compatible personal computers, three printers, and a Hewlett-Packard plotter. The goal was, and continues to be, the exploitation of the new technology to capture and hold student attention. Observation of freshman students and discussion with colleagues in the sciences suggested that direct student involvement in laboratory work provides more than simple augmentation and enrichment of the lecture portion of freshman science courses. It engages the students more fully in those courses, by providing structured hands-on activity that constitutes both a representative introduction to the activity of practicing scientists and also a setting for interesting concrete illustrations of basic theory. The latter aspect of laboratory work is credited by scientists with promoting improved intuitive understanding of the conceptual aspects of their introductory courses.

The Department's new personal computers provided mathematicians for the first time with a means to design similar activities in freshman calculus. These efficient machines seemed to be ready-made laboratory apparatus for exploring the concepts of calculus in concrete new ways. Such exploration had the potential to foster conceptual understanding through *personal discovery* of the many key ideas whose traditional abstract discussion—punctuated with a few simple examples worked out on a chalkboard—seemed each year less and less adequate to build the requisite insight, intuition, and confidence for successful post-calculus study and work. Using computers, beginning students of calculus could solve calculus problems with the same kind of speed and efficiency that researchers obtain on large-scale problems with more powerful hardware. Moreover, practical numerical and graphical programs could greatly extend the range and complexity of calculus problems treatable in a first course. Students could thus acquire valuable experience in solving realistic problems by techniques actually used in applied fields. Thus the challenge was to design laboratory activities that would both motivate and complement lectures about the theory and algorithms of calculus, and also provide hands-on experience with the use of current technology in applying the subject.

Two alternative types of activities were considered: running packaged software and working directly with program code. There was hesitancy to limit the students to use of programs with hidden code that made the underlying algorithm and the method of implementing it invisible. This reflects an uneasiness about the educational impact of simple "button pushing" on unsophisticated students, an uneasiness that seems fairly significant in the general mathematical community and may account for some of the reluctance among mathematicians to incorporate widespread computing activities into the undergraduate curriculum. There seems to be considerable fear that mindless *automated calculation* has no more value to learners than the much-denounced mindless *hand calculation* that comprises so much of traditional calculus instruction. Indeed, that fear in many cases goes farther—to question whether there might not be some inherent eye-hand-mind link between rote computation and conceptual understanding that could be undermined by elimination of the former.

Examples of such concern can be found in John Kenelly's report [11] of fears expressed at *Calculus for a New Century* that computer algebra systems might promote a "black-box syndrome" in which

> students would use mechanical systems without understanding the underlying concepts... There was also concern that students' mathematical maturity upon entering calculus was inadequate to use these systems productively.

Similarly, William Boyce [2] expressed the view that

> one must have some *specific* information about some particular functions. It is probably not necessary to know how to evaluate $\int \sec^5 x \, dx$, but one should certainly know $\int \cos x \, dx$.

Both remarks reflect concern that pervasive use of calculators at lower levels has degraded rather than upgraded conceptual understanding, factual computational knowledge, estimation skills, etc. Such concern was recently stated in graphic terms by Edward Effros in a letter to the editor of the New York *Times* [5]. He asserted that at best computers

> transform the student into a spectator, rather than a creator... Anyone who has taught calculus knows how to recognize those who are failing an exam: simply look for the students who are pressing the buttons on their calculators.

Although publications like [6] correctly point out that no scientific evidence exists to establish a link between use of calculators in elementary and high schools and the diminishing student calculation and estimation skills, the suspicion that such a connection does indeed exist seems widespread. Merely pointing out that no proof of damage exists may be insufficient to induce the general mathematical community to embrace computer-enhanced instructional methods. More positive evidence about observed *benefits* of creative use of new technology is probably necessary to allay such concerns.

In any case, reluctance to risk possible erosion—rather than enhancement—of the quality of calculus instruction contributed to the decision to have students work directly with program code. Another component of that decision was the fact that no single calculus package of that era contained anything like the full range of activity that the faculty considered desirable. Manipulating multiple packages with their differing command structures involved a significant fraction of the effort required to use a single computer language to generate the kinds of programs envisioned. Like many other first forays into computing in calculus, this project started with the language that was supplied with the computers—IBM BASICA (Microsoft GWBASIC). The plan was to introduce relevant syntax as the need for it arose, and give the students a simple program that would illustrate the central idea of each algorithm. The teaching assistant in charge of the laboratory would then present a flow chart for the full-blown version of the algorithm, and assign coding of the full program as the first part of the assignment for the next week. After they succeeded in coding the algorithm to produce a working program, the students would run it in various situations and interpret their output.

The initial results fell far short of the lofty goals of the faculty committee that designed the course. The lack of structure of the primitive version of BASIC in use predictably led to near-perfect examples of "spaghetti code"—unstructured programs whose underlying algorithms were just as effectively hidden as any package's! In addition, many students spent more time debugging their code than trying to understand what the algorithms were actually doing, or interpreting results of running the programs. Little wonder, then, that the head of the University's computer science program sent a letter of protest that use of unstructured BASIC in the course was undermining the computer scientists' efforts to teach structured, logical programming.

The skills students did manage to develop in generating programs in BASICA were of little or no value in subsequent courses, because although the latter typically do involve work with computer code, that almost always is in the form of calls to externally compiled subroutines whose local variables assume user-supplied values. BASICA had no such subroutines or non-global variables. A switch to Pascal was briefly considered, but rejected because its complexity would have required too much attention to technical programming matters.

Meanwhile, the inventors of the original BASIC, John Kemeny and Thomas Kurtz of Dartmouth, were preparing True BASIC for microcomputers. It is a modern, fully-structured language that retains the ease of use of the original BASIC, and its English-like syntax makes it easy to perceive the nature of computational algorithms. As soon as it was released in 1985, it was adopted as the language for the computer-enhanced calculus sequence. The laboratory period activity also changed. Students were given a complete working program to experiment with and modify. That made it possible to discuss the associated mathematical algorithm in detail, and allowed the students to develop independent programming knowledge gradually through simple modification of the initially-supplied code. The latter was logical, and natural for mathematical computing. After five successful years of working with True BASIC, Department faculty who have been involved in the course can identify with Professor Kemeny's assertion in his AMS-MAA invited lecture at the 1988 annual meeting in Atlanta [10] that

> an algorithm should be presented as a computer program. This leads to deeper understanding of algorithms, allows exploring variants of the algorithm, and allows the students to use the algorithm to solve serious problems.

The algorithms of calculus arose as systematic schemes for carrying out calculations, and indeed the name of the subject reflects that. As suggested above, True BASIC programs can effectively highlight the mathematical nature of the computational aspects of calculus.

As the logical evolution of the primitive dialects of BASIC that students meet in elementary and high-school study, True BASIC is easy for students to understand and work with immediately. Its structure requires programs to be logical in order to run, and focus on the logic of program code serves to fill some of the void left by the recent de-emphasis of proofs in elementary calculus texts. The similarity of True BASIC's code for most mathematical processes to the syntax of mathematical English is well illustrated by the way functions with multiple formulas are defined in the language. For example, the function f defined by the two-part formula

$$f(x) = \begin{cases} x \sin \dfrac{1}{x} & \text{for } x \neq 0 \\ 0 & \text{for } x = 0 \end{cases}$$

is definable in True BASIC by the following code.

```
def f(x)
   if x <> 0 then let f = x*sin(l/x) else let f = 0
end def
```

Current Organization of the First-Year Course

Math 120-121 is a two-semester sequence, both parts of which carry five credits (compared to four credits for the standard main-track calculus course). A weekly one-hour laboratory period uses [9] as its manual, and is taught by a teaching assistant. Those assistants have included European and Asian graduate students, as well as undergraduate American students who have completed the course as freshmen. Two of the 28 weekly laboratories deal with the mechanics of the True BASIC language. Each term, thirteen weeks are devoted to work with computing algorithms designed in some cases to motivate and in others to illustrate theory to be presented or already covered in the lecture portion of the course. Students are given disks that contain all the programs whose listings appear in [9], and during the lab period they call up those programs, study their code to understand the algorithms of the programs, run the programs, and analyze the output in relation to the algorithms. At each lab except the last one, an assignment is made at the end of the period, to be completed and handed in at the meeting one or two weeks later. The assignments usually consist of exercises from [9] that range from simple computational or graphical illustrations of the programs worked with in the lab, to modifications of the latter to fit slightly different problems, to interpretation of anomalous results produced by round-off or overflow errors. Sometimes, quizzes are given in the lab period to test student understanding of the algorithms under study.

The first semester begins with a unit on function tabulation. One of the assigned exercises involves two algebraically equivalent formulas for the same function, which because of round-off error give different values for large x. The next lab investigates one- and two-sided limits for functions like the one above. This is approached by using a slightly modified function-tabulation program to get an idea of whether a limit exists. A subsequent program calculates how close, δ, the variable x must be to a limit point a in order for $f(x)$ to be within a prescribed tolerance, ε, of $\lim_{x \to a} f(x)$. Experience with such numerical calculations seems to help students develop a solid intuitive understanding of the essence of the limiting process. Attention then shifts to approximation of roots of equations. The first laboratory on this topic discusses the Babylonian square root algorithm, which provides a simple preview of Newton's method as well as an interesting indication of how digital devices can quickly produce accurate approximations of square roots. The following week, the laboratory constructs the bisection method of approximating the root of an equation $f(x) = 0$ for a continuous function f as an application of the intermediate-value theorem for continuous functions. The closely-related method of false position is presented as a simple example of how a slight variation of an algorithm can dramatically affect the efficiency of computation. Next, numerical differentiation methods are discussed and used to study the differentiability of functions such as the one defined above. Just as results from the limit-calculation lab indicated that the above function has limit 0 at the origin, so those from the numerical differentiation lab suggest that the function is *not* differentiable at $x = 0$.

At this point, the lecture portion of the course is considering optimization and curve sketching. In the laboratory, a simple introduction to True BASIC graphics, from the point of view of automating hand plotting of graphs, is followed by a laboratory on numerical optimization of continuous functions over closed intervals. This not only sets the scene for the next lab—which is devoted to graphing first continuous and then more general functions—but also lays the groundwork for the laboratories on definite integration by using the device of subdividing a given closed interval into small subintervals. Students construct their continuous function plotter in segments, one of which is a subroutine for computing the extreme values of the given function on the input interval $[a, b]$ of the x-axis. Another module of the program asks for the user to select on the basis of that computation the *y*-axis interval of the plotting window. It is then a short step to extending that continuous function plotter to handle the more general functions —like the one graphed below—that are commonly encountered in the curve sketching sections of standard calculus texts.

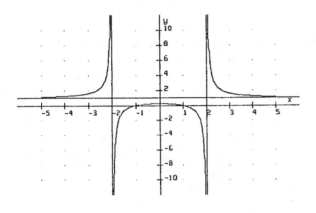

Figure 1

Screen dump of the graph of $y = \frac{x^2 - 1}{x^3 - 4}$
over the interval [-5, 5], computed from the True BASIC program GLNGRAPH running on a Macintosh computer

As a follow-up to the graphing unit, the students work with a lab that shows how the technique of implicit differentiation can be used (via Euler's method) to approximate implicitly defined functions. The numerical instability of differentiation means that overflow error can and does occur in this, and the students are asked to analyze exactly what has gone wrong in such a situation.

The first semester reaches its climax in a unit on definite integration, which is introduced by working with Riemann sums, principally midpoint sums. The midpoint approximation is used to find areas under curves, and the arc lengths of such natural functions as $f(x) = \sin x$ and $f(x) = x^2$. After watching many midpoint sums seem to converge to what they now think of as the definite integral of the corresponding function over an input closed interval, the students consider the question of how accurately the midpoint scheme approximates the definite integral. The following program is produced by modification of the basic area approximation program. Using it, the students discover that the error in the midpoint approximation is proportional to the square of the step size. While many texts include an analysis of the error of the trapezoidal approximation (which is obtained by *averaging the values of the function at the endpoints of the subintervals)*, one rarely if ever sees an indication of the accuracy of the

conceptually similar midpoint rule, which uses the values of the function at the *average of the endpoints of the subintervals*. (Note that the midpoint *mx* is incremented by repeated addition of *h*, which can lead to round-off error that can itself prove instructive in working with arc lengths).

```
REM Program RIEMANN to compute Riemann sums and error
def f(x) = x^2
input prompt "Endpoints a, b of interval      ": a, b
input prompt "Maximum number of subintervals ?   ": U
input prompt "Successive approximations how close
    to stop ? ": E
print
print "n", "Mn(f)", "Error/(h*h)"
print
let S = 1000
let n = 10
do while n <= U and abs(S - T) >= E
    let h = (b - a)/n
    let mx = a + h/2
    for j = 1 to n
        let M = M + f(mx)
        let mx = mx + h
    next j
    let Mn = M *h
    print n, Mn, abs( (1/3 - Mn)/(h*h) )
    let n = 2*n
    let T = S
    let S = Mn
    let M = 0
loop
end
```

Several things about this program are worthy of mention. First, it is simple minded enough for beginners to understand clearly just what it does. Indeed, except for the condition on |S–T| – which is not part of the first area approximation program the students use – it simply automates the computation of Riemann sums that standard introductory calculus texts show and have exercises on for small values of U. Second, the DO WHILE loop calculations continue until one of two things happens: either the specified upper limit U is reached or the specified degree of accuracy is achieved. The following is a typical output screen obtained by running the program.

```
Ok. run
Endpoints a, b of interval ?   0, 1
Maximum number of subintervals ?   5120
Successive approximations how close to stop? 5e-7
```

n	Mn(f)	Error/h*h
40	.333281	8.33333e-2
80	.3325	8.33333e-2
160	.33333	8.33333e-2

320	.333333	8.33333e-2
640	.333333	8.33334e-2
1280	.333333	8.33339e-2

These results provide striking evidence that the midpoint approximations $M_n(f)$ converge to the limit $1/3 = \int_0^1 x^2\,dx$. They also indicate that the error in the midpoint approximation is proportional to the square of the step size h, with the constant of proportionality in this case recognizable as approximately $1/12$. Finally, the error results show some deterioration in accuracy for large values of n – those beyond 320.

After running such programs several times, students quickly start to recognize that convergence of sequences entails successively computed values being arbitrarily close together. When they reach the theorem that the limit of the n-th term of a convergent series must be 0, the students have much of this kind of concrete computational experience to draw upon. One can even introduce Cauchy's criterion meaningfully, because students who have repeatedly encountered it for convergent sequences in the special case $m = n + 1$ are inclined to find it quite natural and believable.

This illustrates some of the potential of direct work with computer code for promoting enhanced conceptual mastery of calculus. Such experience can foster intuitive understanding of an important aspect of convergence that formerly was considered too sophisticated even to mention in an introductory course. In this case, computer calculation seems able not just to substitute harmlessly for hand computation but to build good intuition that until now has been widely regarded as dependent upon significant amounts of hand computation. Consider the following quote from [5], for instance.

> The only way to get a sense of mathematical objects, whether they be numbers, derivatives, or tensors, is to handle them without the aid of calculators.

The second semester begins by continuing to exploit the midpoint rule, this time to estimate values of the natural logarithm. Students modify their simple area calculation programs from the previous course for this purpose. There follows a lab on the trapezoidal rule, in which numerical second differentiation is used to obtain estimates of its accuracy. The next lab treats Simpson's rule, and the students learn how to make both these rules into external libraries that can be called as external subroutines by driver

programs. This kind of activity provides an accurate preview of the process of accessing compiled libraries, which is an important part of later work in both mathematical and scientific computing. A lab follows that uses the midpoint approximation for numerical investigation of improper integrals. Of special interest to the Physics Department is a lab that covers both the Euler and improved Euler (second-order Runge-Kutta) methods for approximating the solution of first-order initial value problems.

Attention then shifts to numerical investigation of sequences and series of constants, and Taylor approximation of transcendental functions. Students recognize that convergence of sequences underlies much of the year's preceding work, and thus come to the discussion of convergence with unusually sharp intuition. As suggested above with reference to the Cauchy criterion, they are well prepared to absorb fundamental theoretical ideas that have traditionally been major stumbling blocks to freshman calculus students. The year concludes with two units on graphing parametric equations and polar coordinate curves (including conic sections). These are much less sophisticated conceptually than the earlier work of the second semester, but they afford an opportunity to review graphical programming in True BASIC, which students who join the sequence in the second course (for instance, students with some AP calculus credit) welcome.

Math 120-121 is advertised as a demanding, honors-level course, but the computing component attracts a significant number of non-honors students. Annual enrollment is currently about 150, out of roughly 600 students in first-year calculus. The course is given in small (25 to 30 students) sections, compared to the 100-to-150 students in large-lecture, main-track calculus, which now has one less section per term than in 1983-84. The sequence is strongly recommended for all Mathematics and Physics majors, and it constitutes an acceptable alternative for Math majors to the freshman computer science course that now uses C as its programming language.

During the first six years of the course, the main text was [1]. It is lean and has many challenging problems, but puts little emphasis on computational aspects of calculus. Starting with the Fall of 1990, there was a change to [3], which provides a better fit to the main-track syllabus and which also gives some attention to computational matters.

Second-year Follow-up Course

Math 220-221 is a two-semester, four-credit, honors-level course that started in 1987-88, with [7] as text. Its goal is to employ the computer skills developed in the first-year course in working with topics from multivariable calculus and vector analysis, elementary linear algebra, and introductory differential equations.

Unlike Math 120-121, there is no formal scheduled laboratory period attached to the course. Instead, students can take advantage of the Department's personal computer laboratory (or one operated by the Computer Center) to access locally produced software written in True BASIC to solve assigned problems. The programs include both numerical computation and 3-D graphics programs that call True BASIC's compiled 3-D Graphics Library. True BASIC's accessibility makes it possible for freshmen with AP calculus credit to take Math 220 successfully. For more about this sequence, refer to [8].

Appraisal of the Sequence

The four semesters of the enhanced calculus sequence cover most of the content of five standard semester courses, and also supply significant computational and theoretical enhancement. Nearly all faculty who have been involved in teaching it have found it to provide superior preparation for upper division work in mathematics, engineering, and the sciences. It has been difficult to obtain quantitative measurements of the sequence's impact on the students, most of whom are highly motivated, bright volunteers for courses advertised as honors-level and ambitious. The fact that they have done well is not surprising, because they generally excel in their course work. And if most students in standard courses worked as hard as the ones in the enhanced sequence, it is unlikely that there would be so much complaint about the success rate in calculus instruction. But some benefits have been observed that seem directly traceable to the nature of the enhanced sequence.

First, students who initially experience limits numerically tend to develop solid intuitive understanding of the limit process. The computer can indeed capture the element of change in that process, and students who watch sequences unfold in real time on a computer screen tend to acquire a good feel for convergence. Moreover, watching the results of algorithms whose code is available on another window of the same screen provides immediate linkage between the underlying algorithms and their computational implementation. As already noted, the simple code used to express the algorithms can foster unusually clear insight into mathematical concepts like convergence. Students whose idea of the definite integral derives from Riemann sum evaluation programs like the above naturally come to think of the definite integral as the limit of such sums. The

fundamental theorem is then correctly recognized as a *theorem* (and not the *real definition* that stubborn mathematicians want to hide for a while, as beginners often conclude from traditional courses). The fundamental theorem provides just one computational tool for evaluating definite integrals—on a par with numerical methods. Students are quite comfortable with problems that require evaluation of integrals such as $\int_0^1 e^{-x^2} dx$, which they recognize as something for which a numerical approximation algorithm is suitable.

Computer methods can significantly expand the range of problems that an elementary calculus class can discuss. Students can find the arc length not just of semicubical parabolas but of *ordinary* parabolas as well! It is also easy to calculate that the sine curve has length approximately 3.82019 over the interval $[0, \pi]$. In this connection, the lab on arc length calculation leads the students to discover that the Riemann sum used to define the arc length formula

$$L = \int_a^b \sqrt{1 + f'(x)^2}\, dx$$

is sometimes just as accurate a means of calculating the arc length of the sine curve as the latter formula itself. This serves to reinforce the idea that different algorithms can be equally efficient in some circumstances.

Results of final examinations given over a five-year period by the same instructor provide some confirmation of faculty perception that the laboratory activity went better after the switch to True BASIC as the language of the code. The examinations reflected a consistent level of expectations from a fairly homogeneous population of above average, highly motivated students. In the two years that BASICA was used, 56 students averaged 64.56 per cent on such examinations, with a standard deviation of 15.10. (For the complete data, see [8].) In the succeding two years, the average on comparable examinations was 70.55, with a standard deviation of 15.24. This represents an improvement of roughly four-tenths of a standard deviation after the adoption of True BASIC. The same instructor taught main-track calculus the succeeding year to more than 200 students. He drew upon his experience in the enhanced sequence to introduce as much optional computing activity with packaged software as was feasible in a four-credit course. He also made a conscious effort to make the final examinations as comparable as possible to those he had used in the enhanced course. The students averaged 64.92% on those examinations, with standard deviation 18.29: results that very closely parallel those in the enhanced sequence before the adoption of True BASIC.

An external evaluation of the enhanced program was conducted in April of 1988 by the MAA Subcommittee of CUPM on the First Two Years of College Mathematics (now renamed CRAFTY—Committee for Reform and the First Two Years). The subcommittee visited Storrs as part of its information gathering about innovative approaches to the teaching of calculus. It conducted an intensive series of meetings with faculty of both the Mathematics Department and client disciplines that require calculus of their students. In addition, outlines, texts, and examinations were scrutinized. In the case of the enhanced sequence, faculty and graduate and undergraduate students involved in the lectures and laboratory were interviewed. The ensuing report to the mathematics faculty and the University administration about the Department's lower division programs included the following remarks about the enhanced calculus sequence.

- "The enhanced calculus program ... has the potential to put the University in a leadership role with respect to the teaching of calculus."
- "we were particularly pleased with the effort being made in designing the enhanced calculus sequence (Math) 120-121, 220-221."
- "Questions in the lab manual reinforce the concepts being taught in the calculus lectures."
- "... the committee would like to see the number of sections increased."
- "It is particularly interesting to the committee to see the successful development of an innovative calculus that could become one of the model courses for other schools trying to reform the curriculum."
- "...our recommendation is that this form of calculus be gradually expanded to a broader range of students. We feel that many students enrolled in Math 110-111 could benefit from the approach of the enhanced courses. Laboratory sessions affording students the opportunity to test basic concepts with concrete numerical examples could be worthwhile."

Shortly before the subcommittee's site visit, a group of mathematicians from the University of Massachusetts at Amherst also visited Storrs. After attending a session of the computer laboratory and interviewing students, faculty, and the teaching assistant, they returned to their campus to study the question of how best to introduce computing activity into elementary calculus. In the Fall of 1988, they started an experimental project using [9] along with locally-generated notes. That work has continued during 1989-90.

Future Directions and Conclusions

The experience of the enhanced sequence has convinced many of the faculty who have taught it that the nature of calculus courses and instruction will inevitably change in response to the ubiquity of personal computers on college campuses. Numerical methods are soon likely to become an *integral* part of all calculus sequences, replacing some closed-form techniques that have traditionally been prominent. The techniques most likely to disappear or to receive only theoretical discussion are those that originally were created to address the fact that direct numerical hand computation of many quantities was too expensive. Traditional procedures for optimizing continuous functions on a closed interval, for example, may survive only as simple examples of important theorems. While it will probably continue to be important for students to understand the graphical content of the second-derivative test, we may not be so concerned with having them use that test to find the extreme values of complicated functions on closed intervals. Simple, fast, numerical computations on the computer that can yield highly accurate results in seconds may become the practical tool for that purpose.

Similar remarks apply to some techniques of formal integration, although an argument can be made for many of them on the grounds of their later importance. (One can, of course, turn that around and argue that such methods be left for those later contexts, where they are better motivated and more likely to make a lasting impression than they currently do in freshman calculus—months or even years before they are seriously used.) Integration by parts will surely endure, but integration of even powers of sine and cosine will likely be de-emphasized. Much material lies between those extremes, of course, and its fate will be decided over a period of years by a gradual process of evolution.

A key question is how much integration of the computer can help thin the bloated content of traditional calculus sequences. Most of the complaints currently voiced about the lamentable stress traditional courses put on hand computation algorithms are accompanied by calls to exploit the power of the computer to supplant that misplaced emphasis by concept-centered activity. It is far easier, of course, to call for such emphasis than to develop strategies that will succeed in teaching the theoretical essence and relevant applicability of calculus to science, engineering, and daily life. The computer does have realizable potential for illuminating theory by supplying a wealth of quick examples whose numerical and graphical patterns are easy for beginners to discern and understand. But such activity takes time. A weekly computer laboratory *added on to* an existing 4-credit calculus sequence has produced encouraging results at the University of Connecticut. But it is at best questionable how attractive an approach to reform is that will raise teaching loads by at least 25 per cent.

In response to the challenge by the MAA subcommittee that it incorporate the benefits of the enhanced calculus program into its main-track sequence, in 1989-90 the Department started a pilot program to integrate computer laboratories into two experimental sections of first-year calculus. With support from the National Science Foundation's Instrumentation and Laboratory Improvement Program and the State of Connecticut's High Technology Project and Program Fund, a new laboratory with 30 Macintosh SE/30 and Macintosh II computers was set up. Outlines have been prepared that appear to make it feasible to cover the traditional four credits of freshman calculus in two weekly hours of lecture, a weekly computer lab period, and a weekly problem discussion hour. (That is the organization of the University's four-credit freshman chemistry class.) Each semester, one large-lecture section of main-track calculus was replaced by three small sections—two taught by teaching assistants and one by a faculty member. Students were assigned randomly from the entire pool of first-year calculus enrollees to all sections, regular and experimental. In two of the small sections—one faculty-taught and the other given by a teaching assistant—one hour of traditional lecture was *replaced* by a weekly computer laboratory period that used techniques developed in the enhanced sequence to teach certain traditional topics in a computer-based way, through student experimentation and discovery. Students in all the small sections took common examinations that did not involve computer calculation, but did have heavy conceptual emphasis. In addition, common final examinations are administered by the Department to *all* main-track calculus students. Examination results will be carefully analyzed, and there will be tracking of the students of all the small sections in subsequent courses—to determine whether their performance differs from that of their counterparts who were taught differently. It will likely take considerable time before the efficacy of of the computer-integrated approach can be assessed with confidence, but the initial results, which are shown in Table 1, are encouraging.

As is to be expected, there was some variation in performance from section to section, but the overall totals show that students in the computer-laboratory sections had a mean final examination score 8.2% higher than those in the traditional sections. That represents about four-tenths of a standard deviation, which is just about three grade levels

Table 1. Final Examination Scores (in per cent)
First Semester Main-Track Calculus, Fall, 1989

Type of Class	Number of Students	Mean	Median	Standard Deviation
Computer Lab1	13	71.73	67.0	13.85
Computer Lab2	12	64.33	67.5	20.40
Totals	**25**	**68.18**	**67.0**	**17.34**
Traditional	19	67.79	70.5	19.65
Traditional 2	86	66.03	64.5	20.54
Traditional 3	112	54.02	55.0	20.34
Totals	**217**	**59.98**	**62.5**	**21.36**

(ordinary C vs. an ordinary D: the University uses plus- and minus-suffixes on grades). Such a difference appears potentially significant, but in view of the small numbers of students in the experimental sections further experimentation will be needed before definitive conclusions can be drawn. (During the course of the semester, by the way, students in the small computer sections consistently scored higher than those in the small traditional section on the common midterm examinations given to the three sections, although the results on the final examination were much closer.)

A frequent question is what the ultimate impact of the calculus reform movement will be on the nature of calculus instruction. It may well be that no monolithic standard will replace the current remarkably uniform approach to calculus. Indeed, the very number of dedicated teachers who are pursuing such a wide range of new approaches —many of which seem to have significant merit—suggests a different outcome. Several parallel new modes of modern calculus instruction could emerge and flourish. Computer algebra systems offer much symbolic processing power as well as computational ability comparable to that of numerical and graphical software packages. The relatively inexpensive new super calculators and graphics calculators can bring numerical, graphical, and in some cases even symbolic-processing tools to the calculus classroom. Sophisticated new software packages provide more user control and power than their predecessors, and thoughtful instructors are continuing to devise ever more effective ways of taking advantage of them. Finally, those concentrating on making calculus courses more relevant to other fields by focusing on mathematical modeling both as motivation for and application of the study of calculus can significantly improve the impact of future calculus courses.

If we look to history as a guide to how technology may impact on calculus instruction, we see that the first approaches to computer enhancement of mathematics instruction differed in degree rather than in kind from what has been described here at Connecticut. Numerical computing was quickly recognized as something that could both save time and illuminate structure. The same is true of computer graphics. Evolution of that approach to calculus enrichment continues at Dartmouth [4]—which pioneered the creative use of computing in mathematics instruction—as well as at the Universities of Connecticut and Massachusetts, all of which are moving in similar directions whose roots go back a generation or more. Their experience suggests that direct work with simple, logical program code can offer significant conceptual rewards.

One appeal of this approach to integration of computing into calculus lies in its perceived conservatism and consequent safety. Those wary of the "black-box syndrome" can involve their students in computing, but by focusing on the algorithmic processes the computer follows can avoid superficial button pushing, mouse clicking, and inputting of data. This provides a built-in defense against critics' contentions that banishing hand calculation may be self-defeating. It is thus possible to get one's feet wet in modern technology, without committing in advance to radical restructuring of the body of material students must study and master. Undeniably, the price of upgrading calculus instruction includes a willingness to work harder. But the work is not unpleasant, and the rewards can be quite satisfying. Many of those who experience student enthusiasm toward calculus for the first time in years become optimistic that calculus instruction *will* improve, and that a common feature of that improvement will be thoughtful utilization of a computing laboratory.

References

1. Robert A. Adams, *Single Variable Calculus, Rev. Ed.,* Addison-Wesley, Reading, MA, 1986.

2. William E. Boyce, "Calculus and the computer in the 1990's," *Calculus for a New Century,* MAA Notes No. 8, Mathematical Association of America, Washington, DC, 1988, 42-43.

3. William E. Boyce & Richard C. DiPrima, *Calculus,* Wiley, 1988.

4. Richard H. Crowell & Reese T. Prosser, "Computers with Calculus at Dartmouth," to appear.

5. Edward Effros, "Give U. S. math students more rote learning," Letter-to-the-Editor, New York *Times,* February 14, 1989.

6. *Everybody Counts: A Report to the Nation on the Future of Mathematics Education,* National Academy of Sciences Press, Washington, DC., 1989.

7. James F. Hurley, *Intermediate Calculus,* Saunders, Philadelphia, PA, 1980.

8. James F. Hurley, "A computer-enhanced lower-division curriculum," *The Mathematics Curriculum: Towards the Year 2000,* Proc. Theme Group 7, 6th International Congress on Mathematical Education, Curtin Univ. of Technology, Perth, Australia, 1989, 259-267.

9. James F. Hurley & Charles W. Paskewitz, *Computer Laboratory Manual for Calculus,* Wadsworth, Belmont, CA, 1987.

10. John G. Kemeny, "How computers have changed the way I teach," AMS-MAA Invited Address, Atlanta, 1988, *Academic Computing,* May/June, 1989, 44-45, 59-61.

11. John W. Kenelly & Robert C. Eslinger, "Computer algebra systems," *Calculus for a New Century,* MAA Notes No. 8, Mathematical Association of America, Washington, DC, 1988, 78-79

Project CALC: An Integrated Laboratory Course

David A. Smith and Lawrence C. Moore
Duke University

Project CALC is a joint effort of Duke University and the North Carolina School of Science and Mathematics (NCSSM), a state-supported residential school for high school juniors and seniors. We are developing a three-semester calculus program based on a laboratory science model; the first two semesters will be implemented at NCSSM. The key features of the course are real-world problems, hands-on activities, discovery learning, writing and revision of writing[1], teamwork, and intelligent use of available tools.

Our intended audience includes *all* students of calculus. We know that many students have no intellectual goals in mind when they enroll in calculus, but we have to assume that someone has set such goals for them as part of the justification for requiring calculus as a prerequisite for something. For almost all students, we assume that the primary goal (whether their own or imposed by someone else) is that they should become *intelligent consumers* of calculus tools and concepts. A very small percentage of students beginning calculus in college will become mathematics majors, perhaps even mathematicians, but neither we nor they know who these students are at the time. Furthermore, we think it is important, even for students ready to grapple with mathematics on its own terms, to learn first how and why it is used by the rest of the world.

The acronym in our name stands for Calculus As a Laboratory Course. Our computer lab is central to the learning experience, not a peripheral "add-on." We provide software "tools" and somewhat structured environments in which student teams carry out open-ended experiments that lead to discovery of fundamental principles of mathematics and its representations of the real world. As much as we can, we use these experiences to drive the classroom and out-of-class activities as well.

We taught the first two semesters of the Project CALC course for the first time to two sections at Duke in the 1989-90 academic year. In the Fall of 1990, seven sections of Calculus I and one of Calculus III were offered at Duke; these were taught by six faculty members and two graduate teaching assistants. Four sections of an adapted Project

CALC course were offered at NCSSM, and our materials are in use in experimental sections at a number of colleges and universities around the country.

Why a Laboratory Course?

Our students come to us with some very firm beliefs about mathematics. These beliefs have little to do with anything communicated directly by teachers or textbooks, but they have a lot to do with the indirect messages that have been hammered home in 12 years of school mathematics.

Teachers and textbooks can alter student *behavior* on a temporary basis (to get past the next test), but they have little impact on student *beliefs*. Here are some of those beliefs, expressed in words the students would not use themselves:

- Only "special" people can really understand mathematics; attempts by "ordinary" people are counterproductive.

- The main purpose of academic mathematics (i.e., of making students take math courses) is to provide a series of hurdles (some low, some high) that one has to get over in order to be admitted to study in a variety of other disciplines.

- Mathematics has *no* connection with anything else, real or academic. This is especially true of the so-called "word problems," which are merely a difficult topic designed to stymie "ordinary" students like themselves. (Possible exception: Some of mathematics looks a little like what they see in a physics class, but they *know* that this too has nothing to do with the way things really work.)

- The fundamental object of mathematics is the *formula;* synonyms include "function" and "equation." If the word "expression" were in their vocabulary, it too would be a synonym - but it isn't.

---

[1]We have written elsewhere at some length about the importance of *writing* as a learning tool in mathematics: G. D. Gopen and D. A. Smith, "What's an Assignment Like You Doing in a Course Like This?: Writing to Learn Mathematics," *The College Mathematics Journal* 21 (1990), 2-19.

• Successful performance in mathematics requires setting aside any attempt to *reason* and replacing that activity with mechanical adherence to *rules*.

• The object of a problem is *the answer*. If you get the answer, it makes no difference how you get it. If you don't get it, the secondary object is to write down enough symbols to get partial credit. The only ways to *know* if an answer is right are (a) it's in the back of the book or (b) teacher says so. (This is actually a logical deduction from the observation that the rules are completely arbitrary and can be changed at any time by the teacher or the textbook author.)

• Reading and writing have nothing to do with mathematics. Assignments can be completed and tests passed without actually reading the book, and it would be unfair for the teacher to ask that anything be communicated in complete sentences.

This list could go on, but the point is this: We must acknowledge and deal with these attitudes if we hope to induce students to *learn*, to *understand*, as we ourselves understand mathematics. We can change their behavior, but only they can change their beliefs.

Much of what our students have actually learned from school mathematics - more precisely, what they have invented for themselves - is a set of "coping skills" for getting past the next assignment, the next quiz, the next exam. When their coping skills fail them, they invent new ones. The new ones don't have to be consistent with the old ones; the challenge is to guess right among the available options and not to get faked out by the teacher's tricky questions. At Duke, we see some of the "best" students in the country; what makes them "best" is that their coping skills have worked better than most for getting them past the various testing barriers by which we sort students. We can assure you that that does not necessarily mean our students have any real advantage in terms of understanding mathematics.

We have to induce students to invent real mathematics by making sure that that's the only kind that will work. "Work" in this context means "lead to good grades." Our students will go to great lengths to get good grades; we have to use that to our (and, ultimately, their) advantage.

Why do we use a laboratory approach? Because students actively involved with computer and calculator projects (and, to some extent, with pencil-and-paper projects) can generate for themselves a large amount of evidence about a small number of crucial concepts. Close attention to this evidence enables them to discard wrong ideas and replace them with correct ones - "correct" because the ideas are their own, and they *work,* not because the book or the teacher said so. What forces the close attention is the process of writing and revising, and laboratory projects provide a natural vehicle for written reports. What gets students to actually *do* the writing and revising is the grading scheme, in which we put primary emphasis on written expression of logical thought.

What We Do and How We Do It

As implemented at Duke, our course meets for three 50-minute periods in a classroom equipped with one computer for instructor demonstrations.[2] Each section (maximum of 32 students) splits into two lab groups; each group has a scheduled two-hour lab each week. Each lab team (two students) submits a written report almost every week; after receiving comments from the instructor, the team revises and resubmits the report for a grade.

Our laboratory (the first of an eventual four) accommodates up to 16 students working at eight IBM PS/2 Model 30-286 computers. The student stations are part of a Novell network that also includes faculty, secretarial, and classroom computers; the file server is a Zenith 386 Model 25 with a 150-megabyte hard disk. The principal software packages used in the labs are EXP (for technical word processing), *MathCAD Student Edition* (for numerical and graphical computation and for discovery experiments), and *Derive* (for symbolic and graphical computation). We have also used *FEEDBACK,* a public domain program for exploring discrete dynamical systems.

We limit lecturing to brief introductions of new topics and responses to demands for more information. Teams of four work in the classroom on substantial problems that lead to written reports approximately every other week. In weeks with no report, each student has a week-long assignment of computations that will be used in the next team project. As with the lab reports, each project report is revised and resubmitted. One class period each week is "group office hours"; the instructor responds to student problems but does not initiate new material.

---

[2] For the details of this installation, see D. P. Kraines and D. A. Smith, "A Computer in the Classroom: The Time is Right," *The College Mathematics Journal 19* (1988), 261-267.

The classroom often functions as a laboratory also. With a computer in the classroom, we can carry out "group experiments," whether planned in advance or in response to student questions. We try to avoid "show and tell" demonstrations, which may be entertaining, but usually have little lasting impact. Rather, our in-class computer use requires active involvement of the students, for example, in selecting parameters or examples and in guiding the course of the exploration.

We also have the students do non-computer experiments in the classroom. For example, they measure the period of a doorknob-on-a-string pendulum (for comparison with analytic and numerical approximations), the height of a bouncing ball and the time until it stops bouncing (to illustrate geometric series), the lengths of their arms and of the blackboards (for studies of the normal distribution), and the balance points of plane figures (for comparison with integral calculations). They take great interest in these activities, and their theoretical calculations become more meaningful when they can compare them with data they *know* are real.

What's in the Course?

We are often asked, "What topics do you leave out of the syllabus?" The only sensible answer we have found is "All of them." The SYLLABUS is part of our problem with calculus; therefore, it is not likely to be part of the solution. Our course is driven by "prototype problems" from other disciplines—physics, biology, chemistry, economics—that students can recognize as important, at least for someone, if not personally for them. These problems are selected to lead into the need to develop most of the usual "topics," but as necessary tools for solving the problems, not as ends in themselves

In our view, the *raison d'etre* for the study of calculus is differential equations. Thus, that's where we start. The need to describe information about rates of change leads to the definition of the derivative. But some rates of change are inherently discrete, and we move back and forth frequently between discrete and continuous models.

The course has a single axiom: Every initial value problem has a unique solution. This statement is intuitively obvious (after the students have seen enough direction fields, both discrete and continuous), and it's also false. We tell them that it's false and that we will take up the conditions that make it true when that becomes important. It hasn't yet. The only real problem we have with our single axiom is that it contains five words that students do not know how

to use correctly. Much to our students' surprise and dismay, mathematics and language are intimately linked.

A number of threads, both mathematical and non-mathematical, run through our course and give it structure. For example, the entire first semester (and part of the second) centers around "slope equals rise over run"—in many different guises. In addition to the obvious ones, this leads to early introduction of Euler's method, which reinforces approximation of derivatives by difference quotients and simultaneously provides a tool for generating solutions of initial value problems. Thus, students are enabled to see their unique solutions, graphically and numerically, long before they know formulas for any of them.

A second (closely related) thread is *scaling,* both of independent and dependent variables, both linear and nonlinear scaling. This leads us, for example, to log-log and semilog graphing and the detection of power and exponential functions by looking for straightness. The first introduction of the chain rule arises from proportional scaling of the independent variable; the general case arises by considering "instantaneous scaling."

A third thread is the role of inverse problems. Problems come in pairs; one of the pair is easier, and the other is interesting. We learn something about the interesting one by working on the easier one first, then turning our results inside out. Studying derivatives to get at antiderivatives is an obvious example; here's a less obvious one: "Partial fractions" is not a topic in our course, but we need to solve the logistic differential equation. Our approach is to suggest that a fraction with two linear factors in the denominator is the "answer" to a problem they have already solved—what problem? There are two possibilities, one of which (the more obvious one) leads nowhere; the other leads to a solution by "guessing" the form of the "problem."

Our non-mathematical threads are the prototype problems. We start with population dynamics, which leads naturally to exponential and logarithmic functions as our first objects of study. We return to this topic several times, introducing immigration, logistic constraints, and superexponential growth. All of these concepts are readily introduced and supported by real data, so students can see at once that we are dealing with important problems of the real world. When they discover that the growth of world population is superexponential, with an asymptote before 2030 AD, they get a very vivid picture of the population problem. Eventually, the discrete logistic model (equivalently, an Euler solution of the continuous logistic model) leads us into the modern study of chaos. Thus, our students find that some of the content of the course has been discovered in their own lifetimes.

Sample Projects

We reproduce here some of the lab and classroom projects that trace our "biology" thread from the start of the course through the middle of the second semester.[3] The following text is an in-class and take-home project from the first week of the course:

We consider the growth of fruit flies in a favorable environment: unlimited food and no predators. In the table below, we list simulated (but reasonable) population data for one laboratory colony. In this case, the function we are considering assumes only integer values. (We do not count pieces of flies!) However, we will allow our approximating function to assume fractional values. When we interpret our estimates using our approximating function, we will have to remember not to be too impressed by a prediction of, say, 788.025 flies on day 20.

Day Number	Number of Flies
0	111
1	122
2	134
3	147
4	161
5	177
6	195
7	214
8	235
9	258
10	283

1. What can we say about the rate of growth of the population? Is it constant?

2. As we have noted in the text, biologists argue that, for biological populations of the type we are considering here, the rate of growth should be proportional to the population. How can we test to see whether the data in hand supports this theory? What is the best choice for the proportionality constant?

3. Decide which (if any) of the following functions has a rate of change proportional to the function itself. For each

one that does, find the constant of proportionality.

(a) $f(t) = t^2$ (b) $f(t) = 2^t$ (c) $f(t) = 10^t$

4. Can you find a function $f(t)$ whose rate of change is proportional to $f(t)$ and that "fits" the data given above?

In the third week, we take up the fruit fly problem again with the following lab project. This is our first use of Euler's method, although the name is not used. Later we abstract the method, name it, and connect it to other uses of approximation by differentials. At this point, all the formulas are provided, and students have only to explain them. When Euler's method becomes one of their working tools, we expect them to provide the formulas.

---

1. In Project 1, we studied the growth of a fruit fly population in a favorable environment. In particular, we assumed that there was no limit, in terms of the resources available, to the population size. In that setting, we discovered that the rate of growth was proportional to the population; this led to the study of exponential functions as candidates for modeling this population as a function of time.

In this lab, we examine the problem of modeling the population as a function of time in a setting where there is a maximum population M that the environment will support. Suppose, as before, that the fruit fly population as a function of time is denoted by $P(t)$, where t is given in days. We assume (as biologists often do) that the rate of growth of P is proportional to the product of $P(t)$ and $M - P(t)$. In symbols, we are assuming

$$(*) \qquad \frac{dP}{dt}(t) = c\, P(t)\, [M - P(t)] .$$

Does a relation of the form (*) seem plausible? Why or why not?

2. Suppose we decide to model the fly population with the following values for the parameters:

(i) an initial population of 111 (as in Project 1),

(ii) a maximum population of $M = 1000$,

(iii) $c = 0.000098$ or $9.8 \cdot 10^{-5}$.

To see that the choice of c is reasonable, find the quantity

---

[3] See Priming the Calculus Pump: Inovations and Resources (MAA Notes No. 17, T. W. Tucker, ed.) for another description of Project CALC with sample materials illustrating our "physics" thread.

$c[M - P(t)]$ when $P(t)$ is small (i.e., when the limitation on resources available presumably has a negligible effect on the rate of growth), and compare this rate to the rate of growth in Project 1.

3. Our present difficulty is that we do not know how to find a formula for a function $P(t)$ that satisfies all the conditions above. For the present, rather than search for such a formula, we will *approximate* the function and its graph.

Suppose we are interested in the growth of the fly population over a 100 day period. We approximate the continuous growth by a sequence of n discrete population jumps. Suppose we decide to use $n = 20$ such jumps to start with, so each jump corresponds to a period of 5 days. We set $t_0 = 0$ and $\delta t = 100/n$ (the length of each jump). (We have been denoting this time increment by Δt in the text, but *MathCAD* does not have a capital delta available, so we make do with the lower case one.) Then we set $t_1 = 1 \cdot \delta t$ (which is 5 for $n = 20$ and $\delta t = 5$), $t_2 = 2 \cdot \delta t$, and, in general, $t_k = k \cdot \delta t$ for $k = 0, 1, 2, ..., n$.

If we approximate $\frac{dP}{dt}(t_0)$ by the difference quotient, $\frac{P(t_1) - P(t_0)}{\delta t}$, we are led to an approximate value $AP(t_1)$ of the population at time t_1 by solving the equation

$$\frac{AP(t_1) - P(t_0)}{\delta t} = \frac{dP}{dt}(t_0) = c\, P(t_0)\, [M - P(t_0)]$$

Show that this gives

$$AP(t_1) = P(t_0) + \delta t\, c\, P(t)\, [M - P(t)].$$

Since we know everything on the right side of this equation, we may compute $AP(t_1)$. Now assume that at t_1 the population is $AP(t_1)$. (Very likely this value is not the same as $P(t_1)$, but this is our "current guess.") Then, at this time, the rate of growth is

$$DAP(t_1) = C\, AP(t_1)\, [M - AP(t_1)].$$

Show that this leads to an approximate value at t_2 of

$$AP(t_2) = AP(t_1) + \delta t\, c\, AP(t_1)\, [M - AP(t_1)].$$

We may continue one step at a time with, in general,

$$AP(t_{k+1}) = AP(t_k) + \delta t\, c\, AP(t_k)\, [M - AP(t_k)].$$

4. Load the *MathCAD* worksheet LAB3. Scroll through the worksheet, and check that the calculations entered are the same ones we discussed above.

5. In general, we expect that the more time steps we use, the better an approximation to the true population function we obtain. Try four or five different values for n between 20 and 200. What happens to the approximating function $AP(t)$ as n increases? What happens to the approximation to the growth rate, $DAP(t)$?

6. Estimate the time at which the fly population is increasing most rapidly, and estimate the size of the fly population at this time. Experiment with several different values of n in determining your estimates. Which value of n seems to yield the best estimates? How accurate do you think your estimates are?

The *MathCAD* worksheet for Lab 3 includes formatted graph boxes already set up to plot AP and DAP as they are computed from the Euler formulas. Thus, students are quickly able to see that a "real" function could arise from a starting point and a growth rate, even if no formula for the function is available. They discover that the logistic model has its maximum growth rate when the population is about half its maximum—but only approximately, because Euler's method isn't that accurate. Later, after the chain rule is available, they obtain and study the second derivative and show that the inflection point occurs at exactly half the maximum population. The next project is the laboratory assignment for the ninth week of the course. Here we are introducing first-order *vector* initial value problems and the closely related parametric representation of curves (equivalently, vector-valued functions). Our initial objective is to show that Euler's method works just as well in this setting—indeed, that we are still talking about "slope equals rise over run."

This lab also sets us up for studying Lissajous figures in Lab 11, pendulum motion (with reduction of a second order differential equation to a first order system) in Lab 12, and projectile motion in Lab 13.

During 1968-69 the United States was swept by a virulent new strain of influenza, named Hong Kong Flu for its place of discovery. We may follow the spread of the disease through the population of New York City by looking at the statistics on the weekly totals of "Observed excess pneumonia-influenza deaths," that is, on the number of such deaths in excess of the average number to be expected from other sources. Starting with a somewhat arbitrary Week 1, the numbers of these excess deaths are listed below

WEEK	EXCESS DEATHS
1	14
2	31
3	50
4	66
5	156
6	190
7	156
8	108
9	68
10	77
11	33
12	65
13	24

If we assume that the number of excess deaths in a week is proportional to the number of cases of flu during an earlier week, these figures reflect the rise and subsequent decline in the number of new cases of Hong Kong Flu.

We would like to model the spread of such a disease in order to be in a position to predict what might happen with similar epidemics in the future. In order to do this, we divide the total population, numbering approximately 7,900,000 for New York, into three groups:

(i) those who are susceptible to the disease (for a new disease like the Hong Kong Flu this includes essentially everyone at the start),
(ii) those who are infected with the disease,
(iii) those who have been removed from both the susceptible and the infected groups.

This last group includes those who have recovered from the disease (and who have thereby acquired immunity) and the small number who die.

The numbers of individuals in each group, $S(t)$, $I(t)$, and $R(t)$, are clearly functions of time; our goal is to model these functions. In what follows, N represents the total population; i. e., $N = 7,900,000$.

How do these functions change? In particular, what governs the rates of change of S, I, and R. Ignoring other minor changes in the population by birth, travel, unrelated deaths, etc., the principal way an individual is removed from the susceptible group is by becoming infected with the disease. What factors are involved in the rate of this infection? Clearly these factors include the number of individuals in the susceptible category, the number of people in the infected category, and the amount of contact there is between them. If we assume each infected individual has

a fixed number, β, of contacts per day that are sufficient to spread the disease, then the number of new infectives generated by one already infected is β times the proportion of the contacts that are with susceptibles. If we assume a homogeneous mixing of the population, the fraction of these contacts that are with susceptibles is $\frac{S(t)}{N}$. Putting these assumptions together, we have

(*) $$\frac{dS}{dt} = -\frac{\beta}{N} I(t) S(t)$$

The quotient $\frac{\beta}{N}$ appears so often in what follows, that it is worthwhile to introduce a single symbol α for it:

$$\alpha = \frac{\beta}{N}$$

Using this notation, we write Equation (*) as:

(1) $$\frac{dS}{dt} = -\alpha I(t)S(t)$$

When we turn to the rate of change of $I(t)$, we need to consider the movement of individuals from the susceptible group to the infected group (as we have done above) and the movement of individuals from the infected group to the recovered group. We assume that a fixed fraction λ of the infected group will recover during any given day. Thus we have

(2) $$\frac{dI}{dt} = \alpha I(t)S(t) - \lambda I(t)$$

As discussed above, individuals pass from the infected group into the recovered group, so we have

(3) $$\frac{dR}{dt} = \lambda I(t)$$

Thus we have three quantities of interest governed by a system of three differential equations. This is again an initial value problem; we need to specify three initial conditions. These are:

(I1) $S(0) = 7,900,000$ (everyone is susceptible at the beginning),
(I2) $I(0) = 10$ (a small initial infected population),
(I3) $R(0) = 0$.

In the lab, we approximate solutions to this initial value problem by using Euler's Method. We start with experimentally determined values for β and λ:

$$\beta=.6, \quad \lambda=.34.$$

We model the epidemic over a period of 13 weeks (91 days) with a time step of a half-day.

Load Lab 9, and scroll through the worksheet. Notice that the three equations are written as one vector equation. For example, the second of these equations specifies that the number of infected individuals I_{k+1} at time t_{k+1} is the rate of change of I at time t_k (namely, $\alpha I_k S_k - \lambda I_k$) times δt plus the value of I_k at time t_k. (See Equation (2) above.) The worksheet is set up to graph I as a function of t.

1. Experiment with different values of β and λ. Explain the observed changes in the graph of I in terms of the differential equations (1) - (3) and in terms of your intuitive understanding of the problem. Notice that β and λ are global variables defined just under the graph of I.

2. Experiment with a larger number of time steps. How does this change the observed solution I(t)?

3. Graph S and R. Explain the shapes of these graphs in terms of your intuitive understanding of the problem.

4. Return β and λ to their original values. We want to investigate the extent to which the model given matches our data for the Hong Kong Flu epidemic. Recall that our assumption was that the number of new flu cases in a week was proportional to the number of deaths reported in a later week. Using the model we have constructed, calculate the number of *new* flu cases for each of the 14 weeks and compare the graph of this data over time with the graph of the "excess deaths" data. Experiment with changes in β and λ to see if you can bring the model into closer agreement with the observed data. For your use, the "excess deaths" data is listed at the bottom of the worksheet as the variable D_j.

In your write-up, discuss the assumptions that went into this model; which ones are most questionable? What might you do to obtain a better model?

We turn now to an in-class/take-home project from the second week of the second semester of the course. The logistic growth equation, used a semester earlier to model a laboratory colony of fruit flies, returns as a model for a human population. The same human population, that of the U.S. before World War II, is studied early in the textbook to determine whether "natural" growth alone or in combination with an immigration term is sufficient describe the recorded census data; the results are not spectacular. Now we are prepared to deal with separation of variables and antidifferentiation to find a formal solution that can be compared with the data. The students know from their chain rule exercise that the "maximum" parameter in the model can be estimated from observation of

an inflection point. They do not know about partial fractions yet, so they have to invent that technique to solve the problem. (See our comments earlier on the "inverse problem" thread. Except for those who will later study Laplace transforms, the two-linear-factors case is the only partial fractions technique students will ever need. In our view, more elaborate techniques should not be presented until they are needed, nor should they be presented to clearly inappropriate audiences.)

In this project you will use the logistic growth equation to develop a model for the population of the United States as a function of time. Recall that the differential equation is

$$\frac{dP}{dt}(t) = k P (M - P).$$

1. Use the graph in Figure 4 on page 6-17 of the text to estimate the date and population when the United States population was experiencing its most rapid growth. Use this to obtain a value for M. [The referenced figure shows U. S. census data from 1790 to 1940; an accompanying table shows the same data numerically.]

2. Use your value for M and the data in Table 1 on page 6-17 of the text to obtain a value for k.

3. Use your values for M and k to define a function to approximate the U. S. population, and use this approximation to estimate the population at five census years between 1850 and 1980. Look up the census data, and compare it to your estimates.

4. If you wanted to predict the U. S. population in the year 2000, would you use a logistic growth model? Why or why not?

The choice of 1940 as a stopping point for the data is a deliberate deception. Students read in our textbook about the remarkable accomplishment of the Belgian mathematician, P. F. Verhulst, who used the U. S. census data up to 1840 to predict the population 100 years later—and was off by only 1%. He was very lucky, of course; his real accomplishment was to invent the logistic model and show how to use it. Our students are, in a sense, repeating his experience, but they have the added luxury of being able to see the inflection point in the data.

Parts 3 and 4 of the project allow them to see that the 150 year run of approximately logistic data was abruptly ended by the post-war baby boom. Many of them realize right away that their estimate of "maximum supportable population" is below the 1980 census figure, and they all discover that the model would be inadequate for any predictions beyond World War II.

At the same time U. S. population is being studied in the classroom, we have a lab project on world population. Here is the text of the handout.

Once again we return to the study of the growth of populations; this time we study world population growth. We have reasonable estimates over the past millennium; of course, the more recent data is more reliable.

Date	Population (millions)
1000	200
1650	545
1750	728
1800	906
1850	1171
1900	1608
1920	1834
1930	2070
1940	2295
1950	2517
1960	3005
1976	4000
1987	5000

1. Load *MathCAD* LAB 2A. When you press the [F9] key, *MathCAD* will read in this data. Use the graph box given to plot the population as a function of time. Does it look like exponential growth? How can you check? (This is easy with *MathCAD* ; think back to last semester when we were trying to decide if a given set of data represented exponential growth.)

In 1960, observing the same trend you just discovered with *MathCAD*, Heinz von Foerster, Patricia Mora, and Larry Amiot proposed a "coalition" model of population growth, to incorporate the idea that the human population has formed a vast coalition against Nature, eliminating things that eliminate us, thereby enabling longer lifespans, decreased infant and child mortality, and greater fecundity. Hence, instead of the ratio of growth rate to population being constant, as in the "natural" model, they proposed that that ratio should be an *increasing* function of the population, say kP^r, for some constant k and some positive constant (presumably small) r. In differential equation form, their proposed model is

$$\frac{\frac{dP}{dt}}{P} = k\,P^r$$

or

$$\frac{dP}{dt} = k\,P^{1+r}$$

For r close to 0, this should give a growth rate close to that of the exponential model; thus, the model incorporates the essential idea of "natural" growth while allowing for gradually increasing "productivity." Is this reasonable?

2. Recall from earlier labs that we can approximate the derivative of a function known only by a table of values by using symmetric differences:

$$P'(t_m) \approx DP_m$$

where

$$DP_m = \frac{P_{m+1} - P_{m-1}}{t_{m+1} - t_{m-1}}$$

Display a log-log plot to see if it is reasonable to assume that DP_m is proportional to a *power* of P_m.

Find a value of r that seems to work best. (The variable r is a global variable near the bottom of the page. When you load the file, r is set to 0.) You can test your value of r by plotting $\dfrac{DP_m}{P_m^{1+r}}$ against m to see if it is "approximately constant". The second page of LAB 2A is set up to allow you to study this ratio and experiment with choices of r. When r is too small (e.g., $r = 0$, as we have seen already), there will be a definite trend of the ratios in one direction; when it is too large, there will be a definite trend in the other direction. An appropriate value of r for the model will not show a definite trend in either direction. (Be sure to check the vertical scale with each choice of r.) Print the sheet showing the graph that corresponds to the value of r you decide to accept.

3. What do solutions of the differential equation

$$\frac{dP}{dt} = k\,P^{1+r}$$

look like? This is another example of a separable differential equation and can be solved by the method described in the text. Show that the solution equation can be written in the form

$$(*)\qquad P(t)^r = \frac{1}{rk\,(T - t)}$$

where T is a new constant that depends on r, k, and the constant of integration. What happens to P as t approaches the fateful time T? The time T is called "doomsday," and the coalition model is also known as the "doomsday model." Show that for any function $P(t)$ satisfying $(*)$, $\ln(P)$ is a *linear* function of $\ln(T - t)$.

4. The calendar date T of doomsday is an "unobservable"

parameter of the model - we can't wait until t = T! On the other hand, it is obviously important to know whether there can be such a date, i.e., whether this model can possibly provide a good fit to the observed data. Load the *MathCAD* worksheet LAB 2B, which will read the same historical data and provide a graph box in which you can study whether ln P can be a linear function of ln(T — t). Enter your own guesses for T, and study the resulting graphs. (The variable T is a global variable defined under the graph box.) Devise a strategy for making the plot straighter, and print the worksheet when you have made the plot as straight as you can. What do you conclude about the reasonableness of the doomsday model and about the best value of T? Did your choice of T depend on your choice of r?

5. Discuss the philosophical and practical implications of what you have determined from the historical population data.

We have described the details of this experiment elsewhere,[4] including the *MathCAD* figure that shows the linearity of log of population versus log of time left until Doomsday when T = 2025A.D. This discovery generates a lot of interesting discussion, orally and in writing. Many of the first reactions are superficial or evasive; some take comfort in their well-established belief that math has nothing to do with the real world. Most eventually come to a realization that the math they just did says something very important—not that population becomes infinite in 35 years, but that historic superexponential growth is a *real* problem, and we're about to see some drastic effects of it.

The following text is one of the more perceptive responses to part 5 of the assignment.

> "If the doomsday model is correct, the practical implications are that we must do something about world population growth quickly. Given the assumed rate of growth in this model, there may not be enough resources to support the human population within 30 years. When you think about it, this is a very short amount of time to accomplish the dramatic social change necessary to curb high rates of population growth. The highest rates of growth (well over 3% a year) occur in the world's poorest countries. These areas already support the greatest numbers of people on the poorest

soils (and therefore have the least sustainable agriculture and the lowest agricultural potential). Such populous countries as China and India only have to experience a slightly higher-than-expected rate of population growth to significantly affect the rate of world population increase.

> "High population numbers are at the root of other, critical environmental problems such as deforestation, food shortages, excessive energy consumption, toxic wastes, and an increase in greenhouse gases leading to global climate changes. The list of problems is long and very depressing. Unfortunately, we don't see many governments taking positive actions to curb environmental excesses, and very few which address and support population control measures (the U.S. no longer supports the U.N. programs on family planning, for instance, and there are actually governments which are encouraging population increases, e.g. Iran). Even if our current projections of world population increase misrepresent the true rate (by overestimating the r value) or the true carrying capacity, the problem of overpopulation remains with us.

> "One current theory (based on the demography of developed nations) is that world population will stabilize when most of the nations of the world have reached a relatively high standard of living (which is equated with wealth in this model). This is projected to occur by the end of the next century at a population size of around 10 billion. Unfortunately, the acquisition of that standard of living requires increased energy and resource use to the exclusion of most other life on earth, and may even exceed the earth's capacity to sustain most life-forms. It's not a very wholesome picture, and there are certainly no easy solutions (as the saying goes, "there is no free lunch"). In fact we think that the dooms-day model is very appropriately named.

> "THIS IS DEPRESSING!!!"

We hasten to add that this is *not* typical freshman writing, although we might see this as a standard we would like our students to work toward. In fact, one of the two authors of the text above is a graduate student in the School of

---

[4] D. A. Smith, "Learning Software MathCAD 2.0" *UME Trends 1,2* (May, 1989). Also see D.A. Smith "Human Population Growth Stability or Explosion?" *Mathematics Magazine 50* (1977), 186-197.

Forestry and Environmental Sciences. We had two such students in the course, and both elected to take the second semester, even though their respective programs required only one semester of calculus. Their interaction with the freshmen was mutually beneficial: They learned a lot about routine manipulations and calculations from the freshmen, who were much better "prepared" for college-level mathematics in the conventional sense, and they conveyed to their teammates a sense of seriousness about and importance of the scientific enterprise.

Our final example is a lab from the middle of the second semester on the chaotic dynamics of the discrete logistic iteration or, equivalently, of an Euler approximation to a solution of the logistic differential equation. In addition to the obvious connection with the previous study of logistic growth, there is a connection with supply/demand models studied in the first semester; in that context, the continuous model always converges exponentially to an equilibrium price, but a similar discrete model becomes unstable if suppliers overreact to previous prices. Still another connection appears when students discover in this lab that they need to sum a geometric series; the behavior of successive prices in the discrete model is precisely that of partial sums of a geometric series.

This lab differs from most of the others in two respects. First, we use a special purpose software package that is not used in any other lab; second, in lieu of a written report, we have a "fill-in-the-blanks" worksheet for recording and submitting results.

Our software is a public domain program called FEED-BACK; it draws web diagrams and bifurcation diagrams for discrete dynamical systems. Students have little difficulty learning to use and practicing with a new program in the same lab session in which they use it for productive work, as long as the operation of the program is reasonably straightforward.

Our reason for not having a written report is that this lab falls right before Spring Break. The time frame required for double submission of written reports makes it difficult to assign them right before vacations and impossible at the end of each term. The results we have seen from single-submission worksheets convince us that the written reports generate much better learning, understanding, retention, and ability to make connections with other parts of the course. Thus, we retreat to the worksheets only when we are forced to do so, and we try to limit their use to topics that are interesting but not central to the course.

The primary purpose of this lab is to discover Feigenbaum's constant, which we describe as a "universal constant" that appears frequently in many apparently different contexts associated with chaotic dynamics. A secondary purpose is to discover an attracting cycle of a logistic iteration of an order that cannot arise from bifurcation. Part 1 is "practice" with the new software, the details of which are not important here; we start with part 2.

2. This step requires teamwork. **Read the rest of this paragraph carefully before beginning.** Generate a bifurcation diagram for $F(x) = r x (1 - x)$, using values of r from 2.9 to 4. (This will show you the dynamic generation of Figure 15 in the text. The entire figure takes more than six minutes to draw, so you don't want to repeat it if you can get everything you need on the first try.) What you are looking for are the r-coordinates of the bifurcation points, i.e., the values of r at which the single limit value x "splits" into two values, then at which each of those "branches" splits again to make four branches, and so on. From your reading, you should already know the first bifurcation point (record it on the result sheet), and you should know about where to look for the second and third. As the diagram is being generated, one member of the team will watch for the next bifurcation, and the other will watch the values of r change. When the bifurcation watcher says "now," the r-watcher writes down the current value of r. Do this for 2, 4, 8, and 16 branches. Keep in mind that each new bifurcation comes quicker than the one before. After you find the fourth bifurcation point, you have one more thing to look for: a value of r at which the "chaos" suddenly settles down, and a 3-cycle appears. (Precise determination of r is not so critical for the 3-cycle case.)

3. We now use the Zoom feature of *FEEDBACK* (We omit the details here.) To check your estimates of the bifurcation points, place the left edge of the zoom box over the approximate bifurcation point and read off the approximate r-value. After you have checked the bifurcation points for 2, 4, 8, and 16 branches, set the left edge of the zoom box just to the left of a bifurcation into 16 branches. Accept this and graph this portion of the bifurcation diagram. Use this portion to estimate the r-value for the bifurcation to 32 branches.

4. Now you will check and refine your result for the second bifurcation point using the web diagram. [Again, we omit program details.] Enter a value of r slightly smaller than the one you recorded for the bifurcation from two branches to four. Run the iteration continuously with trace on until it seems to have "settled down"; then pause, clear the screen, redraw the function graphs, and continue iteration. You

should see a clear picture of a two-cycle. Repeat with a value of r slightly larger than your recorded bifurcation value. You should see a clear four-cycle. Try to refine your value of r until you can't really tell whether you are looking at a two-cycle or a four-cycle; that's what happens precisely at the bifurcation point. Record your best estimate of the bifurcation value r on your result sheet.

5. Compute the distances between the first and second, second and third, third and fourth bifurcation points, and fill these in on the result sheet. Now compute the *ratios* of the first distance to the second and of the second distance to the third. What do you notice about these ratios?

6. Make a conjecture (on the basis of the very flimsy evidence at hand) about the next distance ratio, and then about the distance to the next bifurcation point (i.e., the value of r at which the first 64-cycle appears). Fill in your conjecture on the result sheet.

7. You have seen that bifurcation points emerge faster and faster as r increases, and you now have a good idea of *how much* faster. Calculate the smallest value of r by which you think a bifurcation of *every* order will have appeared, that is, by which a 2^n cycle for *every* positive integer n will have appeared. Choose an r slightly larger than this, in a range where the bifurcation diagram suggests a very large number of limiting x values. Make a continual web diagram for several hundred iterations of this case. What do you see? Repeat with a value of r close to 4. What changes?

8. In step 2, you estimated the location of a 3-cycle. Confirm (or refine) your estimate by constructing a web diagram that actually shows a 3-cycle. Print this screen, and attach the printout to your result sheet.

9. (Optional explorations if you have time remaining) In the logistic bifurcation diagram (see Figure 15 in the text again), the "upper" and "lower" branches come back together again and appear to overlap. Create a new window that starts just to the left of the second bifurcation (emergence of 4 cycles) and ends about where the overlap occurs. (Leave the range of x from 0 to 1.) Construct a new bifurcation diagram (essentially a "zoom-in" on the old one) in this window. What do you see? Can you identify an approximate location for a 6-cycle? Confirm it with a web diagram. Use the "zoom" feature to examine other interesting parts of the bifurcation diagram.

Appraisal of Our First Semester

We taught this course to 42 students in the Fall of 1989-90,

33 of whom continued in the Spring. (These numbers represent about 10% of the students in the standard calculus track in each semester.) Our students were initially selected at random from among those signed up for the standard course. During the first few weeks, students had the opportunity to switch sections, and approximately one-third of our original students did; they were replaced by an approximately equal number of students transferring in from regular sections. Most of these students had taken a calculus course in high school but had not done well enough to place beyond the first semester course.

Students found the first few weeks of the course stressful. Predictably, they were nervous about working in teams and about writing reports. However, we failed to anticipate that they would have such a difficult time with the concept of a *function*. These were, after all, very good students from very good secondary schools, and they should have been ready for a course in calculus. Their notion of function was synonymous with that of *formula;* they found it extremely difficult to think of a function in terms of its graph or in terms of data. Because of this early difficulty, we plan to revise our introductory materials to give students more experience with thinking about functions from a variety of points of view.

On the other hand, the students took to the laboratory easily. They accepted *MathCAD* and EXP readily and were soon using both programs for projects in other course. *DERIVE* was introduced in the second semester and was learned just as easily; we also had no problems using *FEEDBACK* for a single lab.

In addition to the clear pedagogical advantage of working in teams of two, such teams significantly reduce the level of frustration with software; it always helps to have someone else to discuss problems with. For the same reason, we found it important to have an instructor readily available for help with software or clarification of what was expected of the student. We are convinced that eight workstations is close to the optimal number for one lab instructor to keep track of.

We discovered a number of "things that work" by serendipity rather than by conscious design. For example, since our students worked in teams both in the lab and in the classroom, some of the first friends they made at the university were fellow calculus students—and their first joint activity was discussing calculus.

Another example: The laser printer in the lab was a greater asset than we had anticipated. Not only is it quiet and reliable, but it also makes the work the students submit look important. This, in turn, encourages them to spend more time on the write-ups. Some of this time went into cosmetics, but a considerable portion went into deciding what they wanted to say about what they had done. We had originally planned to allow use of the laser printer only for final drafts, but now we have them use it for everything unless we lack an appropriate driver for the software being used.

Student Evaluation of the First Semester

By the time students evaluated the course at the end of the semester, the consensus was that they had worked hard and that it was a good course. The following are typical quotes from their evaluations:

> • The projects were a very positive approach to calculus and taught people to really work together and exchange ideas. The lab pulled things together.

> • I liked the idea of the teacher going over our reports, giving suggestions for improvement, and resubmitting them. This way I was able to learn from and use the teacher's comments rather than not learning until after the fact—and after I received a grade.

> • ... wonderful job helping me understand what the equations, symbols, constants, etc., mean and how they relate ... for the first time I feel as if I have actually learned something.

Numerical ratings on student evaluation forms are notoriously unreliable, but we will report them anyway. In the following table, we abbreviate the questions asked and report the average ratings for both sections combined. After the first two questions, responses were on a five point scale, with "1" meaning "poor" or "not at all," and "5" meaning "excellent" or "extremely." It is not unusual for students at Duke to rate most of the faculty as "above average" (the Lake Wobegon effect), but it is unusual for an entire class to rate a calculus course as any better than "fair" (2.0).

Average number of hours per week outside class and lab	7.5
How demanding, intellectually?	Difficult to Very Difficult
How well presented?	3.3
Instructor's style, enthusiasm	4.0
Instructor's approachability	4.4
Grading fairness	4.2
Clarity of expectations	4.0
Overall rating of instructor	4.2
Text materials	3.1
Class discussions and projects	4.1
Lab projects	4.2
Increased ability to discuss and apply concepts	3.8
Increased knowledge	3.6
Increased interest	3.3
Overall rating of course	3.8

Acknowledgements

Major funding for Project CALC has come from a National Science Foundation curriculum development grant (1989-93) and an NSF-ILI grant (1989-92) to equip three-fourths of the labs and classrooms we will need as the program expands to include all calculus taught at Duke. NSF also provided a planning grant for 1988-89. In addition, the Project has received support through a major grant to Duke from the Howard Hughes Medical Institute, the purpose of which is to improve access for women and minorities to careers in science and medicine. During our preliminary phase (1988-89), we were assisted by the publishers of our selected software packages, Wadsworth & Brooks/Cole, Addison-Wesley, and the Soft Warehouse. Our network software and communications cards were provided at no cost through a grant to Duke from the Novell Corporation.

Calculus With a Symbolic Computing Lab Component

Doug Child*
Rollins College

Symbolic computing programs have the potential to revolutionize calculus courses. It seems that such a revolution will take place, but what its form will be and whether its effects will be beneficial remains to be seen. The discussion in this document reflects our version of the revolution and is based upon three experimental calculus courses held at Rollins College during the last two years. The experimental courses contained a computer laboratory component which used a mathematical computing environment under development by the author. This environment consists of the mathematical symbolic computing program Maple, a graphical interface, simple animation and object-oriented drawing tools, and a HyperCard**-style exploration delivery system that replaces menu systems with a more flexible, user-modifiable format. This computing environment serves as a powerful assistant for both the doing of and the learning of mathematics, and aids in an attempt to diagnose and repair student deficiencies, as we teach an enhanced mainstream calculus course.

Project description

Use of a computing laboratory component by the Rollins College Calculus Project is driven by the goals that we have set for our mainstream calculus course. These goals are to:

- Teach students to read and write mathematics
- Teach concepts so that students can apply calculus to problems which they have not seen before
- Teach mathematical problem-solving
- Change passive students to active learners of mathematics, especially in the 'classroom'
- Give students a feel for 'doing mathematics'
- Teach students to use powerful computing software as an assistant in the problem-solving process

Our course remains a traditional mainstream calculus course in which the methods of teaching are greatly changed by the use of a computing laboratory based upon the author's computing environment. However, some important changes in the curriculum have taken place. We have added

- more formal discussion of basic logic and problem-solving strategies,
- discussions of how to use symbolic computing programs to solve mathematics problems correctly,
- better integration of models of important mathematical objects like 'paths' and sequences,
- better balance between numeric and closed form methods of solving integration and summation problems, and
- the use of iteration to develop models of many important mathematical processes.

The author's computer environment, which is called Calculus T/L, has been designed to help achieve the goals and implement the changes stated above. It has been designed to teach students to read and write mathematics by helping them write complete solutions to calculus problems. Much attention has been given to using careful representations of mathematical objects and optimizing the expressiveness of the program. The solutions created using Calculus T/L contain graphic, symbolic, numeric, and textual information organized on an electronic blackboard. Calculus T/L operates at several distinct levels of detail. Solutions may contain steps like Differentiate f, Find the zeros of Df, Evaluate Df(c), or steps like the first two can be broken up into the pencil and paper size steps that students are normally asked to write. Calculus T/L provides context sensitive help in the form of collections of problem-solving operators that may be applied to the given problem. This feature enhances the operational semantics of mathematical objects used to form solutions of problems. Some topics that are fundamental to the understanding of calculus can be viewed from three or four levels of detail. The student spends time at a given level based upon his needs for understanding. The capability to let different students spend different amounts of time on different activities is important. Our calculus courses draw students with a broad range of mathematical skills. It is important that students can spend as much time as they need in order to master important classes of problems.

*Supported by The Sloan Foundation and by The National Science Foundation

The mainstream calculus course at Rollins serves mathematics and science students as well as other liberal arts students. Classes are restricted to twenty students and meet five hours a week. Total class time during a semester is tending toward being broken evenly into computing laboratory and discussion sessions. The laboratory component is taught in one hour and fifteen minute sessions. Students work in teams of two and prepare lab reports that are due at the next class meeting. Most computer lab sessions are preceded by pre-lab exercises that prepare students for effective use of lab time. Some lab assignments require students to return to the computing lab to do additional work. The discussion sessions contain very little lecturing and are primarily used to probe concept understanding and to work (frequently just set-up) a variety of problems.

Students spend laboratory time discussing mathematics in pairs. After several weeks most students have adjusted to taking an active role in learning mathematics in our computing laboratory. Students in our current courses are required to write and speak much more mathematics than in previous courses. A computing lab is an excellent vehicle for getting students to talk to one another about mathematics. It is important to have an instructor in the lab to facilitate student discussions. Working computer models provide lots of mathematical information for students to interpret and to write about. Students are taught to make observations and then to seek explanations, to make comparisons, and to reorganize calculus topics to obtain a more global view.

The nature of the questions that students are expected to answer is different from that of our previous courses. Less emphasis is placed upon questions that ask students to compute something. Students can compute limits (as an example) without understanding, at any desirable level, what a limit is. We rephrase computational questions in several ways. For example we frequently say "Explain how to compute X," or "Discuss X," instead of saying "Compute X." For many kinds of problems, we provide a collection of English-like commands for students to use. These commands are also understood by Calculus T/L, so students have practice in answering such questions. A different type of problem asks students to write an algorithm for solving a class of problems. A class of problems for which this method has been successful is the class that begins "Use asymptotes to graph ..." We introduce these problems in the usual way, discuss relevant examples, and assign homework problems. During an ensuing computer laboratory, we ask students to develop and test a strategy (method or algorithm) for solving this class of problems. The results are very revealing; there is a strong correlation

between students who can write an effective algorithm and those who can do these problems in the traditional way. By asking students to create an algorithm, the instructor directs student learning energy toward a higher, more abstract level than that obtained by asking students to sketch several graphs. Students must consciously look for heuristics and develop strategies, thus reorganizing declarative knowledge into a form that is useful for solving problems based on the information.

Many students do not seem to approach learning mathematics the way mathematicians do or even the way mathematics instructors think students do. It is important to address this issue early and often in the calculus course. We use highly structured models of problem solving processes until students have developed habits more appropriate for learning mathematics. These models help to communicate to students useful methods of thinking about certain problems. In addition, we stress the importance of mathematical objects, concepts, object classification, and pattern matching in the reading and writing of mathematics. Students seem to have little or no understanding of these important matters. An apparent benefit of our approach is that students write solutions containing proper symbols in appropriate places.

This section concludes with a brief discussion of several of the more important lessons learned in the last three years.

The use of powerful computing tools in a class does not make its students smarter or improve their mathematics backgrounds. It is important to develop accurate models of your students' mathematical knowledge. Spending time with your students in a properly equipped computing laboratory is an especially effective means of achieving this goal. We are also designing tests that focus on non-computational questions and look for knowledge we consider to be necessary for learning calculus. We do not claim to understand what is "necessary," but believe that an appropriately designed computer environment used in a extensive laboratory component creates a setting in which we can experimentally learn how to help our students learn calculus.

Converting students from passive learners into active learners is hard work. Looking for examples and exploring mathematical concepts are foreign ideas to most students entering college. It is important that instructors be sensitive to this issue. Many students expect to be told exactly what mathematics they need to learn and exactly how to solve the requisite collection of problems. An important aspect of this conversion is an open discussion with students about the conversion that is being attempted and potential benefits for them. Students have been well trained to be passive in the mathematics classroom and to strive to obtain answers to a relatively small number of different classes of problems. Breaking this pattern causes discomfort that can usually be ameliorated by frank and open discussion.

Computer activities, designed to introduce an idea and to let students explore it, take more time than traditional methods of presenting the same idea in the classroom. So, in-class computer activities probably should be reserved for those topics that are very important and can be better done in the computing laboratory. On the other hand, precious course time can be saved for interesting problems and important ideas by asking students to practice routine, but necessary problems, as computer-aided homework problems. Our experience has been that coverage of calculus topics is slow at the beginning while we focus on strengthening student backgrounds, but picks up dramatically as the course proceeds. In fact there is some evidence that our students are learning how to learn mathematics. Toward the end of calculus I last Fall, we had more success getting students to read sections in the text and master their problems than in previous years.

Example laboratory activities

The examples included below were chosen to demonstrate a wide range of activities that are used in the laboratory component of our experimental calculus courses.

The meaning of f(x+h)

Let f be the name of a known function. What does it mean to **substitute x+h into f** ? *What will a student learn about calculus if the student does not understand substitution?* Each year a significant number of students come to our calculus courses with a non-functional understanding of the idea of substitution. This is a remedial topic that is dealt with only after student difficulties have been observed.

Computer algebra systems can easily produce f(x+h) for the functions studied in calculus. Many students obtain the necessary functional understanding by trying examples until they correctly anticipate computer solutions.

However, each year we have several students who need something more. We use object oriented drawing facilities to provide the extra sensory dimension of touch for these students. Suppose the function f defined by $f(x) = x^2 + x$ is being considered. First, what is f(2)? How is it formed? Students form $f(\ ) = (\ )^2 + (\ )$ on the electronic blackboard,

create three copies of 2, and drag a copy of 2 into each parenthesized opening. Then students consider f(x+h) in a similar manner. This technique has proven effective for the few students who need it.

Additive property

A discussion of the additive property can be very beneficial near the beginning of a calculus I course. It attempts to anticipate many common 'algebraic' errors, to help students understand the meaning of f(a+b), and to help students learn how to think about problems involving quantifiers. The basic question is

• For a given function named f, is f(a+b) = f(a) + f(b) for all meaningful choices of numbers a and b?

The discussion begins with computer examples of one function, f: x -> 3 x, that does possess the additive property and one, f: x -> 3 x + 1, that does not. Both examples conclude with justified conjectures. Students are then turned loose with a model in our symbolic computing environment to look for as many functions as they can find that have the additive property. The model stresses how one should think about such problems as students make and verify conjectures.

The lab exercise leads students to conclude that the additive property is rarely satisfied by functions known to students. The impact is much greater than if students are just told the result. This activity has proven to be quite effective.

Derivative problems

"Compute the derivative" problems are relatively easy. Most calculus students appear to master them quickly, although some have difficulty properly using the chain rule. But, there are difficulties with teaching students to do these problems. Typically half of our students at Rollins already know how to differentiate when they arrive in our calculus courses. It is awkward to spend much class time helping the other half. Students may not understand differentiation problems as well as instructors think they do. Why do students have so much trouble learning implicit differentiation? Perhaps, students can be taught to understand differentiation problems more thoroughly. We have had great success teaching implicit differentiation. Actually, students seem to be able to learn it themselves. The author spent only fifteen minutes of class time on implicit differentiation in last Fall's calculus I course. All students except for one correctly worked test questions concerning implicit differentiation. We have students do other activities that may help explain this success. For example, we have an early lab session whose main purpose is to make sure students understand composition. After all, *if a student does not understand composition, what will the student learn about calculus?*

SAMPLE PROBLEM: Differentiate x sin(x+2) wrt x

GOAL: Apply derivative problem-solving operators until the original problem has been rewritten so that all derivative subproblems are derivatives of basic functions (i.e., functions that can be differentiated in one step).

There are just a few problem solving operators for these problems. This is a measure of the difficulty level of this class of problems.

> Apply derivative of a scalar product.
> Apply derivative of a sum.
> Apply derivative of a product.
> Apply derivative of a quotient.
> Apply derivative of a composition.
> Apply derivative of a basic function.

The method of problem solution can be described as the recursive application of properly chosen (by pattern matching) problem-solving operators until only subproblems of size one (i.e., derivatives of basic functions) exist. Then differentiate all basic functions.

Students select an operator to apply to the current version of the problem. If the student chooses *Apply derivative of composition* and there isn't such a thing in the current representation of the problem, then an appropriate message is displayed. Otherwise, all derivatives of compositions are expanded using the chain rule. A goal of this activity is to accurately model the way experts solve this class of problems.

Students start a problem like this one by writing it down and then choosing a problem solving operator based upon the structure of the problem. If a student correctly chooses to apply the product formula, then the development of a solution proceeds like a pencil and paper solution.

$$\frac{d}{dx} [x \sin(x + 2)] =$$

$$\sin(x + 2) \frac{d}{dx}[x] + x \frac{d}{dx} [\sin(x + 2)]$$

If a student chooses an incorrect problem-solving operator, then a message stating that no derivative of the chosen form exists in the current problem is displayed and no problem step is taken. A complete script to solve the problem above looks like:

> Apply derivative of product.
> Apply derivative of a composition.
> Apply derivative of a basic function.

This script is a higher order representation of the solution in the sense that it represents the solution of many different problems. We are experimenting with alternate representations of solutions to many types of calculus problems. Our goal is to improve the transferability of problem-solving knowledge to other kinds of problems.

We want students to practice pattern matching so that they can read and construct symbolic arguments (including those involving implicit differentiation). The method outlined above allows students to develop strategies without concern for execution errors. We want students to look at a derivative problem and think, for example, that it is a *derivative of the quotient of two compositions*. We think that many students use lower level, less transferable methods to understand differentiation problems. Students who need practice executing individual problem solving operators can do so without affecting the flow of our calculus course. A computer is a tireless teaching assistant. Most students have to struggle to learn how to solve derivative problems that require the use of more than three problem-solving operators.

Students are tested by asking them to write programs that solve derivative problems as well as the traditional computational problems. The first type of problem seems to be very helpful in developing transferable knowledge.

Mean-Value Theorem

First, students are asked to observe that, for an example function named f and a closed interval [a,b], if one draws a chord through (a,f(a)) and (b,f(b)) and makes a moveable copy of the chord which remains parallel to the original, then the copy can be dragged into position as a tangent line to the graph of f at (c,f(c)) where c is between a and b. Such a function is said to have the mean-value property on [a,b].

Students are asked to find several other functions with the mean-value property on [a,b]. They are also asked to find two continuous functions and corresponding intervals [a,b] such that the function does not have the mean-value property on [a,b].

Students use a model that draws the graph of f over [a,b], draws the chord, makes a moveable copy, proceeds to differentiate f, sets up the equation Df(c) = slope of chord, solves the equation for c, and draws the appropriate tangent line.

The model makes it easy for students to try out various functions. It combines graphics, object oriented drawing, and symbolic computation to introduce students to this important property.

Word Problems

Max/min word problems are good examples of problems for which a powerful computer environment (1) leads to improved problem-solving methods and (2) provides instructor control over the level of detail contained in problem discussions.

Let's consider a standard calculus word problem.

The Problem:

Given a circle of radius R, find the rectangle of largest area that can be inscribed in the circle. We encourage our students to subdivide the problem into the following six parts.

1. Draw a picture
2. Obtain function to maximize. Specify domain.
3. Graph function. (Does the graph look reasonable?)
4. Differentiate f and find zeros of Df.
5. Make conjecture about solution.
6. Verify (corroborate) conjecture.

The use of a powerful computer environment greatly changes how students solve such a problem. The environment allows the instructor to focus student efforts. We let students use the full power of Maple to perform step 4, which is computational. The corroboration steps 3 and 6 are important ones that our students never performed before. Having students carefully draw pictures using drawing commands of the form **Draw rectangle with upper-left corner (a,b) and lower-right corner (c,d). and Draw circle with center (cx,cy) and radius r** helps some students see how to perform step 2. Note that we do not allow students to "prove" things using graphs. Verification steps require application of appropriate definitions and theorems.

Sequences

A computing environment can be very useful for helping students learn about sequences. Calculus T/L contains several tools that students use to examine the behavior of sequences. It also contains an interactive textbook style introduction to convergence of sequences and a model problem-solving method that encourages students to consider examples in order to gain insight about constructing proofs of convergence using a rigorous definition.

Sequence tools and the behavior of sequences

The tools we use to study the behavior of sequences include a simple table tool, a point plotting tool, and a sequence picture tool. The figures below demonstrate the use of each tool for the sequence { 1/ln(n) } . First, define A by A(n) = 1/ln(n).

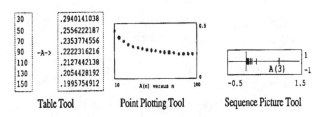

| Table Tool | Point Plotting Tool | Sequence Picture Tool |

The point plotting tool and the sequence picture are designed to draw the next point or line segment upon request instead of drawing all of the objects at one time. The sequence picture above was generated by initializing the sequence index to 3 and repeatedly multiplying it by 10 until $3*10^{15}$ was reached. Students get a sense of the 'rate of convergence' by the way they must choose the next terms to consider (students learn to choose appropriate subsequences) in order to visibly move toward a limit. Our computer environment also contains an animation tool that brings to life the notions of convergence and divergence.

Before convergence is formally considered, students are asked to describe the behavior of various given sequences as the term index grows without bound. All answers are expected to be precisely written in English. They are then asked to find examples of as many different kinds of sequence behaviors they can. Global concept questions like the preceding one are good questions for helping students organize their understanding of a topic.

Series

Students can use functions to represent the difficult series concept. Once appropriate functions are defined, they can be viewed numerically, graphically, and symbolically for improved understanding. A Calculus T/L dialogue for creating series follows. Each time students begin a series problem, they follow the steps below. The series is determined by the choice of A(n) and by the indices used in the definition of S.

Start with a sequence A(n), where A can be thought of as a function from the positive integers to the reals.

- Define A by A(n) = 1/n.
        A: n-->1/n

Define a sequence of nth partial sums of A(n). This sequence is a series derived from A(n).

- Define S by S(n) = sum(A(i),i=1..n).

$$S: n \longrightarrow \sum_{i=1}^{n} A(i)$$

Note that the boxed objects above are Calculus T/L representations of function objects as they appear in the electronic black board. The remaining information and commands appear in a document used to examine series.

There are several difficult ideas here. One is the idea of a sequence as a special kind of function. Another is the definition of a series as a sequence derived from another sequence. A third is the notion that S is a function of the upper limit in a summation. Bringing these ideas to life as functions that can be evaluated, graphed, and symbolically manipulated is helpful to students. A point not to be overlooked is that each time students begin to consider a series, they read (at least they have the opportunity to read) the definitions that precede the define commands above.

Summary

The laboratory activities described above are designed to do the following:

- Prepare students for new concepts and theorems,
- Teach symbolic manipulation,
- Practice routine problems,
- Use computer models to help students build abstractions,
- Diagnose and address student weaknesses,
- Teach improved methods of solving problems,
- Teach students how to "do mathematics", and
- Construct complete, multi-view solutions to problems.

Conclusion

A properly designed mathematical computing environment can lead to remarkably improved methods for teaching the concepts of calculus as well as for solving calculus problems. A computing laboratory can be a place where students gain insights about mathematics that can not be achieved by using textbooks or traditional classroom knowledge delivery methods. It can be a place where students talk about mathematics, a place where students can experience the excitement of doing mathematics, observing important relationships, making and testing conjectures. It is a place where students (and other casual users of mathematics) can learn to solve problems in new and exciting ways. A mathematical computing laboratory is also an ideal home for experiments involving the learning of mathematics using new teaching methods. Given the current sad state of mathematical education in our country, such experimentation may be the most important kind of laboratory activity that can occur in the nation's calculus classes of today.

Calculus and *Mathematica*:
A Laboratory Course for Learning by Doing
D. Brown, H. Porta, and J. J. Uhl
University of Illinois

Introduction

In two years, *Mathematica* has already revolutionized the desk top of mathematicians and scientists all over the world. Pencils and writing paper have begun to disappear as more and more scientists begin to rely on *Mathematica* as their calculating companion of choice. But we believe that the real impact of *Mathematica* will be in the classroom.

Everyone recognizes the dismal quality of the currently available software for mathematics teaching. Writing in 1984, Jacob T. Schwartz explained the situation:

> *While video games have grown into a five billion dollar industry, computer-aided instruction (CAI), which is supported by many of the same technological trends (including cheap microprocessors and high quality graphics), is still a largely static area. Two main reasons for this difference can be asserted.*

> *(A) CAI systems developers have not made effective use of the exciting, fast- paced graphics...*

> *(B) The cost of creating high quality courseware is still very high and because of this, most currently available courseware is mediocre. The cure for this problem will require a substantial up-front investment in improved courseware writing tools ...*

> *Presently, most CAI courseware designers are attempting to build highly constrained miniature systems ... Even though the commercial pressures that lead to this ... are clear, emphasis on today's machines is a very poor way of exploring the true potential of CAI technology. To prepare for the possibilities that the next few years will open up, we need to "lead" the rapidly moving hardware "target" substantially, by writing software and creating courseware not for today's machines, but for systems that will be typical five years or a decade from now. We also need to take courseware very seriously and strive to create "classics" that can be used for years into the future.*

Enter *Mathematica*. Through the medium of *Mathematica* Notebooks, *Mathematica* becomes a powerful system for producing electronically alive graphically active mathematics courseware. *Mathematica* without Notebooks is a superb mathematics processor; *Mathematica* with Notebooks is an unparalleled system for producing mathematics courseware.

Bringing *Mathematica* into the classroom.

There is much excitement in the mathematics community about bringing *Mathematica* into the classroom. Nearly all the grant applications from four year colleges for the NSF instrumentation program specified *Mathematica*. Diverse universities and colleges plan to use *Mathematica* in many courses at all levels.

Too often, however, these plans do not go beyond the simple addition of a laboratory section to existing courses. This sounds good in principle, but in fact this can be a step backwards. In the classroom, *Mathematica* is not a quiet guest. It can singlehandedly destroy a standard calculus course because eighty per cent of the questions usually asked in a calculus test or assignment can be answered by simply evaluating *Mathematica* instructions. Answering standard calculus questions with *Mathematica*'s help teaches little more than a bit of typing. *Mathematica* can be brought into the classroom successfully at the same pace as new courses are designed.

Too often, however, these plans do not go beyond the simple addition of a laboratory section to existing courses. This sounds good in principle, but in fact this can be a step backwards. In the classroom, *Mathematica* is not a quiet guest. It can singlehandedly destroy a standard calculus course because eighty per cent of thequestions usually asked in a calculus test or assignment can be answered by simply evaluating *Mathematica* instructions. Answering standard calculus questions with *Mathematica*'s help teaches little more than a bit of typing. *Mathematica* can be brought into the classroom successfully at the same pace as new courses are designed.

Mathematica Notebooks

Very simply, *Mathematica* Notebooks allow fully word-processed text to be inserted in the middle of active code. This constitutes a new medium of communication that combines the advantages of a standard word processor, the advantages of an enormously powerful easy-to-use computer algebra system and superb graphic capabilities. If the Notebooks are used to their full potential, the result is an electronic text in which every calculation and graph is alive and can be reexecuted with different parameters. Instead of a single dead printed example, each example is infinitely many examples. With *Mathematica* Notebooks, the reasons for an upcoming calculation can be discussed, the calculation can be executed and the meaning of the result can be assessed in one single medium. With *Mathematica* Notebooks, calculations and plots can be done in context with virtually no need for a printed supplement.

We realized early on that the medium of *Mathematica* Notebooks had the potential to change forever the way undergraduate mathematics is taught. And it was for this reason alone that we began the Calculus & *Mathematica* project at Illinois.

What is Calculus & *Mathematica*?

Calculus & *Mathematica* is a collection of *Mathematica* Notebooks that brings students through calculus of one variable (a multivariable sequel lies in the not-too-distant future). The format is that of problems and solutions. Each notebook opens with Basic problems which introduce many of the new ideas in the material under study. Second comes the Tutorial problems which introduce techniques and applications. Full electronically active solutions are provided to each Basic and Tutorial problem. Closing each notebook is a section of Give it a try problems. Here no solutions are given, but the student through word-processing and calculation adds his or her solution directly to the notebook and electronically turns in the completed notebook for comments, suggestions and grading.

Presentation of Calculus & *Mathematica*

The course is taught as a laboratory course. Lectures are held to the minimum. The students themselves asked us to spend less time on lectures and more time in the lab. We obliged and noted significant improvement. Now our typical week allows for about 30 minutes of lecture and the rest of time for the lab.

The computers are arranged in a large U formation and placed the professor at the top of the U looking over the students' shoulders. From this vantage, the professor can monitor the work of each student and quickly spring to the aid of a student in trouble. The professor can be in direct contact with the student at the optimal teachable moment. The Calculus & *Mathematica* professor is taken out of the spotlight in front of the students and is released from the role of curator of the dogma and arbiter of truth. Instead the professor is placed in a strong supporting role behind the students, pushing them instead of pulling them. Teaching in this way is particularly satisfying because the students learn by working. One of our students remarked that ours was the only class in which he did his homework. When asked why, he replied "Because homework is the class."

The correct viewpoint for *Mathematica* courseware

Traditional courses proceed by releasing measured amounts of theory. Then the theory is illustrated by the exercises. The professor is the curator of the dogma and arbiter of truth; the exercises are the catechism. Mathematics becomes a cult, not a human scientific endeavor. Courses based on *Mathematica* Notebooks do not have to be this way.

In fact, courseware based on *Mathematica* Notebooks ought to be based on the principle:

Calculation and plotting set up theory which in turn sets up more calculation and more plotting which in turn ...

This is the way research in mathematics is done and this approach brings mathematics to life. In this way students experience the enchantment and excitement of a calculation in progress. Furthermore because of *Mathematica*'s graphic and calculational power, students learning this way make computations that lead to a mathematical experience far richer than students taught the traditional way. Operating a traditional course based on printed material in accordance with this principle is nearly impossible because the required hand calculations are beyond the reach of the students and some of their professors.

Here are two examples from the calculus & *Mathematica* electronic text:

Example: Natural Logarithm.

Conventional wisdom has it that a student must have studied the integral before they can study the natural logarithm. This approach is not historically accurate and calculations can get around this dogma.

Here is the introduction to the natural logarithm in the Calculus & *Mathematica* electronic text. At this point, the students have seen the chain rule and derivatives of the algebraic functions.

Logarithms: natural and otherwise.

(The dark lines are *Mathematica* input; the lines immediately below the dark lines are *Mathematica* output. The other lines are text mixed between the calculations. In this printed medium, the spirit of the electronic course is blurred because students read the text interactively by executing the *Mathematica* commands as they progress through the lessons. Here we are forced to compromise by showing the electronic text after it has been executed.)

Basic Problem 1.a

Set f[x] = Log[b,x]. Explain why f'[x] must be given by f'[1]/x for x > 0

Answer:

We know f[x y] = f[x] + f[y].

Clear [x,y,f]

On the left, we have:

left = f[x y]

f[x y]

On the right, we have:

right = f[x] + f[y]

f[x] + f[y]

Differentiate both sides with respect to y and set them equal to each other.

derivedequation = (D[left,y] == D[right,y])

x f'[x y] == f'[y]

Set y = 1, and solve for f'[x]:

Solve [derivedequation/ .y-.1, f'[x]]

$$\{\{f'[x] \to \frac{f'[1]}{x}\}\}$$

Done. That wasn't too bad.

Basic Problem 1.b

Compute the derivative of Log[10,x], the logarithm base 10.

Answer:

Let f[x] = Log[10,x]. The last problem says f'[x] = f'[1]/x. So we try to see what the limiting value of (f[1 + h] - f[1]/h is as h closes in on 0.

Plot [(log[10,1 + b] - Log [10,1])/h,{b,-.001,.001}

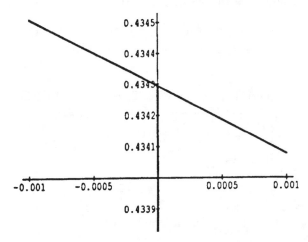

It seems that f'[1] is about .4343. To get more accuracy we can look at:

N[((Log[10,1 +b] - Log[10,1])/b)/.b->.01]

0.432137

N[((Log[10,1 + b] - Log{10,1])/b)/.b->.001]

0.434077

N[((Log[10,1 + b] - Log{10,1])/b)/.b->.00001]

0.434292

N[(((Log[10,1 + b] - Log[10,1])/b)/.b->.000001]

0.434294

N[(((Log[10,1 + b] - Log[10,1])/b)/.b->.0000001]

0.434294

It seems that f'[1] is about .434294

So f'[x] is about .43294/x.

It is clear that we can estimate the derivative of Log[b,x] for any base b the same way.

Check out derivatives of various logarithms by using the *Mathematica* command **N[D[Log[b,x], x]]** to prepare a table of derivatives of Log[b,x] for b= 2,3,4,...,10.

Answer:

Table [{b,N[D[Log[b,x]]), (b,2,10)]

```
    1.4427      0.910239     0.721348    0.621335
{{2,--------},{3,------------),{4,------------},{5,----------},
     x            x             x            x

    0.558111    0.513898     0.48089     0.45512
{6,---------},{7,----------},{8,----------},{9,----------},
     x            x             x            x

    0.434294
{10,------------}}
      x
```

Basic Problem 1.c

What drawback is present in each of the derivatives that appear in the table prepared for Part b)?

Answer:

One bothersome drawback should now be apparent. For the values of **b** we have looked at in the table, the derivative **D[Log[b, x], x]** contains a weird rounded number in its numerator. Whenever we have a **b** such that **D[Log[b, x], x]** contains an untractable number in its numerator, then precise calculations involving **Log[b, x]** and its derivative are hard to do. What we need is a base **b** such that **D[Log[b, x], x]** is as clean

as possible.

Basic Problem 1.d

What base is natural for accurate calculation?

-Answer:

The base for the cleanest possible derivative for accurate calculations is the base called **e** such that **Log[e, x] has derivative 1/x.**

This number **e** is the natural base for logarithms and **Log[e, x]** is called the **natural logarithm** function.

Basic Problem 1.e

Find **e** to three accurate decimals.

Answer:

Look at:

Table [{b,n[D[Log[b,x],x]]}, {b,2,10}]

```
    1.4427      0.910239     0.721348    0.621335
{{2,--------},{3,------------},{4,-----------},{5,----------},
     x            x             x            x

    0.558111    0.513898     0.48089     0.45512
{6,----------}, {7,------------},{8,-----------},{9,---------},
     x            x             x            x

    0.434294
{10,-------------}}
      x
```

The table shows that

D[Log{2, x], x] > 1/x > D[Log{3, x], x]

Therefore **e** is between 2 and 3.

Now look for **e** between 2 and 3:

Table[{b,N[D[Log[b,x],x]]},{b,2,3, .1}]

```
    1.4427       1.34782      1.2683       1.20061
{{2,--------},{2.1,----------}, {2.2,---------},{2.3,--------},
      x            x             x              x
```

$\{2.4, \frac{1.14225}{x}\}, \{2.5, \frac{1.09136}{x}\}, \{2.6, \frac{1.04656}{x}\},$

$\{2.7, \frac{1.00679}{x}\}, \{2.8, \frac{0.971233}{x}\}, \{2.9, \frac{0.939222}{x}\},$

$\{3., \frac{0.910239}{x}\}\}$

This table shows that

D[Log[2.7, x] > 1/x > D[Log{2.8, x],x]

Therefore **e** = 2.7 to one accurate decimal.

Look for **e** between 2.7 and 2.8:

Table[{b,N[D[Log[b,x],x]]}, {b,2.7, 2.8, .01}]

$\{\{2.7, \frac{1.00679}{x}\}, \{2.71, \frac{1.00306}{x}\}, \{2.75, \frac{0.999369}{x}\},$

$\{2.73, \frac{0.995717}{x}\}, \{2.74, \frac{0.992105}{x}\}, \{2.75, \frac{0.988532}{x}\},$

$\{2.76, \frac{0.984998}{x}\}, \{2.77, \frac{0.981501}{x}\}, \{2.78, \frac{0.978042}{x}\},$

$\{2.79, \frac{0.974619}{x}\}, \{2.8, \frac{0.971233}{x}\}\}$

This table shows that

D[Log[2.71, x], x] > D[Log[2.72, x], x].

Therefore e = 2.71 to two accurate decimals.

Let's go to the well again:

Table[b,N[D[Log[b,x], x]]}, {b,2.71, 2.72, .001}]

$\{\{2.71, \frac{1.00306}{x}\}, \{2.711, \frac{1.00269}{x}\}, \{2.712, \frac{1.00232}{x}\},$

$\{2.713, \frac{1.00195}{x}\}, \{2.714, \frac{1.00158}{x}\}, \{2.715, \frac{1.00121}{x}\},$

$\{2.716, \frac{1.00084}{x}\}, \{2.717, \frac{1.00047}{x}\}, \{2.718, \frac{1.0001}{x}\},$

$\{2.719, \frac{0.999736}{x}\}, \{2.72, \frac{0.999369}{x}\}\}$

This table shows that **e** = 2.718 to three accurate decimals.

Mathematica has a built in constant (called **E**) that evaluates as **e** to any desired precision.

Example: Convergence

In previous lessons the students have been shown how to obtain the expansion of Cos[x] in powers of x via the method of undetermined coefficients but the word "convergence" has never been used. They also know that the object **Series[]** produces the desired expansion directly, so that they don't have to calculate each case by the method explained.

At this point the course directs the students to do many plots of the following type. The students learned about convergence on their own. And they even gave it a name; they called it "cohabitation." (Again in this printed medium, the spirit of the electronic course is blurred because students read the text interactively by executing the *Mathematica* commands as they progress through the lessons. Here we are forced to compromise by showing the electronic text after it has been executed)

Part 1

Plot Cos[x] and the sum of the first three nonzero terms of its expansion in powers of x on the same axes for $-1 \le x \le 1$.

Answer:
first three = Normal [Series[Cos[x], {x,0,4}]]

$$1 - \frac{x^2}{2} + \frac{x^4}{24}$$

Plot [{Cos[x], first three}, {x,-1,1}]

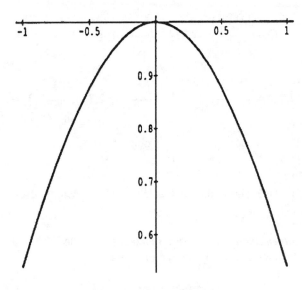

Answer:
firstfour = Normal[Series[Cos[x], {x,0,6}]]

$$1 - \frac{x^2}{2} + \frac{x^4}{24} - \frac{x^6}{720}$$

Plot [{Cos[x], firstfour}, {x,-2,2}]

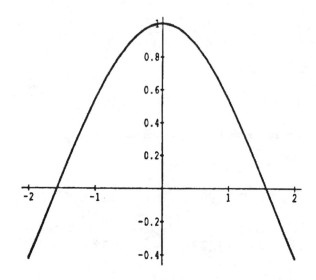

Part 2

Plot Cos[x] and the sum of the first three nonzero terms of its expansion in powers of x on the same axes for $-2 \leq x \leq 2$.

Answer:

Plot [{Cos[x], firstthree}, {x,-2,2}]

Part 4

Plot Cos[x] and the sum of the first four nonzero terms of its expansion in powers of x on the same axes for $-3 \leq x \leq 3$.

Answer:

Plot [{Cos[x], firstfour}, {x,-3,3}]

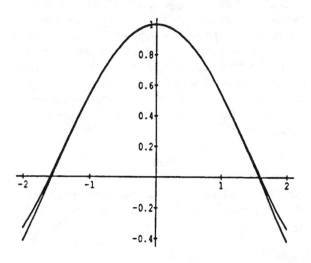

Part 3

Plot Cos[x] and the sum of the first four nonzeroterms of its expansion in powers of x on the same axes for $-2 \leq x \leq 2$.

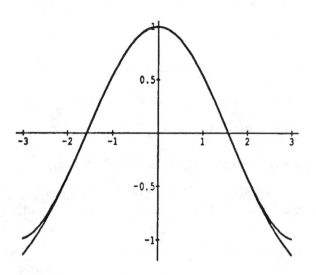

Part 5

Plot Cos[x] and the sum of the first six nonzero terms of its expansion in powers of x on the same axes for -5 ≤ x ≤ 5.

Answer:

firstsix = Normal [Series[Cos[x], {x,0,10}]]

$$1 - \frac{x^2}{2} + \frac{x^4}{24} - \frac{x^6}{720} + \frac{x^8}{40320} - \frac{x^{10}}{3628800}$$

Plot[{Cos[x],firstsix}, {x,-5,5}]

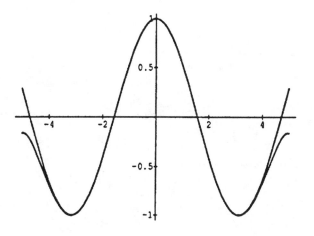

Part 6

Plot Cos[x] and the sum of the first seven nonzero terms of its expansion in powers of x on the same axes for -5 ≤ x ≤ 5.

Answer:

firstseven = Normal [Series[Cos[x], {x,0,12}]]

$$1 - \frac{x^2}{2} + \frac{x^4}{24} - \frac{x^6}{720} + \frac{x^8}{40320} - \frac{x^{10}}{3628800} + \frac{x^{12}}{479001600}$$

Plot[{Cos[x], firstseven},{x,-5,5}]

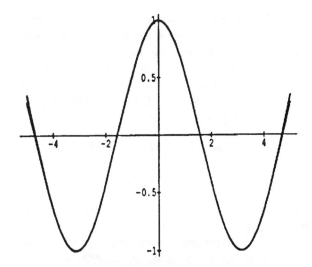

These examples show how calculations and graphics can isolate and explain concepts more satisfactorily than long paragraphs full of vagaries.

Mistakes we made and what we learned

Mistake 1: We thought *Mathematica* syntax would be hard for the students to learn. We were wrong. The syntax may be hard for professors, but it is fairly easy for most students. Through the medium of *Mathematica* Notebooks, the students saw *Mathematica* commands and *Mathematica* routines in context and picked them up very quickly. After the second week, *Mathematica* syntax was not much of a problem.

Mistake 2: We thought a series of prewritten *Mathematica* packages hidden from the students' view would help the exposition. We even went so far as to consider "pull-down menus" for some nasty tasks. But then we realized that "user-friendly classroom interfaces" is a step towards removing calculus from its base position in science and technology. The "user-friendly classroom interfaces" are wrong because they would transfer calculations into video games that can be played only in the calculus classroom. We want our students to be able to use *Mathematica* wherever they find it. Therefore we do not customize *Mathematica* in any way; instead we use *Mathematica* directly as it comes off Wolfram's shelf. Even all the grisily graphics instructions are in the open for our students to use and modify. This decision, more than any other, resulted in our students becoming competent calculators very quickly.

Mistake 3: At first we envisioned teaching the course in

the traditional way using *Mathematica* to solve the problems. In short, we were going to present the theory pretending *Mathematica* was not there and then use *Mathematica* as needed for the exercises. This was wrong. Finally we realized that one of the main uses of *Mathematica* is that of a pedagogical instrument setting up the theory with calculation.

Rethinking the mathematics

As we indicated above, good *Mathematica* Notebook courseware is not produced by merely adapting a good traditional course to the format of *Mathematica* Notebooks. And the difference goes beyond pedagogy. The power of *Mathematica* eliminates the need for certain time-honored mathematical activities. Consider the following three exhibits. In a traditional calculus course, each would be theory-laden. As done below, the theory is replaced by calculation and plotting.

Exhibit: Rational Approximation in precalculus

This is from a precalculus lesson on solving linear equations and plotting during the second week of beginning calculus. By this time most students have little problem with *Mathematica* syntax.

Part 1

Find constants a, b, c, d and e such that the function $(a + b x^2 + c x^4)/(1 + d x^2 + e x^4)$ agrees with Cos[x] for x = 0, 1/2, 1, 3/2 and 2.

Answer:

Enter the points:

**points = {{0,Cos[0],{1/2,N[Cos[1/2]]},{1,N[Cos[]]},
{3/2,N[Cos[3/2]]},{2,N[Cos[2]]}}**

$$\{\{0, 1\}, \{\frac{1}{2}, 0.877583\}, \{1, 0.540302\}, \{\frac{3}{2}, -0.0707372\},$$

$$\{2, -0.416147\}\}$$

Set-up the equation to plug into:

eqn = (y = (a + b x^2 + c x^4)/(1 + d x^2 + e x^4))

$$y == \frac{a + bx^2 + cx^4}{1 + dx^2 + ex^4}$$

Substitute the given points into the equation to get equations involving the coefficients to be determined.

**coeffeqns = Table [(eqn)/. {x->points[[j,1]],
y->points[[j,2]]},{j,1,5}]**

$$\{1 == a, 0.877583 == \frac{a + \frac{b}{4} + \frac{c}{16}}{1 + \frac{d}{4} + \frac{e}{16}},$$

$$0.540302 == \frac{a + b + c}{1 + d + e},$$

$$0.0707372 == \frac{a + \frac{9b}{4} + \frac{81c}{16}}{1 + \frac{9d}{4} + \frac{81e}{16}},$$

$$-0.416147 == \frac{a + 4b + 16c}{1 + 4d + 16e}\}$$

Determine the coefficients by solving the equations:

coeffs = Solve[coeffeqns,{a,b.c,d,e}]

{{a -> 1., b-> -0.454987, c -> 0.0201432, d -> 0.0450125,
    e -> 0.000988357}}

The function we are after is:

**y=((a + b x^2 + c x^4)/(1 + d x^2 + e x^4))/
.coeffs[[1]]**

$$\frac{1. - 0.454987\,x^2 + 0.0201432\,x^4}{1 + 0.0450125\,x^2 + 0.000988357\,x^4}$$

It's a weird looking duck isn't it?

Part 2

How well does the weird duck approximate Cos[x] for -p/2 ≤ x ≤ p/2?

Answer:

Just plot the difference:

Plot [Cos[x] - y, {x, -Pi/2, Pi/2}, Plot Range -> All]

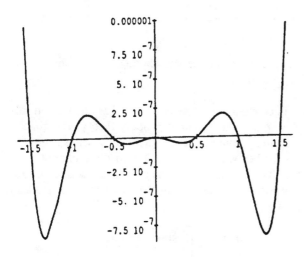

They run within 10^{-6} of each other the whole way.

Exhibit:Approximation of the derivative by the difference quotient:

A question like the following is impossible in a traditional course because its answer would involve careful use of the mean value theorem. In Calculus & *Mathematica*, we ask questions like this very early because the students can answer them and in so doing the idea of what a derivative is is cemented in the mind of the students.

Problem: How well does
    (Sin[x + .0001] - Sin [x])/.0001
approximate the derivative Cos[x] of Sin[x] on [0, 2]?

Answer: Plot both (Sin[x + .000] - Sin[x])/.0001 and Cos[x] on the same axes:

Plot [{(Sin[x + .0001] - Sin[x])/.0001, Cos[x], {x,0,2Pi}]

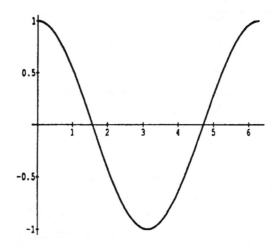

Looks like the approximation is damned good. To get more precise information, plot the difference:

Plot [{((Sin[x + .0001] - Sin[x])/.0001 - Cos[x]}, {x,o,2Pi}]

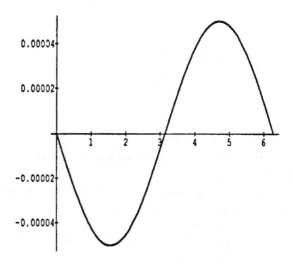

The difference quotient (Sin[x + .0001] - Sin[x])/.0001 and Cos[x] run within 10^{-4} of each other the whole way. For many practical purposes, there is no difference between the difference quotient (Sin[x + .0001] - Sin[x])/.0001 and the true derivative Cos[x].

Exhibit: Approximation by Taylor polyonials

In the original Calculus & *Mathematica* course at Illinois, we taught our students the traditional error term for Taylor polynomials. They replied that they didn't need to know the error term. Instead they used the plotter. Here is how they taught us:

Problem: Find a polynomial that runs within 10^{-7} of $\operatorname{Sin}[x^2]$ for $-1 \le x \le 1$.

Answer: It seems natural to try something like:

guess1 = Normal [Series[Sin[x^2], {x,0,14}]]

$$x^2 - \frac{x^6}{6} + \frac{x^{10}}{120} - \frac{x^{14}}{5040}$$

To test this, look at a plot of $[\operatorname{Sin}[x^2] - \text{guess1}$ and the line $y = 10^{-7}$ for $-1 \le x \le 1$:

Plot [{Abs[Sin[x^2]-guess1],10^(-7)},{x,-1,1},
PlotRange ->{0,2 10^(-7)}]

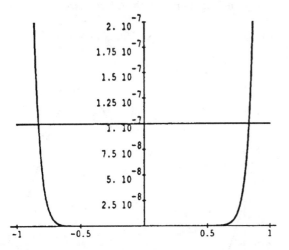

Not close enough; the plot of $\operatorname{Sin}[x2] - \text{guess1}$ breaks through the line $y = 10^{-7}$ Let's try:

guess2 = Normal[Series[Sinn[x^2], {x,0,18}]]

$$x^2 - \frac{x^6}{6} + \frac{x^{10}}{120} - \frac{x^{14}}{5040} + \frac{x^{18}}{362880}$$

To test this, look at:

Plot [{Abs[Sin[x^2] - guess2],10^(-7)},{x,-1,1},
Plot Range ->All]

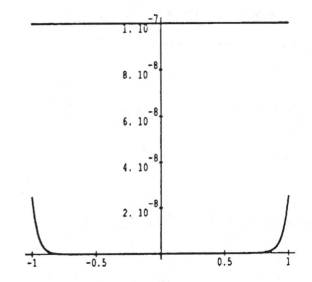

Great. Our second guess is running well within 10-7 of $\operatorname{Sin}[x^2]$ for $-1 \le x \le 1$.

All three examples show how a *Mathematica*-based calculus course can cut through heavy theory and broaden the students' mathematical horizons by dealing with issues far beyond those accessible to the traditional student. The effects of this type of calculus course will reverberate throughout the mathematics and science curricula.

Often asked questions and our answers

-How does Calculus & *Mathematica* fit into the evolution of American calculus courses?

Answer: Calculus is nothing more or less than a course in how to use the tools of differentiation, integration and approximations to make precise measurements. The venerable Granville course had moderate success in this endeavor. Then came the new math and Johnson & Kiokemeister. In the sixties, it became the fashion to study the tools of calculus without actually using them. Instead of using a hammer, we put it on the table and studied it from all angles. By the eighties, the reaction was to pick up the hammer and use it as a rote procedure by driving nail after nail into the same piece of lumber. The tool was used, but nothing was built. Calculus & *Mathematica* picks up the hammer, gets its feel, drives a few nails and then begins to build (tables, ladders, houses . . .)

-Won't *Mathematica* do the same thing to college level mathematics as calculators did to pre-college mathematics?

Answer: We hope so. Many professors and teachers like to blame the poor quality of American students on the calculator. Our view is that uninspired courses taught in a blend of religious and military styles and dealing with singularly uninteresting issues is responsible for the current students' indifference towards mathematics. Lack of interest results in lack of skill. Skill at hand computations and skill at using a calculator or *Mathematica* are similar. Both require knowledgeable management. Hand skills also require clerical ability. Without the calculator, today's math students would be worse not better. -What is the pedagogical advantage of teaching via *Mathematica* Notebooks?

Answer: We again quote Jacob T. Scwartz:

Many conventional academic skills simply amount to the ability to select and apply algorithmic or near-algorithmic procedures rapidly and correctly... Courseware can concentrate on one skill at a time, in a manner impossible for a textbook and hardly available to the classroom teacher, namely by asking the student to handle only that part of a procedure on which pedagogical stress is to be laid, while other aspects of the same procedure are handled automatically by the computer...

This scheme, which combines student interaction with sophisticated computer assistance, has the merit of focusing attention on the key strategic and conceptual decisions needed to handle a problem, while lower level assumed skills are automatically supplied . . . As already said, by relieving the student of part of the burden of low-level arithmetic and algebraic details, we focus her attention on the conceptual content of the material at hand, thus allowing her to comprehend it more richly. This should be of significance to both the strong student, whose progress the computer can accelerate, and the weaker student, for whose inaccuracy in applying the auxillary manipulations which support higher level skills the computer can compensate. Hopefully, as weaker students come to understand how their skills deficiencies impede their ability to handle interesting problems, and as intensive interactive experience pinpoints these deficiencies, they will be motivated to overcome their own shortcomings by systematic review and practice.

Our courseware is developed in this spirit and our experience confirms that Schwartz was right. And we have seen, to our surprise, that the last sentence is on the mark.

-What is the big difference between Calculus & *Mathematica* courseware and the traditional calculus texts?

Traditional calculus texts announce matermatical principle and attempt to reenforce them with glitz, multicolor pictures and propaganda. The text is the bible and the professor is the high priest. In Calculus & *Mathematica*, students learn mathematical principles through calculation and plotting. They don't need a political officer in charge because they recognize truth on their own. One new version of a traditional text advertises that it has about 125 *Mathematica* produced graphics. Calculus & *Mathematica* has the power to deliver 125 or more student-produced graphics in one week of class. The aim of Calculus & *Mathematica* is to make the traditional calculus texts look like a pile of dead Cadillacs piled up in the junkyard for their ultimate disposal.

-Are Calculus & *Mathematica* students handicapped away from the computer?

Answer: Of course they are, but from a realistic point of view students in the traditional courses are handicapped all the time. They shine only on the stock problems of the calculus text. Henri Lebesgue warned us of stock topics when he said:

Unfortunately, competitive examinations often encourage one to commit this little bit of deception. The teachers must train their students to answer little fragmentary questions quite well, and they give them model answers that are often veritable masterpiees and that leave no room for criticism. To achieve this, the teachers isolate each question from the whole of mathematics and create for this question alone a perfect language without bothering about its relationships to other questions. Mathematics is no longer a monument but a heap.

Peter Lax agrees:

Calculus as currently taught is, alas, full of inert material which will remain there as long as the teaching of calculus is controlled by... the group presently entrusted with teaching it.

Away from the computer Calculus & *Mathematica* students can operate on a par with traditionally prepared students, but with the help of the computer Calculus & *Mathematica* can deal with a whole range of problems and situations foreign to the traditional student.

-Are courses based on *Mathematica* Notebooks practical?

Answer: Clearly courses based on *Mathematica* Notebooks are written for the future, but it appears that the future will be on us very soon. At this writing ten universities and high schools expect to be running Calculus & *Mathematica* this spring. In one year the price of a Macintosh capable of running *Mathematica* has fallen by fifty per cent. Addison-Wesley's Advanced Books Program plans to issue a preliminary version of the course for wider scale testing within one year.

References

Peter Lax, *On the Teaching of Calculus* in **Toward a Lean and Lively Calculus**, 61-69, Ronald G. Douglas, Editor, Mathematical Assn. of America, Washington, 1986.

Henri Lebesgue, **Measure and Integral** (Translation of **La Mesure des Grandeurs** edited by Kenneth O. May). Holden-Day. San Francisco, 1966.

Jacob T. Schwartz, *Computed-Aided Instruction* in **Discrete Thoughts: Essays in Mathematics** by Mark Kac, Gian-Carlo Rota and Jacob T. Schwartz, Edited by Harry Newman, Birkhauser, Boston 1986.

MAPLE Laboratory in a Service Calculus Course

Eric R. Muller
Brock University

Introduction

A recent survey [1] of a sample of Mathematics Departments in Canadian Universities estimates that 95 percent of all students taking a first Calculus course do so as a requirement of their major program—a program which is not mathematics. At Brock University this service course has a total enrolment of over 600, and is offered in lecture sections of 60 to 180 students. This twelve-week course is quite demanding; it covers both Differential and Integral Calculus, as well as applications, in four lecture hours and one laboratory hour per week. The Mathematics Department has always devoted extensive faculty resources and substantial time and effort to provide these students with a positive experience in mathematics. Leonard Gillman [2] in his discussion of programs that work has summarized "The fundamental precepts are: challenge your students with difficult problems, demand hard work, and adhere to uncompromising standards; at the same time, constantly promote and bolster their confidence and self-esteem, assuring them that they can succeed and praising each accomplishment, and make yourself available outside of class for sympathetic help and encouragement." The Brock Mathematics Department has made substantial progress in the directions proposed by Gillman.

In the early seventies, the Mathematics Department at Brock introduced computers in the service courses, as a means to motivate students by providing each student with an individualized set of problem data [3]. Since 1982 "student-oriented packages" have been used by students in a laboratory setting for Operations Research and Statistics courses.

This paper discusses the introduction of the Symbolic Manipulator, MAPLE [4], into the Calculus service course, more specifically the use of MAPLE in compulsory laboratory sessions. The reasons for introducing a laboratory are:

(i) to place the student in an exciting and very different mathematics environment;
(ii) to instruct the student in the proper use of a useful and powerful mathematical tool.

It is hoped that the laboratory will also achieve other pedagogical goals such as:

- increasing the student's confidence in doing Mathematics;
- increasing the student's enjoyment when doing mathematics;
- providing more time in lectures for the explanation of calculus concepts and mathematical language by relegating to MAPLE numeric, algebraic, and graphic manipulations;
- presenting more exploratory situations, "what if" situations, and thereby rekindling the spirit of discovery.

There is no single way of achieving these goals. This new environment has much to offer [5], but we recognize that, since learning environments are very personal, it will motivate some students while it will be unattractive to others.

The paper is divided into a number of sections. It begins with a discussion of some of the more practical reasons for introducing MAPLE (or any Symbolic Manipulator) into a Calculus course. How this was achieved at Brock is presented in the next section. Examples of laboratory work and assignments are followed by a section analyzing traditional evaluation indicators. The paper concludes with results of student attitudinal surveys and a discussion of possible future directions.

Motivation

There are two recent developments in computer hardware and software which facilitate the use of Symbolic Manipulators (also called Computer Algebra Systems) in introductory Calculus courses. The first is a concerted effort to make them more "user friendly" while also providing a more general mathematics environment including graphics and other capabilities. For this reason the term *Mathematics Computer Environment* will be used in the rest of this paper. The second is the increase in power of microcomputers at a lower cost thus presenting a more affordable situation. Both of these factors have played a major role at Brock. The first provided the academic and educational drive— we now had an exciting mathematical environment accessible to first-year students—the second helped to produce a budget with a cost per student which the administration could be persuaded to accept. Primarily justified by the needs of this Calculus class of 600 students, Brock purchased 33 Macintosh SEs with an appropriate network.

Among mathematics faculty, the introduction of a Mathematics Computer Environment in service Calculus courses is less controversial than their use in Calculus courses for mathematics majors. Sometimes the controversy stems from a lack of experience with these systems. Unfortunately, the many demands on faculty time make it difficult for them to obtain this experience. Nevertheless the very important departmental decision to include or exclude this environment from a student's undergraduate experience must be made with full knowledge and understanding of the capabilities of these systems as well as a first hand experience in their use. This takes time — how is this understanding to be achieved? Opportunities must be provided since these systems can no longer be ignored. Some smaller systems are already available on pocket-size calculators; faculty must decide whether to forbid, allow, or require their use in class, on assignments, and on tests and examinations.

Many of the faculty in disciplines serviced by Calculus courses require their students to have an early undergraduate experience with computers and generally welcome and encourage the use of technology. Those departments show more interest in mathematical modelling and less concern about the mathematical techniques of solution. They are just as concerned with the proper use of MAPLE as they are with the proper use of any other software; all of those are tools to be used in their area of specialization.

Implementation

In 1988 the author developed a set of laboratory activities using MAPLE for the service course in Calculus. The activities were completed by approximately 100 volunteers registered mainly in the author's section of the course. Laboratories involved fifteen students using MAPLE on terminals from a VAX 11/780. Course evaluations were sufficiently positive to convince the administration that the department was doing something worthwhile and they authorized the establishment of a 30-Mac SE laboratory together with three drivers and the necessary networking. In 1989 weekly one-hour MAPLE laboratories became compulsory for all students in the Calculus service course.

One of the principal thrusts of the implementation was to get early and widespread mathematics faculty involvement. This was achieved by involving three interested

faculty members in the experimental year and all faculty members the following year. Every faculty member in the department served as advisor in at least one laboratory hour per week; those who were unfamiliar with MAPLE and/or microcomputers were provided with a senior student assistant.

The laboratory activities were designed to tie in with the concepts developed in the existing course and were scheduled to follow within a week of their lecture presentation. Some of these activities are described in more detail in the next section.

In the first year, when volunteers registered for the laboratories, credit was given for the work done in the laboratories by requiring one-half the work of the regular class assignments. All the students in the course wrote the same tests and examinations. In the following year, when weekly laboratories became compulsory, the work performed in the laboratories became part of the course assignments. Work done one week was collected the next, marked by senior students, and returned the following week.

A good mathematics educational computer package should require no other language than the standard language and notation used in the course. Existing packages, including MAPLE, have very specific requirements of notation and language. In the laboratory instructions, much of the MAPLE language and notation are provided to the student and a special MAPLE reference card is supplied. At this level, students have enough difficulty mastering standard mathematical language, without having additional notation thrust upon them. We have found that a one hour exploratory introduction to MAPLE is sufficient. Students do not find MAPLE difficult to use.

Laboratory Activities and Assignments

Students are informed early in the course that laboratories involve three types of activities. The first is a preparation for the actual laboratory which requires text and lecture material readings and the completion of straightforward illustrative problems. The second is the actual time with MAPLE in the laboratory and the third is the integration of the first two activities concluding in the presentation of a written assignment. Because marked assignments are returned within one week, they provide regular feedback to the students. The course requires ten assignments excluding an introductory activity and a review laboratory.

Some faculty may argue that a university course should not

remove the responsibility of time management from the student. However, two factors play significant roles in this course. The first is that the majority of students are first year students who have had no experience with time management in a university setting—one course in their regular load of five courses may offer them a model for their studies. The second is that, for the majority of students registered in this course, mathematics is a low priority—a required course specified by their program of interest. Any support that can be provided to help them succeed in mathematics is generally appreciated. Some of the success experienced after the introduction of the laboratories may be due to the continuous monitoring and support provided by the whole laboratory structure.

The in-laboratory activities are evolving continuously. The potential uses of a computer mathematics environment as cognitive technology in Calculus are still to be exposed. Activities are now selected if they have one or more of the following general attributes:

(a) they encourage exploration of Calculus and/ or related concepts;

(b) they probe inductive reasoning and pattern recognition;

(c) they investigate interrelationships between different representations—algebraic, graphic, numeric, etc.;

(d) they involve problems which students cannot solve without the technology.

Activities are not included if they duplicate activities which can be achieved just as easily with pencil and paper. In general this eliminates the standard text book questions. However students are encouraged to use MAPLE to check the answers to all the questions in the text. Here are some examples of in-laboratory activities which meet some of the desired attributes:

Laboratory Activity 1

Consider a polynomial with a number of zeros in a fixed interval, e.g., for x in the interval [1, 2], consider

$$p_5(x) = A(x-1.08)(x-1.19)(x-1.33)(x-1.62)(x-1.97)$$

(the factor A is selected to provide a reasonable looking plot from the graphical package being used). Explore graphically the behavior of p around the point x = 1.54 (say) by plotting on smaller and smaller intervals about this point until the graph appears to be a straight line. Compute the slope of that straight line.

Use MAPLE to compute the slope of the tangent line at x = 1.54 by first graphing the difference quotient function

$$\frac{p_5(x) - p_5(1.54)}{x - 1.54}$$

on an interval including x = 1.54 (MAPLE produces a hole in the graph at x = 1.54). Use this graph to estimate

$$\lim_{x \to 1.54} \frac{p_5(x) - p_5(1.54)}{x - 1.54}$$

Now obtain a more accurate value for this limit using MAPLE's limit command.
Compare this with the slope of the straight line you computed earlier.
Comment on your results.

Now consider the function
$$abs(p_5(x))$$
and repeat the above exploration around x = 1.33.
Comment on your results.

Laboratory Activity 2

This activity is scheduled for the week after Activity 1.

(a) Use MAPLE to get a plot of the difference quotient
$$\frac{\ln(t) - \ln(4)}{t - 4}$$

around t = 4, and get MAPLE to evaluate its limit as t->4. From the graph convince yourself that this limit is correct. Repeat this activity for two other positive (why positive?) integers and three other positive rational numbers. (Recall that rational numbers must be entered in MAPLE as a/b where a and b are integers). From your six activities infer what

$$\lim_{t \to a} \frac{\ln(t) - \ln(a)}{t - a}$$

should be. Get MAPLE to confirm your inference.

(b) The limit procedure is incredibly sensitive and accurate. Use MAPLE and define the difference quotients
$$v = \frac{\ln(t) - \ln(32/10)}{t - 32/10}$$
and
$$w = \frac{\ln(t) - \ln(3.2)}{t - 3.2}$$

Determine
$$\lim_{t \to 32/10} Q \quad \text{and} \quad \lim_{t \to 3.2} W.$$

Explain the results.
(Note: MAPLE will approximate $ln(3.2)$ as
1.16315081 and with this approximation the
$$\lim_{t \to 3.2} \frac{ln(t) - 1.16315081}{t - 3.2}$$
does not exist.)

Laboratory Activity 3

In preparation for this laboratory students are required to complete the following textbook problem.

"A telephone company is laying a new cable between its central exchange on the mainland and its exchange on an island a little way off shore. It has obtained permission to lay the cable along the seashore where it estimates costs to be \$20,000 per kilometer. Across the water it estimates costs to be \$30,000 per kilometer. The geography is as shown on the diagram. Determine the cable layout to minimize total costs.

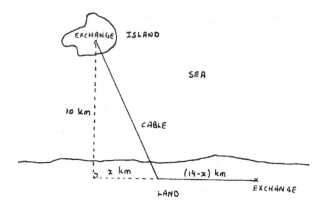

The in-laboratory activity is as follows:

The problem you completed before the laboratory assumes that the cable will be laid horizontally across the sea. Normally the cable is buried in the seabed which has depth variations. Assume that for all reasonable directions from land to island, the sea cross-section between the two can

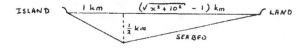

be approximated by the following geometry: Determine the new cable layout and minimum cost.

Laboratory Activity 4

In this course MAPLE is used for finding indefinite integrals. No integral techniques are taught. This is not a major change from what was done previously when far more restrictive integral tables were used. Students are encouraged to always check MAPLE's results by differentiation. A similar approach is used in the introductory study of differential equations. For the definite integral, the difference between an exact value (when MAPLE can do the indefinite integral) and an approximate value (when MAPLE cannot do the indefinite integral and yet can provide an approximation) is underlined.

1. Consider the polynomial

po:=1555(x−1.01)(x−1.23)(x−1.55)(x−1.93).

Use MAPLE to
(a) Evaluate the definite integral
$$\int_{1.01}^{1.93} po \; dx$$
(b) Find the area of the bounded region between po and the x axis (shade the area on the plot you have generated.)

(c) Find the area of the bounded region between po, the x axis and x = 1, x = 2. (Shade the additional area on the plot).

(d) Find the area of the bounded region enclosed between po and the line y = 9.879x − 11.0405 (shade the area on a new plot).

Note: We can save ourselves a lot of typing by using MAPLE's set notation.
        >fsolve(po=y,x);
will find the points of intersection
        >s:=["];
will assign these points to s[l], s[2] etc.
        >int(po-y,x=s[l]..s[2]);
will find the definite integral between the first two points of intersection.

MAPLE computes definite integrals by first evaluating the indefinite integral (if it can) and then substituting in the limits. In cases where the indefinite integral is unknown

MAPLE can be asked to compute an approximation to the definite integral by using the "evalf" command. (It uses some approximating algorithm similar to the Trapezoidal and Simpson's rules discussed in the course).
Consider

$$\int_{-1}^{1} \exp(-x^3)\, dx$$

Results of Traditional Evaluation Indicators

The course data presented are for four consecutive years, namely,

1987 with no laboratories,

1988 with approximately 100 volunteer students in compulsory laboratories (this represents a good cross section of students since they select their section on the basis of their timetable),

1989 and 1990 with all students in compulsory laboratories.

The traditional measures of evaluation are :

(a) Failure rates

The failure rate in 1987 (without laboratories) was twice the failure rate (with laboratories) in 1989. Failure rates in 1989 and 1990 were the same. The failure rate in 1988 for students not taking laboratories was two and a half times the failure rate of those who volunteered for the compulsory laboratories.

(b) Withdrawal rates

No withdrawal rate information is available for 1987. The withdrawal rate in 1988 of students taking laboratories was less than half the withdrawal rate of students not taking laboratories.

The withdrawal rate in 1989 (with laboratories) was two thirds of the withdrawal rate of students not taking laboratories in 1988. The withdrawal rate in 1990 was slightly lower than in 1989.

(c) Average grades

The average grade for all students who wrote

the final examination in 1989 (with laboratories) was five percent higher than the corresponding average in 1987 (without laboratories). In 1988 the average grade for all students who wrote the final examination and who volunteered for laboratories was eight per cent higher than the corresponding average for those who did not have the opportunity to take laboratories in that year. The average grade in 1990 was slightly higher than that in 1989.

These indicators are clearly encouraging. It is important to note that the improvements may be due to factors which are secondary to the use of MAPLE in the Calculus course.

Results of Student Attitudinal Surveys

At the end of the course a questionnaire seeking the students' attitude towards the laboratories was distributed. Students were free not to return the questionnaire and no identification was requested. Approximately sixty percent of the students who wrote the final examination returned the questionnaire in each of the three years. Students who volunteered for the laboratories in 1988 expressed a far more positive attitude towards mathematics and the laboratories than did the students in the class as a whole in both 1989 and 1990.

The following tabulated data for extreme positions demonstrate this change. (The entries are percentages—they do not add up to 100% as the neutral replies of indifference are not recorded.)

(i) Would you recommend laboratories to a friend taking the course next year?

	1988	1989	1990
yes	86	25	41
no	2	46	32

(ii) Student ranking of laboratories versus other modes of learning

	1988	1989	1990
high	29	7	11
low	41	73	68

(iii) Student assessment of laboratories as a learning aid

	1988	1989	1990
good	47	12	15
poor	27	67	56

(iv) Student attitudes towards mathematics

(a) In response to a question of confidence of being able to succeed in the course, the students indicated

	1988	1989	1990
high	52	57	61
low	15	12	12

Of those who had low confidence of succeeding, did MAPLE increase their confidence?

	1988	1989	1990*
yes	5	4	65
no	34	71	35

*In 1990 the question was reworded to commit students.

(b) In response to a question of enjoyment in doing mathematics, the students indicated

	1988	1989	1990
high	41	44	45
low	27	15	15

Of those who had little enjoyment in mathematics, did MAPLE increase their enjoyment?

	1988	1989	1990*
yes	13	14	44
no	33	63	54

*In 1990 the question was reworded to commit students.

Future Directions

In 1989 and 1990, the student questionnaire contained the following question.

You are given the function

$$f(x) = \frac{(3x^2 + 2x)^5}{3x + 1}$$

and the summary sheet of MAPLE instructions. Please indicate whether you would prefer to do the following using pencil and paper or using MAPLE: (a) differentiate, (b) graph for $-5 \le x \le 10$, (c) integrate, (d) find limit as x tends to 1, (e) evaluate f(2).

The results showed the following preferences: (the numerical entries are percentages);

| | Pencil & Paper | | MAPLE | |
	1989	1990	1989	1990
differentiate	58	57	42	43
graph for $-5 \le x \le 10$	15	14	85	86
integrate	37	42	63	58
find limit as x tends to 1	60	49	40	51
evaluate f(2)	72	70	28	30

These responses seem astonishing; it would take an average student less than 10 minutes to do all of these tasks using MAPLE. Nevertheless, the responses support the present course emphasis on (i) the graphing package, which provides a powerful tool for developing good visualization of some mathematical concept, and on (ii) the integration package, which replaced the traditional integration techniques.

At this time there remain more questions than answers:

(i) What are the essential curriculum components of a Calculus course using a Mathematics Computer Environment?

(ii) How are students to be evaluated once a Mathematics Computer Environment is introduced?

(iii) What kind of Mathematics Computer Environment interface do undergraduate mathematics educators wish to recommend?

(iv) Mathematics Computer Environments offer support for far more than Calculus. How are they to be integrated into a multiple-course pattern including College Algebra, Calculus, Discrete Mathematics and Linear Algebra?

The next five years should turn out to be exciting and challenging times for undergraduate mathematics education.

Acknowledgments

Without the support and contributions of all the faculty members in the Mathematics Department this work would have been impossible. Professors John P. Mayberry, Bill Ralph and K. J. Srivastava have contributed far beyond the call of duty. I'm grateful to Mrs. Dorothy Levay who did the data analysis under a Brock Instructional Development Grant, and to Mrs. Barbara Ouellette for typing the document.

References

1) E. R. Muller and B. R. Hodgson "Mathematics Service Courses: A Canadian Perspective" in *Selected Papers on the Teaching of Mathematics as a Service Subject*, R.R. Clements et al. editors Springer Verlag (1988)

2) L. Gillman "Teaching Programs That Work", *Focus* 10 (1990).

3) Auer, J. W. et al. "Motivating non-mathematics majors through discipline-oriented problems and individualized data for each student" *Int. J. Math. Educ. Sci. Technology* 13 (1982).

4) MAPLE, A project of the Symbolic Computation Group, The University of Waterloo. The MacIntosh version is available through Brooks/Cole Publishing Company.

5) Hodgson, B. R., and Muller, E. R., "Symbolic Mathematical Systems and their Effects on the Curriculum" *Notes Can. Math. Soc.* 18 (1986).

First Semester Calculus: A Laboratory Course

Margret Höft*
University of Michigan - Dearborn

Powerful calculators and computers have created new challenges for the undergraduate mathematics curriculum in general, and the calculus sequence in particular. There is no general agreement on how best to teach mathematics with computers, and those who try find themselves on unfamiliar territory. There is however a sense of agreement emerging, that students can benefit from the introduction of technological tools into the curriculum. Many topics can be graphically illustrated with the help of computers, and both teachers and students can be freed from tedious computational tasks in order to focus on important concepts rather than routine calculations. Computers can be used to illustrate and describe mathematical concepts, and they can serve as tools for experimentation and discovery. Moreover, students' interest in mathematics can be stimulated, since access to technology enables them to explore a large variety of examples, and to solve realistic applied problems of considerable complexity.

Goals of the Laboratory Course

In its report *Everybody Counts* [4], the National Research Council addresses the impact of computers on the mathematics curriculum and concludes:

> "Innovative instruction based on a new symbiosis of machine calculation and human thinking can shift the balance of learning toward understanding, insight, and mathematical intuition."

> "Students can explore mathematics on their own, to ask and answer countless "what if" questions. Although calculators and computers will not necessarily cause students to think for themselves, they can provide an environment in which student-generated mathematical ideas can thrive."

Can these ideals be realized in the undergraduate mathematics classroom? If a symbiosis of machine calculation and human thinking is to be achieved in the undergraduate curriculum, then students will have to be introduced to this new way of learning mathematics in their first semester.

The standard approach to first semester calculus teaching, i.e. lecturing and listening, is not likely to contribute to the symbiosis, and alternative teaching methods will have to be explored. How can we teach first semester calculus students to combine machine calculation with human thinking? How do we encourage them to ask "what if" questions? How do we encourage experimentation and an active exploratory approach to learning mathematics? Will we be successful in convincing first semester students to view computation as a means not as an end?

Here is one approach that will *not* work for the majority of students: Make a computer algebra system available on a mainframe computer or in a computer laboratory and suggest to the students that they should use it to do their homework problems. This may be beneficial to an occasional student, but most first semester students need more guidance and specific instruction. If a mathematics department has its own computer laboratories, staffed by graduate student tutors with considerable competency in mathematics, the following scenario might work: The students are given specific assignments to be completed in the lab, and the lab tutors are available for help and guidance. It is likely that the students in this scenario do no more than complete their specific assignments, and therefore will not be motivated to experiment and explore beyond their assigned problems.

At the University of Michigan-Dearborn, we came to the conclusion that faculty supervised laboratory sessions would be the best way to introduce first semester calculus students to technology in the study of mathematics. It was not our immediate goal to achieve a symbiosis of machine calculation and human thinking for all students in just one semester, but the goal was somewhat more modest: *Use technology to enhance the understanding of the concepts of calculus.* A scheduled laboratory session that is supervised by the instructor, will give the teacher an opportunity to interact directly with the student sitting at a computer work station, to look over her shoulder, make suggestions for "what if" questions, encourage experimentation on the spot, and generally foster a spirit of active involvement with the mathematical concepts.

*The course described here was designed and taught by John Fink, David James, and the author. The materials for the laboratories were developed by David James and the author.

In what follows, I shall describe the specifics of a first semester calculus laboratory course as it was taught at the University of Michigan-Dearborn in the fall semester of 1988 and again in the winter semester of 1989. The University of Michigan-Dearborn is a state-supported university located in metropolitan Detroit, and has as its primary mission undergraduate education; 90% of the 7,500 students are undergraduates. The College of Arts, Sciences and Letters enrolls 3,700 undergraduates, the School of Engineering enrolls 1,300, Management 700, Education 700 and Computer Science 350. Essentially all students are commuters from the metropolitan Detroit area, and most are from a working class background. Many are the first in their families to go to college.

The Department of Mathematics and Statistics is an undergraduate department with 140 mathematics majors. Mathematics, engineering, and many science students are required to take three semesters of calculus, followed by a course in differential equations and a course in linear algebra. Calculus is taught in small sections of 32 students per section, and classes meet four times per week for fifty minutes. In the fall semester of 1988 we offered twelve sections of first semester calculus, and in the winter semester of 1989 we offered nine sections. In each of the semesters, three sections were taught as a laboratory course, and these sections were closely coordinated and used the same laboratory materials. The laboratory sections were designed and taught by Professors J. Fink, M. Höft, and D. James; all three are regular members of the department. The materials for the laboratories were developed by M. Höft and D. James.

Because of some scheduling problems the physical setup was slightly different for one of the sections, and what is described here, is what we envisioned to be our ideal situation and what we were able to realize for two of the three sections in the winter semester. After teaching the first three sections in the fall semester, we made one major change in the format of the course in order to integrate the laboratories and the lectures in a more satisfactory way. In the fall semester we had added a laboratory session of an hour and a half to the scheduled four hours of lectures per week, since it proved to be impossible to change the existing schedule of classes. In the winter semester we scheduled three hours of lectures per week, and in addition to those a laboratory session of an hour and a half per week. In the following paragraphs, I shall give a description of the scenario of the winter semester of 1989.

Course Format

The laboratory calculus sections were in design similar to a laboratory course in the natural sciences, where students meet with their instructors not only for lectures and discussion, but also for hands-on work in a laboratory. Students in the laboratory calculus sections met with their instructor in a regular classroom setting three times weekly for fifty minute periods for lecture and discussion, and once a week for hands-on mathematics in a computer laboratory.

A computer with an overhead projection system was available in the classroom, and was used by the instructor during the lectures. Typically, the computer was used in a carefully planned demonstration of a calculus concept, lasting no more than five to ten minutes. Many of the demonstrations illustrated a four step exploratory approach to mathematics: generating examples and conjecturing, then testing, and, finally, proving. Students were encouraged to participate actively in the exploration and make conjectures. The computer might be used spontaneously once or twice again during the lecture in response to a student question, or to complete a lengthy computation, or to display a graph to illustrate the solution of a problem. In the first part of the course we used only the software that the students would also see in the laboratory in order to get them acquainted with the specifics of input and output. Later in the semester, after the students were somewhat experienced with the laboratory software, we occasionally brought in software packages not used in the labs.

The fourth period of the week was scheduled for 80 minutes in a classroom with approximately 30 IBM PCs. In these sessions the instructor typically spent about twenty minutes in the beginning of the class lecturing at the blackboard to introduce the topic of the day, and to provide the theoretical background that was necessary to do the computer exercises. Then the students were given a handout with computer experiments that were designed to explore the topic. The handout also contained a list of homework problems to be done with the aid of a computer, specific instructions for a laboratory report, and what we called "the discovery problem of the week." Students worked in teams of two per computer workstation, but discussion, interaction, and comparing of notes between teams was very much encouraged.

The sessions were supervised by the instructor and one lab assistant. The lab assistant helped with technical problems ("I can't get this computer started", "my printer does not work", "I need more computer paper") and routine mathematical or program problems ("my tangent line does not touch the curve", "I keep trying to graph this function, but the screen stays blank"). The instructor was cast in the role of a roving coach or mobile Socratic tutor, moving from team to team, asking questions, stimulating "what if" experimentation and exploration, and generally encouraging the students to go off on tangents, and try out any conjecture that came to they minds. The instructor's role was to exploit this laboratory setup to teach calculus with the aid of computers.

Equipment and Software

For the computer demonstrations during the lectures we used a Zenith computer with a hard disk that had a variety of calculus software installed on it. A Telex Magnabyte LCD panel was used to project the computer output to a screen via an overhead projector. The supervised laboratory sessions were held in a classroom with approximately 30 IBM PCs. This classroom also had a blackboard and a computer with an overhead projection system in the front of the room, making lectures and computer demonstrations during lectures possible.

The software was chosen with the computer novice in mind. First semester students that have never had their hands on a computer are becoming increasingly rare, but there still are some. (In my section two of the women had never been close to a computer. They teamed up with more experienced students, and once they got over their initial fear, they discovered deep down inside themselves a hacker's soul.) The three instructors agreed that the software should not require any previous experience with computers or, in particular, knowledge of a programming language, and it should require only minimal instruction for use. We wanted one easy to use package with some symbolic manipulation capabilities, and one tutorial package. We chose *MicroCalc* by H. Flanders [2] as the package with some symbolic capabilities, and *Exploring Calculus on the IBM PC* by J. B. Fraleigh and L. I. Pakula [3] as the tutorial. These were the only two packages used by the students. They were installed on the network in the campus computer lab and the students had easy access to them, when they were on campus. Toward the end of the semester, we showed the students other software in classroom demonstrations, but we did not coordinate this, and different instructors used different materials. I used *A Computerized Calculus Tutorial for the First Semester* by K.E.Petersen [S] in several lectures.

Laboratory Work

We scheduled eight lab sessions of 80 minutes each during the semester. There would have been enough time in the semester for nine or even ten sessions, but we did not schedule sessions during the weeks when the students had major tests for this class, and also not during the last two weeks of classes. In these weeks the lab sessions were replaced by regular lectures. To give the reader an idea of the topics covered in the lab sessions, I shall give a brief and incomplete description of six of them, and a more detailed description of two of them (3. and 5. below).

1. Basic skills. first experiments with graphing

The students learned the basics of using the computer and the software: how to make selections from the menus, how to type in functions, how to edit, how to change the viewing window in a graph, how to determine domain and range of a function, and how to decide whether a viewing window is good or bad. They experimented with viewing windows for functions like

$$f(x) = \frac{x^3 - 10x^2 + x + 50}{x - 2}$$

and

$$f(x) = 5\sin 3x - 2\sin(\frac{x}{4}).$$

2. Graphs, tables, and limits of functions

Function tables were introduced. They were used to estimate roots of equations, points of intersection of graphs, and to find limits of functions. The students investigated limits like $\frac{\sin x}{x}$ at zero and $\frac{10^x}{x^5}$ at infinity. They also learned how to graph several functions in one picture, and compared for instance the two functions $y = 10^x$ and $y = x^{10}$, looking at them from close up and from far away. They experimented with various values of a, b, and c in $y = x^3 + ax^2 + bx + c$ until they had three functions which looked quite different from each other close up. Then they enlarged the viewing window to get a global look at the same three functions.

3. The derivative of a function

The computer activities in this session were designed to help students understand the fundamental calculus concept of the derivative of a function. At the beginning of the session the instructor introduced the derivative of a function f at a point c, and discussed general rate of change, and slopes of secant and tangent lines. The computer exercises were divided into three parts, and all were based on *Exploring Calculus with the IBM PC*.

In Part I the students estimated slopes of functions at various points. They first plotted $f(x) = \dfrac{x^3}{8}$ and estimated f′(2) by finding the slopes of secant lines for Δx = 2, -2,1, -1, .5, - .5, ..., .00001, -0.00001. The program they used displayed the function and for each choice of Δx the corresponding secant line. It also computed and displayed the numerical values of the slopes of the secant lines, and listed them in a column so that the students could watch the change in the numerical value of the slope with changing values of Δx. The exercise was repeated to estimate f ′(1), f′(0), and f′($\frac{\pi}{2}$) for $f(x) = \sin(x)$, and also f′(1.5) for $f(x) = 4\sin(1/x)$.

In Part II the emphasis was on realizing that the derivative of a function is again a function. Students used the computer to graph $f(x) = \dfrac{x^6}{10}$ and to approximate f′(x) for x = 0.5, 0.7, 0.9, 1.1, 1.3, 1.5. By hand, they listed the values for x and f ′(x) in a function table, transferred them to a graph and drew a smooth curve passing through the points. This was their first example for the graph of f ′(x), where each function value was the approximated slope of f(x) at x.

Part III was devoted to the exploration of the relationship between the graphs for f(x) and f ′(x). Students used a program where they could choose a function (they could also type in their own choice), then the graph of the function was displayed on the screen, and, very slowly, from left to right, the derivative of the function was drawn on the same screen. Students were encouraged to try to predict the shape of the derivative function by making rough estimates of the slope of the original function. After the students gained some confidence in predicting the shape of the derivative graphs, they used a second program, where the graph of a function (no equation given) appeared on the screen, and the students had to draw what they estimated was the graph of the derivative of this function. They moved the cursor by using the up, down, right, left keys on the keyboard, and could watch the progress of their derivative on the screen. When the student drawing was completed, the program displayed the correct graph of the derivative, the student drawing of the derivative and the function in different colors in one picture, thus giving the students a chance to analyze their mistakes. The student drawing of the derivative was also given a grade for correctness, for instance 70%, 85%, 95%. Computer

homework for the week was to produce a score of at least 90% on a derivative drawing. This was surprisingly easy, once the students had grasped the idea.

There were two challenge problems for the week:

(1) Graph the function f(x) = 2^x and its derivative. Repeat with f(x) = 3^x. Then experiment to find as accurately as possible by graphing, a number c such that the graph of c^x and its derivative coincide. The function c^x is thus unchanged under differentiation. This number c is a very important number in mathematics. Hand in a screen dump for the function that coincides with its derivative, and state what value of c was used.

(2) Students used a program where the derivative at c was estimated by

$$m_{sym} = \frac{f(c + \Delta x) - f(c - \Delta x)}{\Delta x}$$

(They had to figure out how to do this on their own.) They compared the estimates that they got for the derivatives with the secant method to the estimates they got with this new method.

Problems and Questions:

    a. Do both methods provide the same answers? Which method is faster?

    b. Try the new method on y = |x| at x = 0. Write a paragraph to explain why you get different answers in this case. Which answer is correct?

4. Approximations for f(x) and error estimates

Tangent line approximations, e.g., for the function

$$f(x) = \sqrt{x}\ \frac{\sin(x^2)}{x^2 + 2x + 2}$$

for x near a = 1.1,1.3, 1.5 are calculated. Error estimates were done by finding an upper bound M for the values of |f″(x)| near these points graphically. The "discovery problem" of the week introduced quadratic approximations.

5. Newton's method for solving f(x) = 0

The goal of this session was to understand the underlying geometric concepts in Newton's method for finding approximate solutions of f(x) = 0, where f is a differentiable function. At the beginning of the session the instructor briefly reviewed the Intermediate Value Theorem and the Bisection method. Then Newton's algorithm was introduced and some time was spent on discussing the obvious limitations of the method, in particular, what to watch for, if the algorithm is executed on a computer.

The computer exercises were divided into two parts; the first part was based on *Exploring Calculus with the IBM PC*, the second on *MicroCalc*. In Part I the students used a program where they had to supply a function, f, and its derivative, f'. The graph of the function was then displayed on the screen, and after some adjustments on the viewing window, students could obtain rough estimates of all roots. They typed in an estimate x_0 for a root, and the program then used this estimate to display the tangent line at this point, the point of intersection x_1 of the tangent line with the x-axis and the numerical value of x_1. Pressing the spacebar continued the algorithm with a display of the tangent line at x_1 and a numerical value of the point of intersection x_2. A column of x_i- values was generated and the corresponding tangent lines were displayed on the screen. Students solved the equations

$$x^3 + 10x^2 - 50 = 0$$

$$\frac{(x+1)^3}{12} + \frac{x}{5} - 1 = 0$$

$$x - \frac{1}{x} - 6 = 0$$

and found an approximation of $\sqrt{2}$ by solving $x^2 - 2 = 0$. Each time all the tangent lines were drawn, and derivatives had to be computed by hand, making this a slow process. However, the idea was to understand how Newton's method works, and the exercises seemed to accomplish that very well.

In Part II we introduced *MicroCalc* for the first time. Here it was no longer necessary to compute derivatives by hand, an advantage that was immediately obvious to the students. There were no longer any tangent lines either. We used *MicroCalc* to find all real solutions of

$$\frac{x^2 - 5x + 2}{x^2 + 1} = 1$$

and to find the points of intersection of g (x) = 3 cos x and h (x) = x.

Homework:

Use *Exploring Calculus with the IBM PC* to do the following:

(l) Find two real roots of $x^4 - 2x^3 - 2x + 2 = 0$ to six decimal places. First get a plot showing there are two roots and give good initial estimates.

(2) Estimate π to five decimal places by applying Newton's method to the equation sin x = 0. Does it matter what your initial estimate is? Try an initial estimate of 2, then of 1.9, then of 10. For each of the three initial estimates, choose a good viewing window for the graph.

(3) Let $f(x) = -2x^3 + 3x^2 + x - 1$. The equation $f(x) = 0$ has three roots.
    a) Find the middle root to six digits of accuracy.
    b) For some initial estimates, Newton's method will not converge, but will simply bounce back and forth between two numbers. By looking at the graph of f (x), see if you can guess such an initial estimate.
    c) After you have found your answer to b), try initial values which are just a little bit bigger than your estimate, and just a little bit smaller. Explain in words what happens.

The challenge problem of the week extended Newton's method to two equations in two unknowns. This was not covered in the lectures or in the lab; it was up to the students to work out the details.

Challenge Problem:

Newton's method also works for two equations in two unknowns, and for three equations in three unknowns. Recall that to solve one equation in one unknown, Newton's method requires a first estimate x_0, and then it constructs the tangent line to the graph through $(x_0, f(x_0))$. Where this tangent line crosses the x - axis we find x_1, which is an improved estimate (usually) of the true root we are seeking. We repeat the process until the desired accuracy is obtained.

An analogous procedure can be used to solve a system f (x,y) = 0, g (x,y) = 0 of two equations in two unknowns x and y. We start with an initial approximate solution (x_0,y_0) and find the equations of the tangent planes in 3 - space to the graphs of the functions f and g at this point. These two planes can be expected to intersect in a line, which can be expected to intersect the x,y - plane in some point. This new point (x_1, y_1) is an improved estimate (usually) of the true root (x,y) we are seeking. The process is repeated with this new point (x_1,y_1) in place of the original (x_0,y_0) to find still a better estimate, etc.

Problem: The nonlinear system $x^2 + y^2 = 4$, $xy^2 = 1$ has four solutions. Find all four as accurately as you can. First subtract 4 from both sides of the first equation and 1 from both sides of the second equation to get them in the form $f(x,y) = 0$ and $g(x,y) = 0$. Then get a good first estimate for each of the four solutions. You can do this by graphing both these curves (either by hand or by using the implicit function option on the *MicroCalc* diskette), and estimating the coordinates of the four points of intersection. Use Program 5 from *Exploring Calculus* to complete the problem.

6. Local extrema. Points of inflection

The students found local extrema and points of inflection for some functions, where the derivatives would be very difficult to compute by hand. For the functions $y = x^3 + x + k\,x$ and $y = x^4 + k\,x^3 + x^2 + x$, they watched the shape of the curves change for different values of k, and kept track of the number of extrema and inflection points as k changed. By trial and error estimates and graphical feedback from the computer, the students approximated those numerical values of k which produced a major change in the shape of the curve. Then they used their knowledge of derivatives to find the exact value of k by hand.

7. Test your calculus skills

Here the students used software designed to test the understanding of the concepts of differential calculus. They used six different programs from [3]. These programs also related the abstract concepts to some simple applied problems like motion on a line.

8. Integration

Riemann sums, trapezoidal sums, and Simpson's Rule were used to estimate definite integrals. The programs that were used here gave graphic representations of the functions and approximating sums.   The students also found

approximate solutions for $\int_0^x \sin\sqrt{t}\,dt = 1$ and $\int_0^x e^{-t}\,dt = 1$.

The session ended with a tutorial to draw antiderivatives for given functions on the computer screen.

Most students were able to complete the exercises in each lab handout during the 80 minute lab session; some would even get a start on the homework problems, but for all of

them there would be at least something left to do after the session. The students then had to make arrangements to meet with their lab partners in the campus or engineering computer labs to complete the assignments, to write the lab report, and to work on the week's discovery problem. The discovery problems were selected in order to challenge the students, including the better ones. The problems usually went somewhat beyond the material covered in the lectures and required the use of the computer, knowledge of the calculus topics, and some ingenuity.

Student Laboratory Reports

One week after a laboratory session, each team of two students submitted a lab report with solutions to exercises, screen dumps of significant graphs, references to theorems that made a solution of a problem possible, and a solution to the week's discovery problem. The report also had to include a two paragraph description of what was accomplished during the lab session such as, which concepts were studied, which assignments were found difficult and why, how long it took to complete the assignments, where they got stuck and how they got out of it. The rationale for these descriptive reports was twofold. Firstly, the students gained valuable writing experience using mathematical and technical terminology, and they were forced to react on how they accomplished learning of mathematics in the context of a laboratory. They had to think about calculus! Secondly, the instructors received timely feedback - both positive and negative - from the students and could make changes and adjustments in subsequent lab assignments, if students had justified complaints. The lab reports were collected and graded and counted towards the final grade in the course.

Generally, the lab reports were well written and carefully done and contained correct mathematics. Many students used word processors and laser printers to give their reports a polished look.

Evaluation

At the end of the winter semester the students were given a questionnaire to solicit their reactions to the new laboratory course. We modeled the questionnaire after a similar questionnaire in [1], but the questions were changed to reflect the difference between the course format of the Dartmouth project and our project. The questions and responses are shown in Table 1 that appears on the next page.

TABLE 1

Percent Distribution of Responses to Survey

Instructors: Fink / Hoft / James ,
Math 115, Winter 1989
Total Number of Responses: 68

SA = Strongly Agree   D = Disagree
A =Agree   SD = Strongly Disagree
N = Neutral/ No Opinion

	SA	A	N	D	SD
1. Using a computer in the classroom contributed to my understanding of the course material	12	74	10	4	0
2. Using the computer in the laboratory contributed to my understanding of the course material	13	65	12	10	0
3. Using the computer in the classroom enhanced my interest in the course material	13	37	40	7	3
4. Using the computer in the laboratory enhanced my interest in the course material	12	34	34	15	5
5. Using computer graphics helped me to understand derivatives	36	52	7	5	0
6. Using computer graphics helped me to understand integration	18	49	18	11	4
7. Using computer graphics helped me to understand Newton's Method	22	56	7	15	0
8. Overall, the computer labs were a valuable part of this course	15	54	18	10	3
9. The computer labs could be completed in a reasonable amount of time	24	32	21	18	5
10. Learning how to use the computers required little help from others	27	54	6	10	3

The computer in the classroom added a new dimension to the classroom discussions and lectures. I often turned to it spontaneously for a quick demonstration or verification, and the students were motivated to make suggestions on how to pursue solutions of exercises and problems. The students seemed to be more attentive and seemed to concentrate harder when there was something happening on the computer. They also seemed to be more interested in the material, more responsive, and asked more questions than their counterparts in traditional sections usually do. This impression is substantiated by their responses to question 3, with 50% of the students indicating that their interest in the course material was enhanced by the computer in the classroom. The students had few problems learning how to use the computers, and the software turned out to be as user friendly as we had hoped it to be (question 10).

The teams of two students that were formed in the computer laboratories tended to work and study together on their other homework assignments as well. In my section, I became aware of at least seven such regular study groups. One student asked if he could continue to come to the lab sessions after he had dropped the course. He said he enjoyed working with his lab partner so much that he did not want to quit. On a commuter campus, it takes students sometimes a very long time to make friends; computer partnerships seem to help.

We have made no attempt to compare the knowledge level of calculus for the students in the laboratory courses to the knowledge level of those in traditional sections, so we do not know whether the laboratory students learned calculus better or worse or more or less. The final examination did not address material covered in the laboratories with the aid of computers.

We have learned from this experiment that it is possible to integrate the use of a computer into the calculus curriculum, and have the students become comfortable with the computer as a working tool without generating an excessive amount of frustration for the students. For first semester students having difficulties with their first college level mathematics course anyway, a low level of frustration is a very important consideration. We felt encouraged by student response to the course, and in the Winter semester of 1990 we started to teach two pilot sections of second semester calculus using computers. The course format for the second semester laboratory courses was the same as for the first semester courses, three lectures and one lab session per week. However, the software we had been using for Calculus I, seemed inadequate for Calculus II and we switched to *Mathematica* instead. We are still arguing over revisions of the traditional syllabus, and we have no answers yet to the question which topics of traditional calculus should be deemphasized after computers are introduced into the courses.

To implement this laboratory course for first semester calculus has required a tremendous investment of effort and time on the part of the faculty members involved. Anyone contemplating to introduce such a course should be well aware of the commitment of time and energy. However, it has been a long time since I enjoyed teaching first semester calculus as much as I did last year, and I believe that the course format created a group dynamic that stimulated curiosity, and enhanced the learning of mathematics for all students in the experimental sections.

References

[1] Crowell, R. H., Prosser, R. T. *Computers with Calculus at Dartmouth.* Manuscript, 1989.

[2] Flanders, H. *MicroCalc.* MathCalcEduc, Ann Arbor 1988.

[3] Fraleigh, J. B., Pakula, L. I. *Exploring Calculus with the IBM PC.* Addison-Wesley, Reading,MA, 1986.

[4] National Research Council. *EVERYBODY COUNTS A Report to the Nation on the Future of Mathematics Education.* National Academy Press, Washington, D.C., 1989.

[5] Petersen, K. E. *A Computerized Calculus Tutorial for the first Semester.* Harper and Row Publishers, New York, 1988.

The U. S. Naval Academy Calculus Initiative

Howard Lewis Penn and Craig Bailey
U.S. Naval Academy

Introduction

Since 1988 the Naval Academy has been engaged in an initiative to introduce the use of computers into the teaching of calculus. Every student at the Naval Academy takes 3 semesters of calculus. Additionally each student must purchase a Zenith 248, which is an AT clone with an EGA card and a color monitor. Each classroom is equipped with a Zenith and each faculty member has a computer on their desk. The authors have been involved in the production and procurement of software and the production of assignments to use with the course. These assignments are used in conjunction with a standard calculus textbook[9]. Almost all the instructors have used the software and assignments in the class. This project is not strictly a laboratory course since the students complete the assignments on their own in their room. However, these projects can easily be used in a laboratory setting. The large number of faculty, students and machines involved in the project had a great impact on the choice of software. First, the software had to be available on a site license arrangement since it would not be possible to ask each student to spend, for example, $50.00 for a package. Second, since the students would be for the most part working by themselves, the programs used had to be easy to learn and simple to run. Since the students completed the assignments on their own, the number of individual programs had to be kept to a minimum. The Math department purchased site licenses for two commercially available programs, Calculus-Pad and Microcalc. Additionally, the first author, along with Jim Buchanan of the Mathematics Department and Frank Pittelli of the Computer Science Department produced a computer package, Mathematics Plotting Package(MPP). These programs are described later in this paper.

There are several goals that we hope to achieve with the assignments. First we hope to stress the concepts that are covered in calculus. The programs that we use are graphical packages that help students better understand the connection between the analytic equation or function and the graph of the solution to the equation or the graph of the function. Another goal is to have the students work with more applications. The assignments place additional stress on the numerical techniques that occur in calculus. Many typical calculus books introduce the techniques of numerical integration but then go out of their way to avoid problems in the application sections that require a numerical solution. The use of graphics and numeric programs allows the assignment of problems with more complex functions. Another goal of the project is to make assignments that require the students to use techniques from more than one section of the book. Many of the assignments require the students to write their conclusions about the results of a project. One more aspect of this project is to at least occasionally show the students that mathematical equations can be interesting and that their graphs can be beautiful. Above all, we wish to force the students to think about calculus rather than try to memorize a collection of, to them, meaningless formulas.

The problem assignments

The authors have produced a collection of 43 computer and supplemental assignments covering topics from all three semesters of calculus. Each assignment contains several problems. Most of the assignments require the use of MPP. A few assignments use Microcalc or Calculus-Pad. Two of the projects in Calculus III require the students to write a short Turbo Pascal program. This is feasible since each student is required to take a three hour programming course in Turbo Pascal during the freshman year.

Not all the assignments require the student to use the computer. Some of the problems require the students to do work by hand. Several other problems require the students to compute an integral by hand if they can or to use the numerical integration portion of MPP if they cannot do the integral by the techniques that have been covered. This forces the students to decide if they are able to compute the integrals. Many of the problems require the student to turn in a written explanation as well as the graph of a solution. Later in this paper, we will look at several specific examples from the problem assignments.

The computer software

Calculus Pad[4] is an easy to use program. It is rather limited in scope but does two activities very well. The program is at its best when graphing surfaces in three space. Once the initial graph has been produced, the user can rotate the view. During rotation, the graph disappears and the user only sees the coordinate axes. The arrow keys are used to change the view. As these keys are pressed, the coordinate axes rotate in real time. When the correct position is obtained, the return key is pressed and the graph is quickly redrawn. The other useful feature of this program is its ability to compute quickly and draw Taylor polynomial approximations to a function. The program is available from Brooks-Cole for $30.00. The address of the publisher may be found in [1]. A site licence may be purchased for $200.00 with a $50.00 per year renewal fee. The program is free with the adoption of Steward[8].

Microcalc is a comprehensive program covering many of the topics in calculus. The program is available from MathCalcEduc for $425.00 which includes an unlimited site license. One useful portion of the program is surfaces of revolution. The program draws the area to be rotated and then shows the surface as it is generated. Another topic that is covered well by Microcalc is generation of parametric space curves. The program can plot the unit tangent vector and the unit normal vector at the point. The program was selected as a distinguished program in the 1987 EDUCOM/NCRIPTAL awards[3]. A review of the program appears in [2].

MPP is a collection of nine modules integrated into one master program. Eight of the nine modules make use of graphics. The program will run on IBM-PC's and clones with at least 512K of memory and a graphics board. However, the program is designed to be used on machines with an EGA or VGA card and a color monitor. The first module is a function plotter. It will plot up to 6 functions at the same time. These functions may be of the form $y=f(x)$, $x=g(y)$, parametric, polar or data points stored in a file. One of the strengths of the program is the ability to mix easily the types of graphs. One of the modules illustrates the definition of the derivative by repeatedly drawing secant lines. Another is used for finding the roots of a function using Newton's method, the bisection algorithm, or the secant method. Both of these modules zoom in as the points get closer together. There is a numerical integrator that uses Riemann sums with left, right, mid or randomly chosen points, the trapezoidal rule or Simpson's method. The program contains a contour plotter that will plot up to 15 contours for a function of two variables. There are also 2 and 3 variable integrators. These allow integration in any order and polar, cylindrical and spherical coordinates. The 2 variable integrator shows the region of integration. This program may be obtained by sending to the first author either two 5 1/4 inch. disks or one 3 1/2 inch. disk. The author asks that the disk(s) be formatted. In addition to the

program, the documentation[7] and the computer and supplemental assignments[8] will be included.

Specific assignments

The first assignment in Calculus I has the students graph a few functions. One of the functions they are asked to graph is $sqrt(x^2-4)$ from -5 to 5 with a step size of $.1$. When the program plots this function, it gets to -1.9 and reports "undefined function at -1.9." The student is then faced with the problem of how to get the rest of the graph. The solution is simply to plot the function from -5 to -2 using one function input line and from 2 to 5 using a second function input line. The students are then challenged to come up with a single line input which has the same graph as $sqrt(4-x^2)$ on its domain of definition and the same graph as $sqrt(x^2-4)$ on its domain of definition. This assignment should help the student better understand the concept of domain.

Another early assignment concerns the graphing of trig. functions. The last problem asks the students to graph $cos(100\pi x)$ from 0 to 4 with a step size of 0.02. The program draws a line across the graph at a height of 1. The students are asked to explain in writing if the graph is surprising and why the graph is as it appears. Unfortunately, some of the students are not surprised by the graph because they have no idea of what a cosine curve should look like. After a while, someone comes up with the explanation why the graph looks the way it does; that is, the step size is the same as the period of the function. The students are then asked to plot the graph again with step sizes of 0.021 and 0.019. Finally the students are asked to graph the function with a step size of 0.001. Eventually, the students learn that there is no way to get a good graph of this function on the interval given because the computer cannot draw enough points. This exercise is intended to help the students better understand the concept of the period of a function.

The epsilon-delta definition of a limit is one that gives the students much difficulty. Most of the instructors teaching calculus can remember being required to come up with a formula for delta given epsilon. Using a graphing program, such as MPP, this concept can be explained in term of keeping the graph on the screen. The students are told for various epsilons to set the minimum value of y to L$-\varepsilon$ and the maximum value to L$+\varepsilon$. They are then to try to find a value of δ such that if the minimum x is a$-\delta$ and the maximum x is a$+\delta$, the graph does not go off the top or the bottom of the screen. The assignments ask the students to find values of δ for $\varepsilon=1, .1, .01$ and $.001$. Naturally, some of the examples are of discontinuous functions with jump discontinuities of about .005. Figure 1 shows the graph of $x^{2/3}\sin(200x)$ from −0.1 to 0.1 with Ymin = −0. 1 and Ymax = 0. 1.

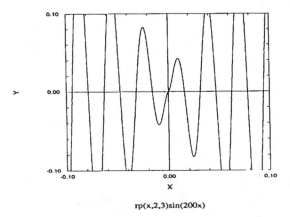

rp(x,2,3)sin(200x)

We can see that for ε=0. 1, δ=0. 1 is not small enough. Now we show the same graph with Xmin=−0.25 and Xmax=0.25. We see that δ=0.25 is sufficiently small to keep to graph from crossing the top or bottom of the graphics window.

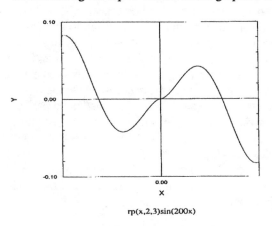

rp(x,2,3)sin(200x)

In Calculus II, there is an assignment to compute

$$\int_{1}^{b} 1/x \ dx$$

for b ranging from .5 to 2 with steps of .1. As well as illustrating the definition of the natural logarithm function, this assignment reminds students what happens when the limits of integration are switched. We assign this project just before the students are introduced to the natural logarithm function. Likewise, there is an assignment to use Newton's method to construct a table of values for the inverse of ln(x). This assignment is made just before the exponential function is covered.

One of the assignments in Calculus III concerns the Maclaurin polynomial approximations to ln(1 + x). The students are required to plot the function together with the 1st, 3rd, 5th, 7th and 9th order Maclaurin polynomials. The students are asked to discuss the behavior of the polynomials on the interval (−1 ,1) and the interval (1,3). The figure below shows the graphs of these functions.

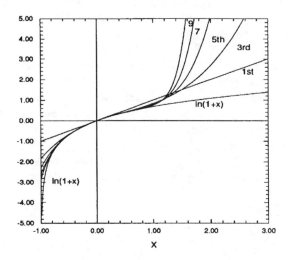

ln(1+x) and the 1st, 3rd, 5th, 7th and 9th Maclauren Poly

We decided to use MPP for this project because the graphics is in EGA mode and each function can be plotted in a different color.

When the topic of inverse trig. functions is covered in Calculus II, there is an assignment to plot the graph of the function, arcsin(sin(x)) from −2π to 2π.

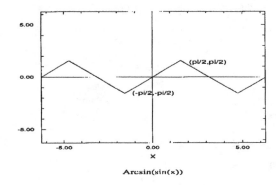

Arcsin(sin(x))

One look at this graph should convince any student that the arcsin is only the inverse for the interval [−π/2,π/2].

Programs like the ones used with this project make the graphing of complicated functions much simpler. Therefore, students can view many more examples as well as more complex examples. This will force those of us teaching calculus to rethink the way we approach the graph sketching using derivatives and asymptotes material in Calculus I. The connection between the analytic and the graphical representation of a function is still very important, but the focus need not anymore be only on using information about the function and its derivatives to sketch the graph. There are several assignments involving these topics. There is an assignment to have MPP draw the graph of a function which is beyond their capabilities to differentiate and then to sketch the derivative by hand. There are assignments to draw a graph and label where the first derivative and the second derivative are positive and negative. There is also an assignment that is intended to show the students that you cannot rely only on one graph. This involves the graphing of the function,

$$(x^{2/3})(x^2-5x-6)/(((x-1)^2)(x+6))+4\arctan(x)/\pi.$$

First we show the graph of the function on a large scale.

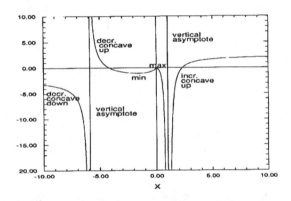

The second graph is the same function plotted near (0,0). It shows that there is an additional point of inflection that does not show up in the broader graph.

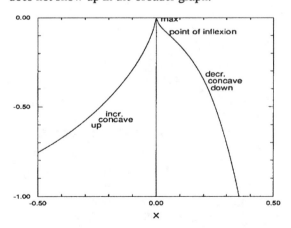

One of the assignments in Calculus II is to plot the functions, $\ln(x)$, $\log(x)$, $\log_2(x)$, $\log_{1.2}(X)$. $\log_5(x)$ and $\log_{100}(x)$ and turn in a written explanation of how the base of the logarithm effects the graph. Here is a graph of some of these functions.

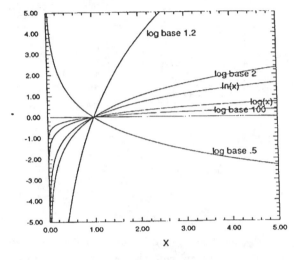

Logarithms with different bases

The contour plotter is used 3 times in Calculus III. It is used first when the topic of functions of several variables is introduced. The second time is when the gradient is covered. There is an assignment to plot the contour lines for a function of the form z=f(x,y), draw by hand the gradient at several points and then use the vector field plotter to plot the gradient. The third use is for the topic of extrema of functions of several variables. For this topic it helps to look at the contour plot produced by MPP as well as the surface plot produced by Calculus Pad. The next two figures show the surface and the contour plot for a function that has a saddle point at the origin but is a local min in both the x and the y direction.

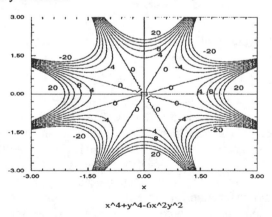

x^4+y^4-6x^2y^2

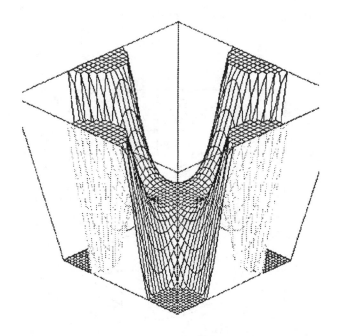

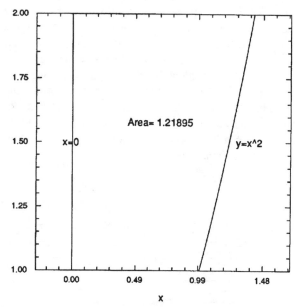

The increased emphasis on applications is illustrated by a couple of examples. The first involves the root of $J_0(X)$, Bessel's function of the first kind of order 0. Bessel's function of the first kind is one of the functions available in MPP. The roots of this function are very important in the study of the heat equation in a cylinder and the vibrating circular drum head. A second example is an application of the integral. The students are told that the weight of midshipmen is normally distributed with a mean of 170 and a standard deviation of 30. They are to use the integrator to estimate the probability that a midshipman will weight between 200 and 250 pounds. The function, normal(x,m,s) is available in MPP so they need only integrate normal(x,170,30) from 200 to 250.

An example of the use of the double integrator is a problem on finding the center of mass of the region bounded by y=1, y=2, x=0 and y=x², to the right of the y-axis, if the density is given by cos²(x²). They are told that only one of the integrals can be worked by hand, they are to compute this integral by hand and use the double integral evaluator to estimate the other integrals. The following figure shows the output for the integral to compute the area. The double and triple integrators are also useful to teach students how to set up multiple integrals. One student when faced with a difficult triple integral set up the problem all 6 possible ways and after some corrections was able to get the same answer from each integration. This student never had any more difficulty doing multiple integrals.

The next couple of examples show how material from more than one section is used to solve problems. Students are asked to find the area between the curves y=sin(x²) and sin²(x) from 0 to the next time the curves cross. The students are to first graph the two functions, then use Newton's method to determine the point where the functions cross and finally use the numerical integrator to find the integral. A side benefit of this assignment is that the students have a better feel for the difference between these functions. A similar problem comes up in Calculus II. The students are asked to plot x^π, π^x, π^π and x^x. Below is a graph of these functions. The students are then asked to find the area between the graphs of x^π and π^x.

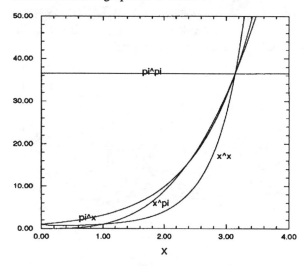

Power functions

The final goal of the project is to show the students occasionally that mathematics can be fun and beautiful. When the topics of parametric equations, polar functions and contour plots are covered, the students are assigned some interesting graphs and are asked to also produce an original graph. These assignments turn out to be a lot of fun for some of the students, and they put a considerable amount of effort into the original graphs. At the same time they need to think about the domain of the independent variable and how closely together the points need to be plotted. The following shows a parametric equation that one of the students submitted.

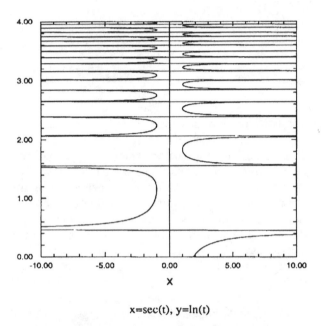

x=sec(t), y=ln(t)

A few years ago a student submited the following polar coordinate graph. Dr. Penn's paper on polar coordinate graphs[5] gives several other examples.

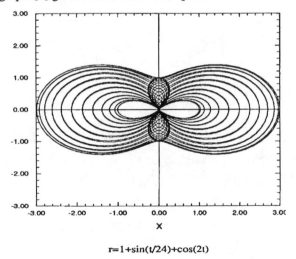

r=1+sin(t/24)+cos(2t)

Here is an example of a contour plot.

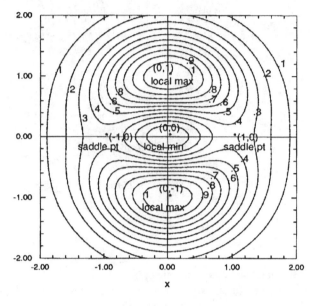

(x^2+3y^2)exp(-x^2-y^2)

A final example that shows the beauty of the graphs, while at the same time requiring the students to think, is the example concerning the epicycloid and the hypocycloid. The students are required to plot the unit circle together with the epicycloid and the hypocycloid generated by having a smaller circle of radius 1/4 roll around the unit circle. The students are then asked how many times the smaller circle revolves around its center. The most common answer is 4. However, after some discussion of the problem, the students reach the conclusion that for the epicycloid, the circle rotates 5 times and for the hypocycloid, the circle rotates 3 times.

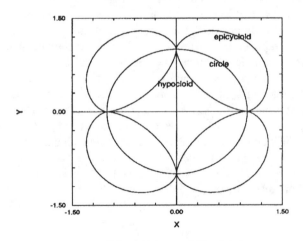

Epicycloid, hypocycloid and circle

Conclusions:

There are both advantages and disadvantages to having the computer assignments as homework. One of the advantages is that the midshipmen have constant access to the computer and the software. Therefore, they can work on the project at the time that is best suited to them. They can use the software for activities other than those assigned. There are times during the semester when 1 hour a week is not sufficient to spend on the assignments, such as when the new functions are introduced in Calculus II. Likewise, there are times when a couple of weeks may go by when there is no need to spend time with the computer. The homework approach allows the instructor to give more than 1 assignment during some weeks and skip weeks at other times. The instructor also has constant access in the classroom and can use the software when the flow of the lecture and the students' questions suggest the need.

On the down side, the students must work the assignments without immediate access to help. Therefore, the number of individual programs must be kept small. There is not as much chance to discuss the outcome of each assignment.

In general the reaction of the students has been very favorable. Most of the students feel that the projects help them better understand the concepts. When every student is required to work the assignments, there are some who do not like doing them. Some students have an aversion to using the computer. The most common negative comment is that the assignments are a waste of time. These comments come from precisely the students who try to learn calculus as a collection of meaningless equations that are memorized. Yet these same students make less of the conceptual mistakes that are so common. The number of times that students differentiate π^π improperly is much smaller since we have been assigning these projects.

Overall, we feel that the use of the computer assignments has had a very positive effect on the teaching of calculus. The assignments work well as homework but would work just as well in a laboratory.

Acknowledgements

The authors wish to thank Jim Buchanan and Frank Pittelli for their contributions to the program, MPP. We also wish to thank all the instructors and students who have used the software and the assignments connected with this project. The work described in this paper has been funded by an Instructional Development Project Grant from the U.S. Naval Academy. We wish to thank the administration of the Naval Academy for their support of the use of computers in teaching.

References

[1] Cunningham, R.S. and Smith, David A. "The Compleat Mathematics Software Database", **College Mathematics Journal**, Vol. 19(1988), pp268-289.

[2] Kozma, Robert B., Bangert-Drowns, Robert L. and Johnston, Jerome, **1987 EDUCOM/NCRIPTAL Higher Education Software Awards,** NCRIPTAL, 2400 School of Education Building, University of Michigan, Ann Arbor, MI 48109-1259.

[3] Leinbach, L. Carl, "Review of Microcalc 3.01", **College Mathematics Journal,** Vol 19(1988) pp367-368.

[4] Norman, Daniel and Vemer, James, **Exploring with Calculus-Pad.** Brooks Cole, Monterey, CA, 1987.

[5] Penn, Howard Lewis, "Computer Drawn Polar Coordinate Graphs", **Collegiate Microcomputer,** Vol 2 (1984) pp 273-279.

[6] Penn, Howard Lewis and Bailey, Craig K. "Mathematics Plotting Package Computer and Supplemental Assignments", Mathematics Department, U.S. Naval Academy, Annapolis, MD 21402

[7] _____, "Mathematics Plotting Package Documentation, Mathematics Department, U.S. Naval Academy, Annapolis, MD 21402

[8] Stewart, James, **Calculus,** Brooks Cole Publishing Company, Monterey, CA, 1987.

[9] Swokowski, Earl W. **Calculus with Analytic Geometry** (Second Alternate Edition), Prindle, Weber and Schmidt, Boston, MA 1988.

Iowa's Computer Labs for Calculus

K. D. Stroyan
University of Iowa

Introduction

Since the early 1970's, the University of Iowa (a public university with nearly 30,000 students) has offered optional computing "laboratories" that run concurrently with calculus 1, 2 and linear algebra. (See Hethcote [1972].) The essential format of these labs stabilized around 1974 and has been very successfully taught by a number of faculty and advanced T.A.'s during the last 15 years. We have expanded our materials with advances in computing, moving from main-frame to mini, then to IBM-compatible personal computers. In 1986 I published a lab manual with software for IBM-compatible or Apple II computers called, *Computer Explorations in Calculus* (Harcourt, Brace, Jovanovich). The linear algebra lab is still taught from notes and locally available software. This is the current status of our BASIC labs as they continue to be offered, but we also have ambitious new plans to use up-to-date computing with both calculus and linear algebra.

Last October we installed a network of 10 NeXT computers as a lab for our special "Accelerated Calculus" course (22M:45-46). Since mid-October, weekly computing "lab assignments" or "electronic homework" have been a required part of that course. Accelerated Calculus is taught to about 10% of our beginning regular (non-business) calculus students, with an enrollment of 100 students in the fall semester. (The BASIC calculus labs remain optional for the remaining 90%.) We are using *Mathematica* software with a combination of prepared "Notebooks" and student programming. Since we received NSF ILI funds, our NeXT lab is dedicated to just Accelerated Calculus, whereas the BASIC calculus labs share public computing clusters with all students. Ten computers are enough for 100 students since our lab is open 50 hours per week. Moreover, we are set up with room for two students to work at each computer and encourage that on many assignments. The discussion and interaction between two or more students working at the computer often seems beneficial mathematically and helps to control the distracting technicalities of computing - one student can type while the other thinks.

My colleague Eugene Johnson is developing *Maple* materials (to use on Maclntosh SE computers in public clusters) for our linear algebra computing lab.

The new plans will greatly expand the scope of our labs, but some of the fundamental lessons we learned along the way form an important springboard for these plans. This article describes both our BASIC calculus lab and the beginning experience with our NeXT/*Mathematica* calculus lab. I explain some of the reasons for organizing things the way we do in the hope that this will contribute to more wide-spread success of such labs and help to speed the development of more advanced materials such as the ones on which we are working. It is time to bring the teaching of calculus into the computer age. Many "square wheels" have been re-invented in this quest, there are pitfalls in "integrating" calculus and computing. I hope this article will encourage newcomers and give them a head start.

Why use computers with calculus?

Many issues and ideas (particularly numerical ones) relating calculus and the computer are only tangential to the main stream of ideas in calculus. On the other hand, typical high school graduates now are more accustomed to everyday use of computers than typical mathematics faculty, so that complete separation of computing and mathematics must seem out-of-date to them. Even a simple program like *Eureka* seems to answer all the big questions of calculus, so why do their professors ignore computing and insist on hand computation of problems contrived to come out in whole numbers?

True rigor in calculus requires a certain kind of sophistication. Euler, Gauss and Cauchy didn't have it. Yet we professors all know how "simple" it is, thanks to three centuries of collective effort and years of our own training. We can get the concepts across, if with less than complete rigor. Computing offers students a chance to "experiment." This is the most important reason to offer computer labs for calculus - computer experiments can strengthen students' understanding of the concepts by giving them a realm they can explore. I also believe that computing is an important mathematical tool which we need to train our students to use as such. Computing can contribute both to learning the conceptual part of calculus and its "real world" applications. This is not to say that we should teach full-blown "programming" courses. The preface of the manual (Stroyan [1986]) that we use in our BASIC lab says it this way,

At the University of Iowa we use these materials in an optional one-hour computer "lab," where students use the computer to explore both fundamental concepts of calculus and approximation procedures arising in calculus. We try to encourage "experimentation." A theory as polished as calculus makes it easy for teachers to forget that it wasn't always known and hard for students to see how it was discovered. The computer offers students unique opportunities to explore ideas in calculus before and during the process of learning the theory.

In order to focus on mathematics we strive to keep programming technicalities to a minimum. Students can readily use our materials without any prior programming experience, but we do expect students to be actively involved in the workings of all the programs (except some of the graphical demonstrations.) Parts of the exercises test this understanding by having a mathematical bug that requires a program modification. This kind of active computer use helps students understand calculus by having to formulate ideas in the computer's terms.

Our materials also teach students how calculus and the computer can interact to the benefit of both. Roughly speaking, calculus is good for qualitative understanding and the computer is good for quantitative accuracy. Using the two together, one can obtain accurate answers to elaborate problems.

Modern scientific software even promises to make computing an important wide-spread part of research in pure mathematics - and out-mode my numerical remarks in the previous paragraph. The spectacular graphics, well integrated with numerics and symbolics in *Mathematica* offers many new avenues of exploration to students. The convenient "Notebook" front end (or editor) provides teachers with a powerful medium with which to guide that exploration. In the 1990's, our students should leave calculus with a good start at their scientific computing education - symbolic, graphical and numerical.

Examples from our BASIC lab

Our lab manual, *Computer Explorations in Calculus* (Stroyan [1986]) contains many exercises at various levels intended for weekly assignments. As remarked above, our current lab is largely numerical (rather than symbolic or graphical) and is based on simple programs locally written in BASIC (for either Apple II or IBM-compatible computers.) We have several graphical programs, such as a "microscope" to see that the linear approximation of a derivative is the line you see under a powerful microscope. We also make microscopic views of the nowhere differentiable function of Weierstrass and draw "Gibbs goalposts" (in the section on uniform approximation.) Despite these graphical examples, the technology directs the students' efforts toward numerical programs. Our exercises have been tested on many students, reviewed and developed informally by the people mentioned in the acknowledgement and reviewed in painstaking detail by 4 experts and the editorial staff of Harcourt, Brace, Jovanovich publishers. There are still many more computing exercises and projects to develop. I hope many of you will try to do so. Ours are only a beginning and we also recommend the book by Smith [1976] to help get you started.

The Table of Contents of our lab manual shows the organization. We begin with two class hours on programming in Part 1 and return for one more lesson on "round off errors" at the beginning of the second semester. Part 2 is a list of weekly exercises that follows the development of most first semester calculus courses. The emphasis is on the fundamental concepts of calculus as viewed from a personal computer. Part 3 emphasizes numerical approximations from calculus and also points out some of the main pitfalls in numerical computation. For example, the harmonic series converges on a floating point computer, but not in our lifetime. (Exercise 3.9.6 asks the students to estimate how long it would take a perfectly accurate computer adding a million terms per second to reach 99. It asks for the answer in units of age of the solar system...)

Table of Contents from *Computer Explorations in Calculus*

Some of our exercises build themes and each semester I assign one larger "project" requiring two weeks. We will describe two of the themes.

Riemann sums

A beginning programming exercise in our lab shows students how to accumulate a sum, $1+2+3+...+n$, $1/31+2/31+...+31/31$, $1+4+9+...+n*n$, etc. It is a mathematically interesting way to learn about loops and the way computers treat equality as replacement.

Before integration is discussed we have a mini-project to compute the arclength of an ellipse. We use this as an application of the uniform derivative, because the polygonal approximation tends to the length on small pieces as the pieces become straighter. (Naturally, this problem cannot be solved in elementary terms even once students learn integral calculus, as Liouville showed 150 years ago. We remind the students of this in the second semester when they are in the thick of integration techniques.)

When students begin integration in the regular class, we calculate Riemann sums. **This reinforces the idea of an integral as a generalized sum,** since students are already proficient at calculating sums with the computer. Almost immediately, students are dissatisfied with the computer's accuracy, so we discuss Riemann's definition and illustrate it with left-, right-, mid-, and random-evaluation points in a simple program. Midpoints win and, in the first

semester, we opt for more conceptual questions as the next step.

Next, we give some exercises that amount to asking students to derive integral formulas. We ask them to use their midpoint rule program to compute the sum of the formula, not to integrate only by rules. These exercises are chosen in the hope that no book has a derivation of a formula of that kind. Plugging in formulas is what we are trying to avoid. The idea is to reinforce the concept of integrals as sums by actually having students evaluate sums.

In the second semester we begin with a section on accurate integration. The first lesson is that summation with too many terms produces a bad answer. (This is because the machine uses fixed length decimals, so that addition of a tiny number to a big number loses a significant portion of the tiny one.)

The final moral of this is that, unlike numerical differentiation, numerical integration (Simpson's rule with the Richardson extrapolation self-check) is a reliable, accurate tool. Applications to log, arctangent, normal statistics, elliptic functions, and the si-function follow. Any one of these could be built into a student project.

The details of this theme have changed in our NeXT lab. Summation is done with a single command. We have many graphical illustrations (the BASIC lab has only one). It ends with a single command numerical integration routine, so the numerics are suppressed. With polish we hope to make the main connection between summation and integration even clearer with the new technology.

The Catenary and Suspension Bridge

Section 3.4 of our BASIC lab manual can be done at several levels up to a full scale student project of calculating the optimal sag in the support of a suspension bridge model.

The first exercise is easy and is used to illustrate the root finding programs of the previous section (cf. Contents above). The implicit formula for horizontal tension of a catenary in terms of sag is simply given and the student is asked to use either the program for Newton's method or the secant method (the latter being easier) to solve and then compute the maximal tension. This problem has some of the features of a real-world problem. It is hard to start these local root-finders. The root is near 1,000 and so a tolerance smaller than .0009 can cause an infinite loop on a six decimal machine (and it happens!) Getting started takes one weekly lab session. (These same accuracy problems arise in *Mathematica*, by the way.)

This is no place to stop and we are appalled that some calculus books omit the derivation of the differential equation for the catenary. The specific catenary is not so important, but **the concept of defining a function by simple (nonlinear) properties easily described by a differential equation is a central idea of calculus.** Computing tends to bring the conceptual idea out ahead of tricks needed to find explicit solutions. (Take the tractrix if you don't like catenaries; both turn out to be important in calculus of variations and differential geometry.) So we prove that the tension is given by the formula we gave in the first exercise by deriving the differential equation of the catenary and solving it. This is a deep application of the "increment principle," which we illustrate with computer "microscopes" and use repeatedly. The catenary equation is later extended to arrive at the equation for a bridge with a light support wire (w) and a flat heavy road-bed (W):

$$dV = \left[W + w\sqrt{1 + \left[\frac{V}{H}\right]^2}\ \right]\ dx$$

This "suspension bridge" equation can be "solved" by separation of variables. In fact, the program *Maple* finds a symbolic implicit solution (slightly different from ours, Stroyan [1986], p.169.) However, implicit solutions are useless. (It may well be that it cannot be solved in elementary terms. For example, one cannot solve $y = x^x$ for x in terms of y in elementary terms.) A project lets students use a simple 4-5 Runge-Kutta program to solve numerically for the tension at a desired sag.

Between the big project and the opening exercise, we ask various levels of optimization questions. 'Find the minimal maximum tension,' etc. The point is that **computing makes the concepts come to the fore, even in the context of "messy" real-world problems,** because the students are setting the problems up and asking the computer to solve them, rather than the professor setting them up and asking the students to plug in.

Our materials also contain many routine exercises. Two of the early surprises (to us) in the computer lab were the following. First, students are often reassured by the computer getting the right answer to obvious problems. For example, limits of clearly continuous functions or derivatives of simple polynomials. Second, students need to learn to explicitly interpret the result of computations. We insist that they write out their interpretations. Students need to be taught that computers are not omnipotent number gods -

they can get the wrong answer, or the right answer may be that the limit isn't converging, etc. Most students want to circle the last line of print-out and not write any comments. Hopefully, our lab students will no longer accept the clerk's reply, 'The computer says...' This is important learning, but modern software will let us do much more.

Examples from our NeXT lab

The conventional order of topics in our BASIC lab (given in the table of contents above) is not followed in our NeXT lab. We are developing completely new materials for our whole Accelerated Calculus course, not just the computer lab that goes with it. Last fall, our computers arrived about 5 weeks into our course. We had a single loaner computer for a while and students' first assignment was to log on and run a Notebook that solved a previously assigned homework problem. The first truly mathematical electronic homework was to make surface graphs and contour graphs of six functions. Students computed partial derivatives for the same functions as a class assignment. *Mathematica* has simple single commands for surface and contour graphs (as well as density plots). The students themselves made the plots. A *Mathematica* Notebook explained what to do and guided the exploration of the various options and pitfalls. Some of the plots are shown below. A follow-up question was: Find a point on each graph with a horizontal tangent plane. One of the functions $z = \sqrt{x^2 + \text{Sin}[y]^2}$ has a non-differentiable conical point which most students discovered themselves.

Students knew from lecture and preliminary Notebooks on the loaner machine that *Mathematica* could find symbolic derivatives. Several of them wrote their own Notebook to check their partial differentiation homework. This was not assigned, just something that they found convenient. This was week 2 of our lab - you don't need to spend much time teaching programming as a separate subject.

Max-min in two variables can solve many interesting problems that one doesn't find in the regular calculus books because the systems of simultaneous equations that arise are intractable. *Mathematica* can solve many complicated systems of equations (although instructors should try them before they assign them.) I handed out solutions of some nice geometric problems with nasty equations in the text materials and planned to make our first "killing" with machine symbolics in our two week old lab. Unfortunately, physical plant decided we needed an electrical upgrade and shut us down for that week. I mention this because some of these kinds of problems are bound to strike any lab every semester. You need some flexibility to deal with such problems.

Our next computing topic was one of our most successful - discrete dynamical systems used to study simple population models. The prepared Notebooks were minimal and the student interaction was maximal. We did experiments and asked students to formulate conjectures in three successive assignments leading up to the local stability theorem for equilibria of nonlinear systems - a nice use of calculus, since students can work out the linear case and calculus linearizes locally. (We certainly aren't the first to think of this topic, James Sandefur has extensive elementary materials and in lecture we followed notes of Frank Wattenberg that were being used at U. Mass.) Two main term projects, one on an economic model and another on real scientific data of harvesting whales (See the paper by Johnston, *et al.* in this volume) built on this experience. We also followed into the study of differential equations from here, using Euler's method in the computing assignments.

Our Notebooks contain an integrated mixture of numerics, symbolics and graphics and all three were used by students in their term papers. It is easy to begin to write such Notebooks with *Mathematica*. *You* solve a difference equation for 20 terms as a list and plot the list with a single command. Both the numbers and graph are right there. In the first semester I wrote about a dozen in six weeks (while keeping up with all my other work). Students can learn the part of *Mathematica's* programming language that you want "as they go" by working your prepared weekly Notebook assignments. I classify mine according to the level of student interaction that they require. Some have flashy animations of microscopes expanding surfaces or dynamical systems moving particles and these are often only used by the students to help cement a concept from lecture with relatively little "programming." This lowest level of interaction still allows the students to re-run the Notebook with their own function or parameter, but the next level requires more contribution from the student to complete the assignment and often teaches a *Mathematica* command or two in the process. The third level is only a "shell program" aimed at reducing the technicalities in a "programming" assignment. The fourth level is a student written Notebook. I believe that each type can play a valuable role in teaching, learning and using calculus.

My students each wrote a *Mathematica* Notebook "from scratch" with their term paper, though the difficulty of the programming varied a great deal sometimes because the project did not demand much computing and sometimes because they lacked experience. One project on optics had a very elaborate and beautiful Notebook. Several students

invented an interesting programming style. Our NeXT network is very convenient and they would open several of my old assignment Notebooks, clip command segments they needed, and paste them into their own notebook. It made me feel extra responsibility for developing good programming style. One of our goals is to make *Mathematica* (or some scientific computing package) a tool our students can continue to use beyond this course.

Mathematica is not a simple language once you get beyond the basics or want to modify a built-in command like "Plot3D" or "ContourPlot" in a seemingly minor way. Much of the advanced syntax is cryptic with dots and dashes where clear connectives would help. The original *Mathematica* book is good to start you off, but is not an adequate reference. Hopefully the new *Programming in Mathematica* book (by Maeder) will help us learn the intricacies of list manipulation, mapping onto lists, contexts and the like. We have developed some intermediate level Notebooks for flows of continuous dynamical systems. Much of that has been hard work for us, but something from which we have sheltered the students so they could concentrate their efforts on learning mathematics. (Essentially we made some new commands like "flowMovie" to animate their differential equations.)

NeXT/*Mathematica* has a few bugs, but they are obscure and minor enough not to be major distractions. The NeXT network has been terrific - a real Unix network, but with a helpful front end. Students like the NeXT software package and use it for many things unrelated to the course.

The most exciting time in our lab was the last two weeks of the semester when students were working on their term papers or "projects." Several faculty passing our lab in the hall remarked to me that they had never seen such excitement and high level student discussion. What was going on? This was a time when students could use the calculus and computing they had learned to work in some detail on a problem that interested them. I put together a list of 15 proposed projects and also solicited original ideas. Students signed up one week, submitted a first draft the next and completed the paper the third. My students have very diverse interests from physics and engineering to music and sociology. My project ideas came from ideas in our BASIC lab, from materials developed at The Five College Consortium in Massachusetts, Duke University, Iowa State University and other places.

Classical physics or engineering problems really excite a small fraction of my students, but repulse another segment. World population growth, epidemics, economic models, harvesting whales, cooling each had their proponents and opponents. Good projects capture student interest, are based on sound use of mathematical ideas from the course and help start the student investigation without giving out the punch line. They're hard to get just right. Many students like "curve fitting," but use it in grotesquely uncritical ways. It is a rich source of student interest, but one where I need to learn more about guiding their exploration.

The down side of the projects is that they are almost all-engrossing, very time consuming and deal with a limited number of mathematical ideas. I advocate two or three per semester, but am not in the camp that thinks the course should be all projects.

Why a computing laboratory?

Why have a separate lab for computing? Why not "integrate" calculus and computing? I believe that there are two major reasons. First, a separate lab can deal with the inevitable technical problems that arise with computing. For example, computing can bring the role of a differential equation as a description of the way a quantity changes to the fore, but we still don't want a main conceptual discussion interrupted with a question like, "Why couldn't I get the printers in room 18 to print the graphics from those differential equation examples?" (The answer, requiring some time to verify, turned out to be that the student did not follow the directions to "boot" the software.)

More than hardware and logistical problems, one must strive to minimize "programming" technicalities, whether using BASIC or *Mathematica*. 'Each minute spent learning how to run the computer is a minute not spent learning mathematics.' The model we developed by 1974 provides students with "shell programs" which they modify for various uses. One does not need to spend much time teaching programming per se - just do it! That isn't any harder with *Maple* or *Mathematica*. (One does need to spend considerable preparation time designing the experiments and shell programs.) On the other hand, our students do program and modify programs to avoid certain mathematical difficulties. We have found that average students must write a few short programs (which compute mathematical things) in order to have the confidence to modify the larger programs.

One must be careful not to let technicalities (or toys, even wonderful ones) distract students from the main objectives of calculus. A separate lab is a natural place to confine these technical problems. We don't wash beakers during chemistry lectures and we can't afford to waste time on the computing equivalent during calculus lectures. Required computer labs could still allow the main lectures on calculus to rely on machine computations. (Use of computers during lectures for demonstrations is really a separate issue. I am discussing things students actively do. I do not believe computer lecture demonstrations are necessary and wonder if they are cost effective, but that would take a whole article to explain.)

Second, a lab provides a good environment for students to work on "open-ended projects." We have described the suspension bridge problem from our BASIC lab and the more extensive efforts in our NeXT lab above. Much recent praise has been given to "open-ended problems," but they are not any easier now for students than they ever were. They can be exciting and worthwhile, but take considerable time. A lab can require a multi-week project without completely interrupting the flow of ideas in the main course on calculus. A host of excellent projects is waiting to be developed and computing can be a key part of it, while a "lab environment" is almost essential.

References

H. W. Hethcote and A. J. Schaeffer [1972], *A Computer Laboratory Course for Calculus and Linear Algebra*, American Mathematical Monthly

David A. Smith [1976], *Interface: Calculus and the Computer*, Houghton Mifflin

K. D. Stroyan [1986], *Computer Explorations in Calculus*, Harcourt, Brace, Jovanovich Publishers.

Accelerated Calculus 1, University of Iowa, Fall 1989, Electronic Homework 1

f[x_,y_] := Sin[x y] Cos[x y]
Plot3D[f[x,y], {x,-2,2}, {y,-2,2}, PlotPoints -> 25]

Accelerated Calculus 1, University of Iowa, Fall 1989, Electronic Homework 1

f[x_,y_] := Sqrt[1 - x^2 - y^2] Cos[2 Pi x y]
Plot3D[f[x,y], {x, -1, 1}, {y, -1,1}, PlotPoints -> 25]

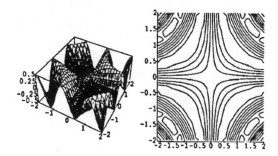

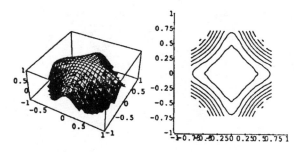

f[x_,y_] := Sin[x]^3 + Cos[y]^4
Plot3D[f[x,y], {x, -Pi, Pi}, {y, -Pi, Pi}, PlotPoints -> 25]

f[x_,y_] := Sqrt[x^2 + Sin[y^2]]
Plot3D[f[x,y], {x, -Pi, Pi}, {y, -Pi, Pi}, PlotPoints -> 25]

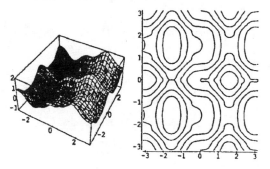

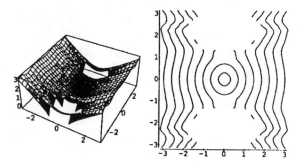

Local and Global Explorations in Calculus Using Computer Algebra

Elton Graves and Robert J. Lopez
Rose-Hulman Institute of Technology

Introduction

Rose-Hulman Institute of Technology (RHIT) is a small, selective, private engineering college where excellence in undergraduate engineering and science education is a 115 year old tradition. Although calculus is the entry level mathematics course at RHIT, lamentations about our calculus and differential equations sequences could easily have been the sole source for the kinds of critiques recently published in *Calculus for a New Century* [1], and *Towards a Lean and Lively Calculus* [2].

Richard D. Anderson [3] states "student learning procedures consist primarily of working textbook problems - usually several or many of the same sort - following the model procedures given in the textbook or by a teacher. Thus students learn various paper-and-pencil algorithms for producing answers to special types of problems." In [4], Susan S. Epp says that "the state of most students' conceptual knowledge of mathematics after they have taken calculus is abysmal." In [5], Peter D. Lax complains that "there is too much preoccupation with what might be called magic in calculus. For instance, too much time is spent pulling exact integrals like rabbits out of a hat, and, what is worse, in drilling students how to perform this parlor trick." Small and Hosack write, in [6], that "too much of the time of mathematics undergraduates is spent carrying out routine algorithmic manipulations (which the students will not long remember.) This is done at the expense of conceptual understanding of the material and an appreciation of mathematical processes .... Today, the major emphasis in college mathematics instruction is placed on imparting specific mathematical facts and algorithms, rather than on understanding and the development of inquisitive attitudes, analytical abilities, and problem solving skills." Paul Zorn [7] suggests "cautious use of [computer algebra systems] for routine computation can help shift calculus's focus back to ideas, where it belongs." S. K. Stein, in [8], proposes that "in whatever course we teach we include a significant number of what might be called 'open-ended' or 'exploratory problems'." And finally, Leonard Gillman [9] concludes his study of mathematics teaching programs that work by declaring "the fundamental precepts are: challenge your students with difficult problems, demand hard work, and adhere to uncompromising standards..."

To rejuvenate our calculus courses in the above spirit, we decided to implement a computer algebra system as the tool of first recourse in teaching, learning, and doing engineering analysis. In order that our use of Maple software running on VAXstation 2000 workstations be more than just an "add-on" we sought a total integration of computer algebra into the calculus and differential equations curricula.

The computers and the Maple software are available to students during class, for homework, and during exams. We saw this as essential for the change of emphasis from manipulative to conceptual. Students are tested in the same way they are taught to learn and perform the calculus. To use Maple in weekly laboratory assignments, and not to use Maple in class and on exams neither fosters a new curriculum nor allows a de-emphasis of manipulations.

We therefore set up a Symlab (short for symbolic, algebraic, and numerical computations laboratory) as a teaching classroom with marker boards, twenty-nine workstations and an overhead projection system. An adjoining workroom received another thirteen workstations, thirteen other terminals, and a printer. Both Symlab and the workroom are open from 7:00 AM until midnight. Within this framework, we organized our course revision around three specific activities, namely, in-class experimentation and practice, regular homework assignments, and special projects.

In-Class

We will give two examples illustrating how pedagogy changes when computer algebra is available in the classroom. The first example is a mundane separable differential equation, culled from a recent text that offered it as an excuse for practicing integration skills. Unfortunately, solving the differential equation $y' = t^2/(1 + 3y^2)$ by obtaining $y + y^3 - t^3/3 = C$ is essentially a stultifying exercise. The **dsolve** command of Maple will find that implicit solution, or the appropriate integrations can be done. But what does y(t) look like? What does the resulting implicit representation of y(t) imply? What is the meaning of such manipulations anyway? Are such manipulations ends in themselves or is there some intellectual content in the activity?

The Maple **dsolve** command will deliver
$$y = [A + B]^{1/3} + [A - B]^{1/3} \text{ ,where}$$
$$A = t^3/6 + C/2, \text{ and}$$
$$B = \frac{\sqrt{4 + 3t^6 + 18\ Ct^3 + 27\ C^2}}{6\sqrt{3}}$$

as the explicit representation for y. Choosing specific values for C, (e.g. 0, 2, 10, 30, 68) and plotting the resulting functions yield the following figure.

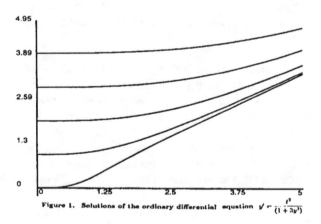

Figure 1. Solutions of the ordinary differential equation $y' = \frac{t^2}{(1 + 3y^2)}$

This figure suggests that, for large t, y(t) is asymptotic to a straight line. This conjecture is verified by computing the limiting value of the derivative y'. Maple delivers the answer $3^{-1/3}$, and the "what if" questions involved are coaxed from the students in a participatory manner, with the students performing the analysis for themselves, as various suggestions are made.

We conclude this example by suggesting that the real question we should strive to have our students ask at this point is whether or not the asymptotic behavior of y(t) could have been predicted from the differential equation itself, and if so, how? Indeed, if y -> a + bt with increasing t, then, noting that y' will approach b and using the differential equation, it is not hard to show that b = $3^{-1/3}$.

The second example of in-class teaching strategies is more mundane, but no less effective. The double and triple integrals of multivariate calculus are effectively introduced by computing the appropriate Riemann sums via

Maple's sum and limit commands. Thus,

$$\int_0^1 \int_0^1 xy\ dx\ dy = \lim_{m \to \infty} \lim_{n \to \infty} \sum_{j=0}^{m-1} \sum_{i=0}^{n-1} x_i y_j\ \Delta x_i\ \Delta y_j$$

can be computed in Maple for a grid with uniform spacing $\Delta x_i = 1/n$, $\Delta y_j = 1/m$, so that $x_i = i/n$ and $y_j = 1/m$. The double sum is obtained as

$$R := \textbf{sum}\ (\textbf{sum}\ (x_i y_j\ \Delta x_i \Delta y_j, i = 0 \ldots n\text{-}1), j = 0 \ldots m\text{-}1)$$

and the iterated integral via **limit** (**limit** (R, n = infinity), m = infinity). The syntax is natural and reinforces the standard notation. The approach allows the student to leave the lecture with a good sense of the definition of the multiple integral because he has experienced this definition in the computation.

Homework

Homework exercises consist of both practice and discovery activities. For example, students discover and articulate the product, quotient, and chain rules for themselves. The product rule yields to experiments with functions like $f(x)$ = exp(x)sin(x); the chain rule, to functions like $f(x)$ = sin(exp(x)); and the quotient rule, to functions like $f(x)$ = $x^3/\sin(x)$. After these explorations, the student is required to formulate precisely each of the three rules.

To test the effectiveness of this discovery approach, we asked the following on an examination. Given the "functions" y = nip(x) and y = bod(x), with derivatives pin(x) and dob(x) respectively, find y'(x) for y(x) defined by a) sin(x) nip(x); b) bod(x)/ ln(x); c) $\sqrt{nip(\cos(x))}$. Over 90% of the students were successful with these problems, indicating that they learned the product, quotient, and chain rules, on their own, from the discovery assignments.

Another discovery assignment has students articulating the effects of varying the parameters *a, b,* and *c* in functions of the form $f(x) = af(x + b) + c$. Such graphical investigations are related to an early treatment of optimization problems. We assign one or two such problems each night, starting with the third day of class. The objective function is formulated and graphed, with the extrema located, graphically, to within two or three decimal places. Thus, students develop an intuition for the concepts of optimization, graphing, and limits. This also considerably reduces the time needed for formal treatments of these topics.

Another activity with conceptual content is having students examine the graphs of $f(x)$ and $f'(x)$ to decide which curve is which. Polynomial functions are no longer a challenge because of the relationship of the extrema of $f(x)$ and the zeros of $f'(x)$. However, in the figure below we have the graphs of $f(x) = \sin(x)\cos(x)/[1 + \cos^2(x)]^{1/2}$ and its derivative.

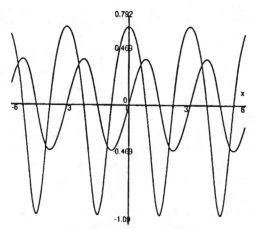

Figure 2. Which is $f(x)$ and which is $f'(x)$?

Each seems to have a zero at an extrema of the other. Hence, one has to examine the slopes to decide that $f(x)$ is the function with extrema of larger magnitude. A variant of this type exercise is sketching $f(x)$ from a graph of $f'(x)$.

Our assignments on finding volumes of revolution by disks and shells require using both methods on the same volume. For many problems one method is "easier" than the other because of the integrations involved. We require our students to use both methods and show that the results agree. In addition, we make liberal use of regions with piecewise continuous boundaries. The strategy of "self-checking" by solving a problem in more than one way is also used with multiple integration. There are six ways to iterate a triple integral. We assign several problems to be iterated all six ways.

We are willing to consign appropriate parts of a computation directly to Maple. To keep the unit on separable differential equations from being merely an arena for practicing integration, we cover population growth models, including the logistic equation. The logistic model, given by $dP/dt = aP - bP^2$, has two parameters and one constant of integration. Computing these constants from three arbitrary data points is an algebraic main event, a show the students can now enjoy by letting Maple solve the three non-linear equations that determine these constants.

We are equally willing to consign to Maple the algebra of solving systems of non-linear equations that arise from the Lagrange Multiplier method. Maple will solve many, but not all, such problems that appear as exercises in most calculus texts.

Projects

The third component of our revised calculus sequence consists of special projects that require the integration of several techniques in the pursuit of a solution. A first such project, based on [10], studies the dependence of a conic section on five points through which it passes. Four points are held fixed and one point is perturbed. The variation of the conic in response to this perturbation is sought. Since the most general form of a conic not passing through the origin is $Ax^2 + Bxy + Cy^2 + Dx + Ey + 1 = 0$, specifying five points through which the conic must pass leads to a system of five equations in the five unknown coefficients A, B, C, D, and E. In particular, the students are assigned the following:

a) Find and graph the conic which passes through the points $(\frac{1}{\sqrt{2}}, \frac{1}{\sqrt{2}})$, $(0,1)$, $(-1,0)$, $(0,-1)$, and $(1,0)$.

b) Find the equation for the conic which passes through the points $(\frac{1}{\sqrt{2}}, \frac{1}{\sqrt{2}})$, $(0,1)$, $(-1,0)$, $(0,-1)$, and $(z,0)$. This will be an equation in z, x, and y. The coefficients A, B, C, D, E will all be functions of z.

c) Let $w = B^2 - 4AC$. What is $w(z)$? Plot w for $-2 \le z \le 3$.

d) Pick a value for z such that $w < 0$. Graph and identify the conic obtained.

e) Pick a value for z such that $w > 0$. Graph and identify the conic obtained.

f) Pick a value for z such that $w = 0$. Graph and identify the conic obtained.

g) What does the value of $w(z)$ tell you about the conic?

This project was inspired by the authors' attempts to

generate data to simulate orbits of comets. Until the above activities were conceived and completed, neither of the authors had ever "observed" the deformations studied in this project!

A second project [11] incorporates related rates, Newton's Method, the angle between two curves, and the law of cosines. The challenge to the students is stated as follows. A certain football play requires the running back to go in motion and run a "down-and-out" pass pattern. The quarterback rolls out to his right and can either pass the ball to the running back or run with the ball. When thrown from five yards to the right and ten yards behind the point where it was centered, the ball follows a trajectory that makes an angle of 72 1/3 degrees with the line of scrimmage. How fast is the distance between the ball and the running back changing when the running back is one yard, and the ball is three yards from the point where the ball is caught?

We assume the path of the running back can be approximated by $R(x) = -2 \exp(\arctan(3x^2 - 62x + 316)) + 8$, the speed of the running back is 15 mph, the speed of the ball is 45 mph, and the ball travels in a straight line.

Assessment

The first two sections of students to have started the Maple calculus have now completed the required five quarters of calculus and differential equations. For these students Maple has been a very positive experience. They have mastered material that has a reputation for being very difficult. They are, moreover, very confident in their grasp of engineering analysis. Many of these sophomores have been delighted to discover, in their engineering classes, that they are unique in their mastery of the Laplace transform, and of second order linear differential equations with constant coefficients. So far, reports from both students and instructors indicate that the Maple students are amongst the best prepared students. However, no formal tests or comparisons have been made, and only anecdotal evidence is available to indicate the gains of implementing more demanding activities in the Maple courses.

Consigning tedious manipulations to Maple does not seem to hinder the development of conceptual strength in the calculus. There has been debate, even on our campus, about the role of "by hand" manipulations in the learning process. The freshest evidence we have of the viability of computer algebra as a fundamental learning and production tool comes from a recent hour exam involving two sections of (Maple) Calculus II. The test was on integration, and the students were not allowed to use Maple. The

integrals were not trivial, and there was a random selection of problems of different types taken from previous final exams. The median score was 90% for the two sections and in one section, the median was 97%. We find such results very encouragingly positive.

Conclusions

In an article [12] on the effect of standardized testing, Professor Lorrie Shepard, University of Colorado, points out "cognitive psychology teaches that children learn better by doing tasks, seeing for themselves how concepts are connected to one another. And those kids .... do better on the skills tests than the ones forced to drill their days away." The parallel with calculus classes is too strong to ignore. We believe our computer intensive environment for teaching calculus is an appropriate response to the problems in calculus courses today. By making our calculus class *be* the calculus laboratory we see clear indications that the conceptual understanding of our students has been enhanced while their manipulative skills have not been diminished. However, our clearest indication of success is the formal appearance of Maple in the engineering curricula at RHIT.

References

1. *Calculus for a New Century*, **MAA Notes**, Number 8, Mathematical Association of America, 1988.

2. *Toward a Lean and Lively Calculus*, **MAA Notes** Number 6, American Mathematical Association, 1986

3. Richard D. Anderson, "Calculus in a Large University Environment," in *Calculus for a New Century*, **MAA Notes** Number 8, American Mathematical Association, 1988, 157-161.

4. Susan S. Epp, "The Logic of Teaching Calculus," in *Toward a Lean and Lively Calculus*, **MAA Notes** Number 6, American Mathematical Association, 1986, 4159.

5. Peter D. Lax, "On the Teaching of Calculus," in *Toward a Lean and Lively Calculus*, **MAA Notes** Number 6, American Mathematical Association, 1986, 69-72.

6. Donald B. Small and John M. Hosack, "Computer Algebra Systems, Tools for Reforming Calculus Instruction," in *Toward a Lean and Lively Calculus*, **MAA Notes** Number 6, American Mathematical Association, 1986, 143-155.

7. Paul Zorn, "Calculus Redux," **Focus**, Volume 6, Number 2, March-April, 1986.

8. S. K. Stein, "What's All the Fuss About?" in *Toward a Lean and Lively Calculus*, **MAA Notes** Number 6, American Mathematical Association, 1986, 167-177.

9. Leonard Gillman, "Teaching Programs That Work," **Focus**, Volume 10, Number 1, January-February, 1990.

10. Robert J. Lopez and John H. Mathews, "Exploring Conic Sections with a Computer Algebra System," **Collegiate Microcomputer**, Autumn, 1990, Vol VIII, Number 3, 215 -219.

11. G. Elton Graves, "With CAS Engineers Can do More than Set Derivatives Equal to Zero," *Proceedings of the 17th Annual Conference Mathematics and Statistics*, Miami University, 1989.

12. Connie Leslie and Pat Wingert, "Not as Easy as A, B, or C", **Newsweek**, January 8, 1990, p. 56-58.

Calculus with Computers at SUNY, Stonybrook

Dusa McDuff and Eugene Zaustinsky
State University of New York, Stony Brook

Introduction

For the past three years, we have been experimenting with the use of computers in teaching first year calculus. The aim of Calculus with Computers is precisely the same as the aim of our standard calculus offering: The mastery of conceptual calculus. Consequently, it employs the same textbook and follows essentially the same syllabus, as our standard calculus course. We see Calculus with Computers as being a "computer-enhanced" calculus course. The student expends a little additional effort in learning how to use the computer. But this effort is more than repaid by a deeper understanding of basic concepts.

Our standard calculus course involves three hours of lecture and two hours of recitation per week. Our experimental Calculus with Computers sections replace one of these recitation hours with a two hour computer laboratory session.

Our software is generally not tutorial in nature. Rather, we use the computer as an additional teaching tool, to enable the student to explore the concepts of calculus by means of explicit examples. For the most part, we try to illustrate calculus concepts graphically, but for some topics (e.g., Riemann sums and series) we also use numerical examples.

The software we use is all menu-driven and is very easy to learn and to use. We have avoided using any software that can in any way be seen as a specialized non-procedural programming language. We believe that the small amount of programming effort we require should be applied to learning to use a standard general-purpose procedural language.

The aims of our use of computers in teaching calculus include:

- To illustrate and explore the ideas of calculus graphically and numerically.

- To develop powers of visualization, especially in 3-dimensions.

- To be able to express *procedural* notions, such as Newton's Method and the definition of the definite integral, in a more natural way by using a programming language.

- To make the study of calculus more enjoyable and stimulating.

We attempt to use the computer in ways that

- Encourage the students think critically about the output of the computer.

- Suggest problems whose answers cannot be obtained by reading data from the computer, but require some thought and calculation.

Our experiments have involved the development of suitable text materials, in the form of weekly worksheets, and also the development of some of our own software. In what follows, we shall describe the main features of the course and of our worksheets and software.

The Calculus with Computers Course

The formal part of Calculus with Computers involves two 80 minute lectures, one 50 minute recitation, and one 2 hour computer laboratory session per week. Recently, we have determined the student's final grade as follows: Final Exam: 40 %, Two Mid-Term Exams: 35 %, Computer Laboratory Assignments: 20 %, and Graded Homework: 5 %.

Our Exams have been slightly shorter and somewhat more straightforward, than those given in our standard calculus sections, to compensate for the additional effort required to complete the weekly computer laboratory assignments.

The principal topics covered in the first semester include functions and limits, the derivative and its applications, the integral and its applications, and the logarithmic and exponential functions. Second semester topics include techniques of integration, infinite series, complex numbers, three-dimensional space, vectors, functions of two variables, partial and directional derivatives, and multiple integrals.

The Computer Laboratory

Our Computer Laboratory is equipped with two dozen stand-alone IBM PC/XT machines with enhanced color graphics. Each pair of machines shares a dot matrix printer. Each machine is provided with a Menu System, which enables a student to load and run the required programs without learning anything at all about how to use either the machine or its operating system.

The Computer Software for
Calculus with Computers

We use three kinds of software for Calculus with Computers. First, there is commercial software, that we purchase or license. Second, there are several programs that we have written ourselves to support topics not supported by our commercial software. Finally, there is the True BASIC Programming Language, that students use to write and run their own little programs.

1. Commercial Software

During the first semester of Calculus with Computers, much student computer laboratory work revolves about a single commercial program: Kemeny's *True* BASIC CALCULUS (TBC) Program, Version 2 (from *True* BASIC Inc., 12 Commerce Avenue, West Lebanon, NH 03784) TBC is a collection of subprograms, which carry out a variety of calculus-related computations, both numeric and symbolic, and plot graphs. The TBC General subprogram begins by asking the user to enter the formula for a function of one variable (using the notation and syntax of BASIC) and then offers the user a list of options to select from a menu. These include printing a table of function values, plotting a graph of the function, computing and displaying the formula for its derivative, and so on. One can easily and quickly plot the graphs of x, x^2, x^3, x^4, each in a different color, on the same graph. Or $f(x)$, $f'(x)$, and $f''(x)$. The screen image can also be printed on the printer. We provide an illustrated handout on how to use TBC General, to get our students started. The program is very easy to learn and very easy and intuitive to use, so the students soon become proficient with it.

As far as it goes, TBC is fine. Unfortunately, TBC is generally restricted to one dimension and, even in one dimension, TBC does not support all of the material in our syllabus. For instance, the TBC *Area* subprogram plots a function, computes a very accurate approximation to $\int f(x)\,dx$ over the interval [a,b], and computes and illus-

trates various approximations to the integral. It does not, however, allow the user to specify the partition, so we use our own programs to enable students to do this.

We also use a second commercial program, *MicroCalc*, (from MathCalcEduc, 1449 Covington Drive, Ann Arbor MI 48103). It also does symbolic differentiation and it does an excellent job of simplifying the results. It plots parametric curves, makes projections of solids of revolution, plots graphs of functions of two variables, among many other things. We acquired it more recently than TBC and, since our worksheets are keyed to either TBC or our own programs, we do not make full use of it.

2. Our Own Programs

Our students also use a number of programs we have written ourselves. These are not the result of an effort to develop our own comprehensive calculus software package. Rather, we wrote all of them to support specific topics in our syllabus, which we felt could not be adequately supported by our commercial software.

We briefly describe two of these programs and, in the next section, we will try to show how we use them and how they relate to our Computer Laboratory Worksheets.

Our first program, SERIES, was written to support our unit on infinite series. Like TBC, SERIES consists of a collection of subprograms which can be selected from a menu. One of its menu selections is Help, which provides a brief on-line tutorial on how to use the program. Figs. 2 and 3 show two of its screen displays.

Our second program, PLOTSURF, was written when our students had no way to plot the graph of a function of two variables. It was intended to be a partial generalization of TBC General to two dimensions. It plots perspective projections of graphs of functions of two variables, level curves, the gradient vector field, and it demonstrates partial and directional derivatives. The last is done by tracing sections on the surface, drawing tangents to these, and displaying their slopes. The user is required to enter the desired function, as a *True* BASIC expression and also the desired domain and eye position.

In writing our programs, we have attempted to make full and effective use of high-resolution color graphics. For instance, as the plot shown in Fig. 5 appears on the screen, the quadrilaterals seen from above are colored bright blue, while those seen from below are colored darker blue.

3. The *True* BASIC Programming Language

During the second half of the first semester of Calculus with Computers, our students begin to do a bit of their own programming. We use True BASIC, which is an implementation of the ANSI Standard for the Full BASIC Programming Language. The students find it very easy to learn and to use.

We offer no formal instruction in programming, since the aim of our use of programming is not learning programming, as such. Rather, we wish to provide students with an alternative way of looking at and thinking about several topics, that they usually find very difficult, by using a structured procedural programming language. Accordingly, we give the students short and heavily-commented programs which they can compile and run. The source code is provided in the worksheets, but the programs can also be called up from files to save the trouble of typing. We first ask that these programs be studied, understood, and run. Next, we ask that simple modifications be made to these programs. The first few programs are concerned with such things as Newton's Method and the computation of Riemann Sums.

The Worksheets

We provide our Calculus with Computers students with weekly Worksheets. These contain some exposition, as well as the problem assignment the student is expected to work out and hand in to be graded.

To illustrate how we use our worksheets and how we try to unify the several components of the course, we now discuss and give examples from several of the Worksheets.

1. Worksheet on the Derivative and Tangents

In the worksheet on derivatives and tangents, the students are first asked to do very concrete things, which illustrate the ideas being introduced in class. For example, they graph a function f together with its derivative f', get a printout and then color the part of the graph of f where it is increasing and the part of the graph of f' where it is positive. They also obtain a graph of a function f together with some of its tangents (whose equations they calculate).

The worksheet ends with some exercises which are designed to illustrate the fact that the graph of a differentiable function looks like a straight line when you look at it on a small enough interval but that a non-differentiable function can be infinitely wiggly. They start by looking at a

polynomial such as $x^7 - 10x^5 + 10x^3 - 2x$ which has a fair number of wiggles over the interval $-2 \leq x \leq 2$ but which becomes straight when plotted over an interval such as $0.759 \leq x \leq 0.761$ or $0.479 \leq x \leq 0.481$. By way of contrast, they end with the following exercise:

Exercise: Plot $f(x) = x \sin(1/x)$ using the plotting area $-b \leq x \leq b, -b \leq y \leq b$ for $b = 1, 0.1, 0.01, 0.001, \ldots$ so that your plotting area is always a square. Do you think that if we extend the domain of f by defining $f(0) = 0$, the resulting function will be differentiable at $x = 0$? What limit would you have to calculate to find $f'(0)$? Hint: use the definition of the derivative. Does this limit exist?

Remarks: We have found that the students do not pick up hints very well, and that it is hard to write exercises which encourage them to explore. However, we try to make the exercises as interesting and varied as we can. The labs are timed so that topics are sometimes treated first in the lab and many of the themes explored in the lab are later taken up in the lectures. Many students found the last part of the above exercise very hard, since they are not used to working from definitions. However, the function $\sin(1/x)$ had been thoroughly investigated in a previous worksheet, so that if they did find the correct formula for $f'(0)$, they immediately recognized it.

2. Worksheet on the Riemann Integral

This worksheet is the first serious application of programming. The previous worksheet is an introduction to programming, in which the students read, run, and edit a very simple program, either on limits or on iteration.

The students are provided with the following program:

```
!Program RAREA to approximate the signed area
!between f and the x-axis using right-hand rule.
DEF f(x) = x*x
INPUT PROMPT "Endpoints of intervals a, b = ?": a, b
INPUT PROMPT "Number of subdivisions?": n
LET h = (b - a)/n
LET z = a + h
FOR j = 1 to n
LET R = R + f(z)
LET z = z + h
NEXT
PRINT "APPROXIMATE AREA IS Rn(F) = "; R*h
END
```

They read it, run it and are shown how to edit it so that it runs a little more efficiently and implements the midpoint rule. They compare the accuracy of the right hand and midpoint rules, and then do the following exercises.

Exercise 5. Change the function to: $f(x) = \sqrt{1 - x^2}$.
(This is coded as SQR(1 − x * x) or, preferably, as SQR(ABS(1− x * x)). Run the program with [a,b] = [0,1]. What does the midpoint rule give when n = 100, 500, 1000?

Note that your results get closer and closer to $\pi/4 \approx 3.14159/4 = 0.78539$. Sketch below the area that this integral measures, and explain this result.

Exercise 6. Finally, let's calculate the integral of the function l/x over various intervals $1 \leq x \leq b$ (or $b \leq x \leq 1$). We will call this integral *L(b)*:

$$L(b) = \int_1^b \frac{1}{x} \, dx$$

Note that *L(b)* is an increasing function of *b*. Also, when

$0 < b < 1$, then $\int_1^b$ is defined to be $-\int_b^1$, so the integral still

makes perfectly good sense − you just get a negative number as a result.

(i) Complete the following tables:

b	2	4	8	10	32
L(b)					

Can you see any pattern in these results? What is their relation to *L(2)*?

b	.5	.25	.125	.065	.03125
L(b)					

What is the relation between these results and those in the table above?

(ii) Find value for *b* by experiment, such that *L(b)* = 1. Of course you won't be able to solve this equation exactly, so find a number *b* of the form #.## with *L(b)* < 1 and such that you get *L(b)* > 1, if you add 1 to the last digit of *b*. Since you have already found that *L(2)* < 1 and *L(4)* > 1, you know that this value for *b* should be between 2 and 4.

Remarks: We have found the Riemann sum to be one of the most difficult ideas in first semester calculus for students to understand, partly because the terminology and notation are so cumbersome. It is perhaps easier to understand as a procedure, expressed in a programming language, than as the usual mathematical definition. We therefore use very simple programming so that the students can see what is being computed, with the aim of making the definite

integral completely concrete and accessible. We ask the students to edit the programs to encourage them to read and understand them. Most students have no trouble with this and many enjoy it. Exercise 5 is designed to remind the students that they really are calculating (signed) areas. The worksheet culminates in Exercise 6 which is both a preparation for a later discussion of the Fundamental Theorem of Calculus, and an introduction to some of the marvelous properties of the logarithm.

3. Worksheets on Iteration and Newton's Method

Following an idea of Devaney, Newton's method was one of the themes of the course that was taught in Fall 1989. It occurred in week 4 as an application of differentiation, in week 7 as an interesting application of our introductory program ITERATION, and in week 12 as an exercise in the use of complex numbers.

In week 7 we apply Newton's method to the cubic equation $x^3 − 2x + 2 = 0$. The students discover the periodic orbit $x_0 = 0$, $x_1 = 1$, $x_2 = 0$, . . . and find that the behavior for starting points between 0 and 1 is much more complicated than one would naively imagine.

The full beauty of this is seen in week 12 when Newton's method is done over the complex numbers. Our students have fun playing with an interactive tutorial program that demonstrates the orbits of points under Newton's method for the cubic $z^3 = 1$. This is depicted in monochrome below. The figure shows the first few points of the orbit of $-1.2875 + i\,(-.15)$, which converges to the root $-\frac{1}{2} + \frac{\sqrt{3}}{2}i$.

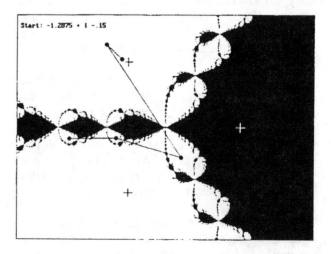

Figure 1

We have found that most students, even those who think they have no interest at all in mathematics, are fascinated by the incredible brightly colored figures associated with such processes. The program displays a number of these figures, as well as giving the interactive demonstration of how they are computed.

The exercises of this worksheet are largely concerned with the arithmetic and geometry of complex numbers. For example, the students are asked to calculate the first few points of an orbit.

4. Worksheet on Convergence of Series

In the second semester, programming is used for numerical calculations with sequences and series. Here is an example to show how we try to integrate the computer work into the course. The students were first asked to investigate the convergence of the sequence

$$s_n = 1 + \frac{1}{2^p} + \frac{1}{3^p} + \cdots + \frac{1}{n^p}$$

numerically, for different values of p. That is, they were asked to make an informed guess as to whether the sequence has a limit and to estimate the limit if they thought there was one. The convergence properties of the p-series were established in the usual way, in the next lecture. Now, knowing the facts, the students returned to the laboratory and made a more serious numerical investigation of the convergence of the p-series using the program SERIES.

For p = 2, 1.4, 1.25 and 1.1 they were asked to find a value of N such that the Nth partial sum s_n is within 1 of the actual sum. For the first three given values of p this could be done easily by trial and error, using SERIES. However, for p = 1.1 the value of N is so large that the calculation must be done by hand.

```
This program calculates the partial sums sM of the p-series:

          1 + 1/2^p + 1/3^p + 1/4^p...

for different values of p. Because the sum of this series from the Mth
term on is well approximated by the integral of f(x)=1/x^p over the
interval {x: x>M}, it is possible to get a very good estimate for the
actual sum {when this exists!}

p-Series for p = 1.1

N         sM        The actual series lies between these values:

10        2.68016       10.544        10.6234
100       4.27802       10.5813       10.5876
1000      5.57283       10.5842       10.5847
10000     6.6834        10.5844       10.5845
100000    7.42217       10.5844       10.5845

     Press Any Key to Make Another Run or Esc to Exit
```

Figure 2

5. Worksheet on Taylor Series

SERIES also includes routines which plot the graphs of Taylor polynomials for certain specified functions such as sin(x), exp(x). This is a typical screen display and it is followed by a sample exercise:

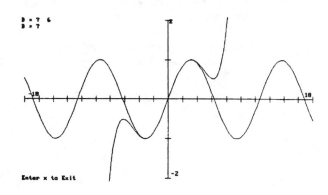

Figure 3

Exercise. Notice that the polynomial
$$P_5(x) = x - x^3/3! + x^5/5!$$
approximates sin(x) very well over the range $-2 \le x \le 2$. What estimate does Lagrange's formula give for the error $|\sin(x) - P_5|$ over this interval? Because the Taylor expansion of sin(x) has no term of degree 6, we have $P_5(x) = P_6(x)$ You can therefore estimate this error more accurately by using the formula for R_6, rather than R_5.

6. Worksheet on Functions of Several Variables

One aim of the labs involving 3-D graphics is to develop the student's ability to interpret the concepts learned in class in terms of the graphical information given by the computer. Here is a typical exercise in which the students use the Perspective, Level Curve, and Gradient Vector Field plotting functions of PLOTSURF

Exercise.   Plot the function $f(x,y) = \dfrac{xy}{x^2 + y^2}$ with

Xmax = 1. Experiment with eye positions and, when you have found a good one, get a printout. Note: This function has rather unexpected behavior because it is not defined at (0,0) and $f(x,y)$ does not tend to a limit as (x,y) --> (0, 0).

(i) Notice that the lines y = x, y =-x, x = 0, and y = 0 are level curves of this function (i.e., f(x,y) is constant along each of these lines.). What constants correspond to each of the above lines? Draw on the printout the corresponding horizontal sections and label them.

(ii) Calculate grad(f) for this function and write this on the printout. If you were on the surface at the point $(0,-1,0)$ in which direction should you move to increase your height above the (x, y)-plane as fast as possible? Mark this direction on the graph.

(iii) At what points (x,y) is $f(x,y)$ a maximum? a minimum? What can you say about grad(f) at these points?

Here are two examples of plots generated by the program PLOTSURF.

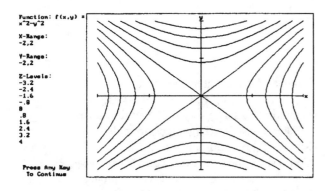

Figure 4

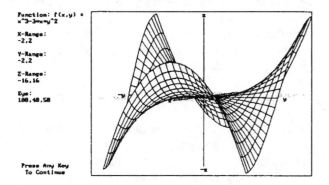

Figure 5

Evaluation of the Project

The number of students taking Calculus with Computers has varied from semester to semester, from about 15 to 50. At the beginning, while we were developing our materials, we did not advertise the course widely. However, now that we have solved many of our problems, we hope to serve about 50 or 60 students each semester.

1. Student Feedback

The feedback from students has been very positive this year but, at the beginning of our experiments, they had some valid and useful criticism. They have always appreciated the graphics but we originally expected too much work from them and many felt that what they learned from the computer laboratory work was not worth the additional effort it required. Some also felt that the laboratory had little to do with the Lectures and Exams. We now expect much less programming and we provide much more detailed instructions as to what is to be done. We have also greatly improved our worksheets and have tried to better integrate the computer component the course, by organizing our Lectures around examples from the worksheets, rather than problems from the textbook.

2. Student Achievement

We have not attempted to design an experiment to measure the performance of our students, as compared with those in the standard sections. Our Calculus with Computers students are self-selected and, as already noted, they take somewhat different exams in the interest of fairness.

We have, however, been able to make an informal assessment of comparative achievement. One of us (McDuff) lectured to both our Calculus with Computers II students and also to several sections of our regular Calculus II students, in Spring 1988. This provided a sample of 128 students, who had essentially the same lectures. They also took final examinations with 6 common problems. The results of this informal experiment are given in the table at the top of the adjoining page.

The percentages are percentages of students obtaining at least 8 out of 15 points for the given question on the final examination.

Student Populations: regular = all regular students, regular66 = all regular students receiving > 66 % on the final, computer = all computer students, computer66 = all computer students receiving > 66 % on the final.

Student Population	equation of plane	directional derivative	power series	vectors	Taylor polynomial	double integral	number students
regular	52 %	65 %	60 %	63 %	79 %	34 %	99
regular66	77 %	87 %	87 %	79 %	94 %	60 %	53
computer	79 %	83 %	62 %	79 %	79 %	76 %	29
computer66	90 %	95 %	81 %	90 %	85 %	95 %	21

The computer students, being self-selected, tended to be better and more motivated. Also, because they had to work more during the semester, there were many fewer very weak students in that class and so it is probably only meaningful to compare the regular66 figures with the computer66 figures. The most striking results occur for the question on the double integral. Both groups of students got essentially the same lectures, and the subject was not treated in the computer laboratory. Therefore, if anything, the regular students should have had a slight advantage since they had more recitation time to go over the topic. However, the question involved some geometry, since the students had to figure out the correct limits of integration for the triangle, and the calculus with computer students clearly did better on this problem.

Conclusions

Attempting to integrate the use of computers into calculus requires a heavy commitment. First, there is equipping a laboratory, installing the software, and maintaining the software installation. Second, there is the production of suitable text materials. This can use up as much time and energy as one has and is the most formidable problem associated with the integration of computing into calculus. Ideally, we would like our own new textbook, since the introduction of computing changes the character of the subject in many ways. Third, there is the problem of writing some of one's own software, to augment what is available. Fourth, there is the problem of staffing the laboratory, so that it can remain open for a sufficient number of hours each week.

We believe that our calculus with computers students are doing at least as well as— and probably better than—the regular students, and that most of them really enjoy the computer component of the course. We have no doubt that graphics and numerical experimentation can significantly improve the understanding of calculus concepts.

The Laboratory Component for an Integrated Course in Calculus and Physics

Joan R. Hundhausen, Department of Mathematics
F. Richard Yeatts, Department of Physics
Colorado School of Mines

The Calculus Reform movement has provided the opportunity and momentum for the initiation of a variety of experimental programs in the teaching and learning of calculus. Ours is a fairly unique project which combines the efforts of two instructors, (one in physics and the other in mathematics) to integrate the subject matter of calculus and physics into a single course for freshmen. The Colorado School of Mines is fertile ground for such a project, since it is an institution specializing in science and engineering; all students are required to complete three semesters of calculus and two semesters of physics. Our work in implementing this course has been supported by an NSF Grant for Curriculum Development (NSF Grant #USE-8813784).

Since calculus has historical roots in the study of physical phenomena, there are many ways in which calculus and physics can be mutually supportive in developing students' understanding of concepts in both subjects. We attempt to seize these opportunities wherever possible, both in lecture-recitation periods and especially in weekly two-hour Laboratory/Workshop sessions. Here the students can work in informal groups and benefit from close interaction with the two instructors as well as a student assistant. The projects that are planned for these Laboratory/Workshop sessions generally parallel the subject matter being addressed in either calculus or physics, and wherever possible, are designed to provide opportunity for integrating calculus and physics in a meaningful way.

GENERAL COURSE DESCRIPTION AND GOALS

A. Emphases and Structure

Our course is designed to foster both a deeper understanding of the concepts of calculus and to promote transferability: i.e., *the students should have the facility for recognizing and applying the mathematics they have learned in calculus to other contexts in more advanced physics or engineering courses.* Thus we place much emphasis upon geometry, modeling, versatility in the use of symbolism and parameters, and of course, problem-solving.

In addition to the two-hour laboratory workshop session, the class meets for four lecture sessions weekly. The class was offered for the first time in the fall of 1988, running for a full academic year. The material of Calculus I and II (4 and 3 credits, respectively) as well as Physics I (Mechanics; 3 credits) is spread throughout the two semesters, with five hours of credit being earned each semester. At the end of the year, students must be prepared to mesh with the mainstream courses in Calculus III and Physics II (Electricity and Magnetism), so we have tried to avoid wide deviations from traditional course material. Students volunteer to participate in the course, and we have attempted to select those with adequate pre-calculus backgrounds. Enrollment has averaged 21 in the fall semester and 16 in the spring semester. All classes meet in a large classroom/laboratory which is equipped with tables accommodating 5-6 students; there are also several laboratory benches in another part of the room. Thus a single facility is used for lectures, exams, and laboratory projects.

Many leading educators both inside and outside the mathematical community have questioned the effectiveness of the traditional lecture-recitation method with students playing the role of passive listeners, often resorting to rote memorization and manipulation of formulas largely unsupported by understanding of concepts or appreciation of applications.

> "All the research protocols, as well as insightful efforts at computer-based instruction, are showing that students must be given more time rather than less, more verbal and phenomenological experience rather than less to master basic concepts and acquire command of lines of reasoning. It is necessary to engage the mind of the learner in the intellectual activity; inculcation, however lucid, has very little effect on passive listeners." [1]

The integration of calculus with physics guarantees the role of applications, and the laboratory workshop setting also provides opportunity for some guided "discovery and exploration" as well as some hands-on experiences in physics. Our intention has been to involve students in the learning of calculus and physics and to persuade them to take a more active role in their own learning.

B. Use of Technology

Where appropriate for enhancing the understanding of a concept or to the performance of routine calculations, we utilize the symbolic, graphical and numerical power of the HP28S calculator. We are able to lend students this instrument for the year if needed, although many students, realizing its usefulness and convenience (in other classes as well) purchase their own. Initially a 1-1/2 hour tutorial enabled students to became familiar with the operation of the calculator. This along with a little additional ad hoc guidance was sufficient for them to use it effectively in graphing, equation-solving, and numerical (including trigonometric and logarithmic) calculations. Later in the first semester, another tutorial introduced them to the programming capabilities of the calculator, enabling most of the class to program the Euler Method for finding and graphing approximate solutions to initial-value problems. Use of the calculator is often incorporated into the Laboratory/ Workshop exercises, and it is always available for use on homework, quizzes, and exams.

C. Assessing Student Progress

In any innovative program, it is essential that student work be examined and assessed regularly. During each semester, four one-hour exams are scheduled; each of these provides the opportunity—and the challenge—of integrating the subject matter of calculus and physics on some problems. In addition, at least one half-hour quiz is given between exams, to help students gauge their mastery of the material. (These do not necessarily cover both subject areas.) Homework is assigned and discussed regularly and sometimes graded, either by an instructor or a student assistant. Close interaction with students in the Laboratory/Workshop and careful grading of the worksheets they produce also enable us to evaluate the teaching/learning progress.

COURSE SCHEDULE AND SURVEY OF TOPICS

In order that readers may appreciate the parallel development of calculus and physics concepts, we include here an informal outline of the course for the first semester; textbooks are also listed. We have generally found the presentation of physical concepts (such as velocity, acceleration, center of mass, work) appearing in the calculus text to be very supportive of the teaching and learning of these concepts, and often assign introductory readings and exercises for these concepts directly from the calculus text. The concepts are then explored in greater depth via assignments from the physics text. Those who are not primarily interested in the integration of calculus and physics will nonetheless recognize the strong emphasis upon applications.

Informal outline of topics in Physics and Calculus

Texts: Giancoli: Physics for &Scientists & Engineers
Johnson & Riess: Calculus with Analytic Geometry
(Field Test Edition)

Semester I (15 weeks [CALCULUS 4 credits; PHYS-ICS 1 credit]

Precalculus Review (inequalities, slopes, trigonometry, functions, graphs, operations with functions)
  LAB/WORKSHOPS: Calculator tutorial;
                 Exploring inequalities;
                 Functions, Graphs, and Models.

Limits; linear approximations; continuity
  LAB/WORKSHOPS: Limits and linear approximations
                 Continuous Functions

Tangent Lines and Velocity
  Velocity and uniform acceleration, from Physics text
  LAB/WORKSHOPS: Linear Motion [2]

Derivatives; implicit differentiation, chain rule; Euler's Method
  LAB/WORKSHOPS: Numerical calculation of
                 derivatives and antiderivatives; the initial value
                 problem

Motion with air resistance; parametric equations;
  Vectors, vector kinematics, from Physics text.
  LAB/WORKSHOP: Euler's Method applied to
                problems with air resistance

Related rates; extreme value problems; derivative tests; mean value theorem.
  LAB/WORKSHOPS: Set-up and analysis of max-
                 min problems
                 Advanced graphing tech-
                 niques using HP28S

Newtcn's Laws and applications; dynamics of circular motion; parametric equations; initial value problems in two dimensions.

> LAB/WORKSHOPS: Analysis of forces;
> Motion on an inclined plane.

Anti-derivatives, Euler's Method, and Area under a curve; unification and interpretation of these concepts; Riemann Sum; Integration; Fundamental Theorem of Calculus

> LAB/WORKSHOPS: Numerical Integration;
> Riemann Sum;
> Elementary Differential
> Equations

Applications of integration; area, volumes.

Descriptions
of LABORATORY/WORKSHOP Projects

Laboratory/Workshop projects fall into three main categories:

A) Pencil and paper exercises which delve more deeply into analytical problems than typical homework;

B) Numerical exercises utilizing the power of the HP28S calculator to study "real world" problems (usually related to physics) or to explore some mathematical concept;

C) Simple experiments which bridge the gap between physics and calculus, often involving the collection and interpretation of data, and appealing especially to the kinesthetic learners.

TYPE A

These exercises may require the student to derive a mathematical formula describing a geometric (right cylinder inscribed in a sphere, polygon inscribed in a circle) or trigonometric (lighthouse beam traveling along a straight shore) situation. This serves to prepare students for the concepts of optimization and related rates, before the derivative is treated. Graphs of the functions describing some of these physical situations are also examined (sometimes with the aid of the HP28S), and significant characteristics such as domain and range, monotonicity, local extrema, and limiting behavior are discovered and analyzed.

Parameters and their special roles in functional relationships are introduced and contrasted with variables, as illustrated in the sample exercises in Part III of this Volume: "Graphs and Functions." The use of physical units and functional notation figures prominently here, and the exercise culminates with the student being required to correlate graphical and analytical solutions to a physical problem.

Other projects in this category would include a study of simple differential equations illustrating the relationships among acceleration, velocity, and distance. Special attention is given to the equation $dv/dt = -kv$, which describes motion under resistive force; the equation $dv/dt = g - kv$ presents a greater challenge for students, but is useful in illustrating exponential behavior and the concept of terminal velocity. The latter can be demonstrated in the laboratory with a bead falling in a column of viscous liquid, such as glycerine. Students appreciate the reference to "real-world" problems.

TYPE B

Some of these exercises use numerical methods implemented with hand calculation and/or the HP28S to illustrate discrete versions of continuous processes such as differentiation, integration, or finding limits. For example, several workshops involve hand calculations to compute a Riemann Sum from graphical or measured data, to find the work done by a force or the distance traveled by a moving object. Another numerically-oriented Workshop exercise examines the ideas of approximation or convergence as the HP28S is used to calculate and graph successive Taylor Polynomials.

Let us describe in somewhat greater detail (see part III of this Volume: "The Initial Value Problem") a Workshop exercise which is particularly suited for reinforcing calculus concepts by numerical practice as well as by meaningful physical examples. In the first semester, after the concept of tangent line and linear approximation have been developed, Euler's Method of integration – or numerically solving a differential equation – is introduced.

After practicing hand calculations to find several "predicted" values of the unknown function, students are guided to program the Euler Algorithm on the HP28S. By repeated use of the simple equation $f(t+h) = f(t) + hf'(t)$ and appropriate initial data, they can observe the calculation of many successive approximate values of the unknown function and also watch the evolution of its graph. Here the students are investigating the motion of a body propelled upward and returning to the surface of the earth, with or without air resistance. In subsequent workshop exercises, the Euler Algorithm is adapted for the purpose of formulating a Riemann Sum for a continuous function whose formula is known; making use of the numerical algorithm in this way helps to unify the central ideas of calculus – differentiation and integration.

TYPE C

"Linear Motion" is a type C exercise which serves to bridge the gap between calculus and physics (See Part III of this Volume.) Based on exercises primarily developed by L.C. McDermott of the Department of Physics at the University of Washington, Seattle, WA, this LABORATORY/WORKSHOP consists of a series of demonstrations (best performed by the instructor, in turn, for small groups of students) which focus on the physical meaning of and distinction between displacement x, velocity $v = dx/dt$ and acceleration $a = dv/dt$, where t is time. In these demonstrations, the students observe critically the motion of ball bearings on connected sections of track (aluminum channel). The students:

- answer pointed questions about the motion they observe;
- draw sketch graphs of x, v, and a versus t;
- correlate the graphs according to the requirements of calculus.

It has been assumed by a number of authors that velocity and acceleration are well understood by students because of their familiarity with such things as speedometers in cars. This "understanding" is often very superficial, however. For many students, the newly-learned concept of differentiation is probably better understood than the concepts of velocity and acceleration. "Linear Motion" provides a focused laboratory experience which is set to refine students' naive understanding to one that is conceptually sound.

In another Type C-Workshop, students determine experimentally the center of mass of an irregularly-shaped metal plate by using a plumb bob. This is followed by a guided mathematical exploration into why the center of mass, based upon the common formulas used in its definition, can be determined in this experimental fashion. The equivalence of the two notions is carefully established via an optimization technique (which students have learned as an application of derivatives) applied to a simple expression involving potential energy of point masses.

The center of mass is then calculated alternatively, via a numerical method involving the use of paper strips that model the shape of the plate; the latter illustrates the concept of the Riemann Sum, and would be classified as a "Type B" laboratory exercise.

SUMMARY AND EVALUATION

The implementation of this course integrating Calculus and Physics has enabled us to observe some of the challenges students face in mastering calculus concepts and applying them in another subject area. On the basis of both anonymous questionnaires and personal interviews, every student in the Integrated Course has agreed that the laboratory experience is valuable in making calculus practical and meaningful. The following comment is typical:

"They (Laboratory/Workshops) give us a chance to go over concepts more in depth and learn the practical applications."

Responses to the question "Did the Laboratory/Workshops contribute to your learning experience?" were heavily weighted toward the positive end.

It is too early in our assessment process to say how successful the LABORATORY/WORKSHOP sessions are in helping prepare students for their chosen disciplines. (Our first group of students are just completing their sophomore year.) However, compared with a control group our students have performed equally well in two subsequent, required courses: Calculus III and Physics II. Furthermore, the ICP students have the following advantages:

- greater experience in applying calculus to real-world problems;
- greater experience in numerical calculation (discrete calculus);
- experience in symbolic manipulation.

This innovative approach to teaching calculus and physics is an investment that should reap dividends for both instructors and students in upper division courses.

Reference

[1] Arons, Arnold B., "Proposed Revisions of the Introductory Physics Course", **Am. J. Phys.**, 57 (8), 681.

[2] Rosenquist, Mark L. and Lillian C. McDermott, "A Conceptual Approach to Teaching Kinematics", **Am.J. Phys.**, 55 (5), 407.

Part III
Projects for Use in Calculus Laboratories

Seeing Calculus Through Weekly Labs
Robert Decker and John Williams
University of Hartford

Introduction

At the University of Hartford, we are in the second year of experimenting with a weekly calculus lab. We have taught five sections of calculus I and four sections of calculus II with a weekly lab. Starting in the fall of 1990, all sections of calculus I will have a weekly lab. The reasons for changing the way we teach calculus have been well documented (see (3), (4), and (5)). Most of our students became at best symbol manipulators without much feel for how calculus is used or the ideas on which calculus is based. As mathematicians we are able to picture ideas from calculus and our students were unable to do this. With the increased availability of good software and graphing calculators, we felt we could get the students to picture the ideas of calculus instead of just manipulating the symbols.

The two key pieces to the puzzle were good, easy to use software and a graphing calculator. The software we use is EPIC (Burgmeier and Kost) and DERIVE (Soft Warehouse). The former is very easy to use; the function input screen allows students to type functions into the computer just as they appear in texts and in the professor's lectures. EPIC provides all the pedagogical support we needed for Calculus I. After the students learn the ideas of Calculus in the first course they start using Calculus more as a tool in the second semester. In the second semester we move DERIVE into a major role. DERIVE is a symbolic manipulation package for the IBM (see the review in (2).

The calculator we use is the Casio fx7000g. We chose it for the price (as low as $60), for the ease of use and the programming capabilities. The calculator is something the students can keep with them wherever they do their work. In all the student evaluations of the course, the calculators were the clear favorite piece of equipment. We require the students to purchase a calculator and then provide them with 8 programs for finding roots, doing numerical integration, plotting differential equations and more.

We are teaching the course in a computer classroom outfitted with 12 networked AT&T 6300's and a Dataview projection system. The students sit two to a computer. We have found the interaction between the students cuts down on the learning time for the software. In addition there are computers on campus which the students can use outside of class.

We have developed 20 labs, 10 for Calculus I and 10 for Calculus II. Here is a brief outline of each:

Calculus I labs:

Lab 0: Warm-up exercises on using the graphing capabilities of EPIC. Forming a function dictionary.

Lab 1: Fit data from a pendulum to the equation $Ae^{-k}\cos(wt)$.

Lab 2: Approximating roots by blowing up the graph.

Lab 3: Discovering the rules of differentiation

Lab 4: Finding roots by using a root finder.

Lab 5: Discovering the connection between the graph of a function and its derivative

Lab 6: Newton's Method

Lab 7: Implicit Differentiation

Lab 8: Area by rectangles

Lab 9: Discover the fundamental theorem of Calculus

Lab 10: Arc Length

Calculus II labs:

Lab 0: Warm-up with DERIVE

Lab 1: Modeling a balloon (description below)

Lab 2: The logistic equation

Lab 3: Integration by parts; discovering strategies

Lab 4: Graphs in polar coordinates

Lab 5: Kepler's law; relating area to time

We fit these labs into the regular Calculus course, about one per week. The students start all the labs in class but need to spend an average of two hours outside of class to complete the labs. We were able to find time to fit in the labs by allowing the labs to cover some of the topics usually covered in lecture. For instance the lab on integration by parts has the students formulate methods for using integration by parts such as driving the x out of $x^n e^x$. This worked well in the first semester where ideas are taught more than techniques. Some topics were dropped such as graphing functions while others were added or emphasized such as numerical techniques and the idea of approximation. More topics were dropped in the second semester with the use of DERIVE. Only substitution and integration by parts survived from the techniques of integration section.

In a previous article (1), the authors discussed the first year of this project and gave a detailed description of the first lab of the first semester. Here is a description of two other labs, both from the second semester and both involving a falling balloon.

Calculus II   Lab I
Falling Objects

Introduction: When an object is dropped, there are two forces acting on it: gravity and air resistance. If there were no air resistance, the object would continue to accelerate, i.e., go faster and faster. With air resistance, the object stops accelerating at some point and reaches what is called "terminal velocity." One assumption that seems to work fairly well is that the force due to air resistance is proportional to the velocity squared. From this assumption, one can write down a differential equation for the velocity of the object (something we will do later in the semester). Then one can solve this differential equation to get the distance of the object above the ground (y) as a function of time (t); the result is

$$y = y_0 - \frac{1}{p} \ln(\cosh(t\sqrt{pg})).$$

In this equation y_0 is the distance above the ground when the object is dropped, and p is a parameter which is determined by the mass m of the object and the air resistance constant k by the equation $p = \frac{k}{m}$. The air resistance constant k is determined by the shape of the object. g is the acceleration due to gravity (g =32 ft/sec).

Experiment

1) Drop a balloon from seven feet and time how long it takes to fall to the ground. Drop the balloon three times and average the results. Assume that the equation

$$y(t) = 7 - \frac{1}{p} \ln(\cosh(t\sqrt{32p}))$$

is a good model for the distance $y(t)$ of the balloon from the ground (y is in feet and t is in seconds). Let T_2 be the time at which the balloon hits the ground.

2) Find the parameter p which corresponds to the balloon. From the experiment we know that $y(T_2) = 0$; graph $y(t)$ and adjust the parameter p so that the graph crosses the t-axis at time T_2. Note what happens to the curve as you vary p.

3) Did the balloon reach its terminal velocity before it hit the ground? Using the value of p determined from a), graph the first derivative $y'(t)$ (which is the velocity). Use this graph to estimate the time T at which the balloon reached its terminal velocity, i.e., stopped accelerating.

Write-up

1) How does p effect the shape of your graph? Sketch a few graphs with different p's.

2) How did you find the p that fits the data? How accurate is your value?

3) If p is increased, does it take more or less time for the object to reach terminal velocity? Why?
(Remember $p = \frac{k}{m}$)

Evaluation

This lab demonstrates some of the features we try to put into the labs. First the use of real data and real situations. We want the students to identify the graph of the function with what they have seen. Second the use of parameters, how do parameters effect the shape of the graph, how do they relate to the actual problem? Third, the use of a model and the limitations inherent in any model. In this model for the balloon, the balloon mathematically never reaches a terminal velocity while in practice we see that it does seem to stop accelerating. Fourth, in the equation above, you cannot find p explicitly; so you must approximate the value of p and discuss the error involved.

For this lab, the students used EPIC, DERIVE, their calculators and combinations of all three.

Calculus II   Lab 10
Differential Equations and More Balloons

Two differential equations are commonly proposed as models for falling objects with air resistance. They are

$$\frac{dv}{dt} = p\,v - 32$$

and

$$\frac{dv}{dt} = pv^2 - 32.$$

One of these equations was used in a previous lab to model a falling balloon. Both of these equations can be solved using separation of variables (which is one reason that they are popular candidates to model falling objects). However, as was shown in class, neither works very well for balloons. A more general form of these equations would be

$$\frac{dv}{dt} = p|v|^n - 32$$

where

$$\frac{dx}{dt} = v.$$

v is the velocity of the balloon and x is the distance that it has fallen. The parameter n is new and must be determined along with p; the goal of the lab is to determine the n and p which best describe the actual balloon.

The problem is that this differential equation cannot be solved as easily as the other two. Thus a numerical technique must be used. This is much like using Simpson's rule to find areas; a formula can't be found which represents a solution, but a graph and a table of values can be generated directly which represents a numerical solution. The solution technique proceeds much like a recursively defined sequence; to get the next value x value as time proceeds you use the x value at the previous time. You decide how much time elapses between consecutive x values; this time increment is called the step size. The

smaller the step size, the more accurate your answer is (just as larger n values in Simpson's rule give more accuracy).

Experiment

1) Drop a balloon 7 feet and measure both when the balloon reaches the half-way point and when it reaches the ground. Repeat the drop three times and average your results.

2) Use your differential equations software to solve the equations given above. Use a step size of 0.1 and start with $n = 2$ and $p = 1$. Solve the equation on the interval $0 \le t \le 2$. Get both graphical and numerical output. Use the window $-10 \le x \le 10$. This should look like the graphs you did on the previous balloon lab. Now decrease the step size by half to see if you get the same solution; if you do then the solution should be fairly accurate. When did the balloon hit the ground for this model? When did it reach the halfway point? Sketch this graph.

3) Now vary n and p to get the best fit for the data. We want the solution to work for the time when the balloon hits the ground and for when it reaches the halfway point. Also find the time the balloon hits terminal velocity for each model. Which n represents the best model? When did the balloon reach terminal velocity for this model? If you were to continue, which n would you try next?

Write-up

1) Describe the effect of varying each of the parameters n and p. Describe the relationship between n and p for the four different models. Explain how the four different final models (for the different n values) compare graphically. Justify your choice of step size.

2) Compare solving differential equations numerically (as in this lab) to solving them by separating variables, integrating, and then graphing the resulting equation as you did on the previous balloon lab. Which is more accurate? Which is easier? Which is more powerful?

Evaluation

In this lab the students get a chance to understand differential equations, a topic which seems to have dropped out of Calculus II. In addition we can discuss how models can be refined. We also get them to think (and write) about different methods for solving problems.

Conclusion

For each of these labs, the students individually write up lab reports which count for a substantial part of their grade.

Our main goal in introducing this weekly lab is to produce students who know when to do what and what it means when they do it. Our students should know when to differentiate, when to integrate, when to approximate, when to turn to the computer and when to work problems by hand. They should know what a derivative is, what an integral is, how good an approximation is, and whether the computer is producing garbage.

We are evaluating the project in two ways. First we are comparing how our students are doing on Calculus I topics compared to other Calculus I students. For the last two years we have given an exam to both lab sections and non-lab sections. The exam has consisted of eight questions, four conceptual questions which we felt our students would be better prepared for and four standard computational questions on which we hoped our student would do at least as well as the other sections. On both semester's exams, the lab students did statistically significantly better on the conceptual part and better but no statistically significantly better on the computational part. So we were able to conclude that our students were gaining understanding without losing computational abilities

The second way we are evaluating is comparing the lab students success in the next two courses, Calculus III and

D.E. 's. Again there was no statistical difference in their performance. We should caution that both of these studies have few students (90 in the first and just 30 in the second) and are continuing.

As a result of our work so far, we will offer computer labs to all our Calculus I students staring in the fall of 1990.

These labs will be published by Prentice Hall in the fall of 1991.

References

1. Decker, R. and Williams, J. 1989. "Calculus as a Laboratory Science", **Collegiate Microcomputer**, Vol. VII, Number 4.

2. McGivney, R., Decker, R. and Williams, J. "DERIVE", **AMATYC Review**, Spring, 1990.

3. *Towards a Lean and Lively Calculus* , **MAA Notes**, Number 6, Washington DC, The Mathematical Association of America, 1986.

4. *Calculus for a New Century: A Pump not a Filter* , **MAA Notes**, Number 8, Washington DC, The Mathematical Association of America, 1988.

5. *Computers and Mathematics* , **MAA Notes**, Number 9, Washington DC, The Mathematical Association of America, 1988.

Computer Laboratory Experiments in Calculus

Bruce H. Edwards, Department of Mathematics
Patrick H. Stanley, Undergraduate Student
University of Florida

Introduction

The Mathematics Department at the University of Florida offers a variety of classes in which computer projects and applications play a significant role. In particular, the first author taught a section of second semester calculus during the fall of 1987, in which students were required to complete weekly computer projects using both numerical and symbolic software. The second author of this paper was a student in that class. Much has been said about the "lean and lively" calculus, as well as the introduction of computers into the calculus curriculum. It is evident from our experiences in this course that one of the most pressing problems in organizing such a class is the need to generate a corpus of interesting and relevant computer projects which are accessible to students. In this paper, we will offer four of our projects that have proven "lively" in the classroom. Hopefully, one outgrowth of this volume of the MAA Notes will be a continuing search and development of computer projects for calculus.

Our calculus section met the usual four times a week, one meeting each week being in the University computer laboratory instead of the regular classroom. Here, each student had their own IBM/XT and software. Throughout the course we used the numerical package CALCAIDE (associated with the Ellis and Gulick Calculus text), and the symbolic program muMath.

Altogether, there were 12 projects during the semester. Four of them are described below just as the students would receive them in the laboratory. The introductory material would be discussed together as a group, with the instructor going around the room making sure everyone was doing things correctly. The students then completed the questions using the relevant software, and wrote up solutions. Projects were due five days after the laboratory meeting.

Our experiences indicate that this kind of laboratory does not require that the course be shortened or quickened. Because the laboratory motivates the calculus so effectively, we had no difficulty covering the entire course syllabus. The laboratory environment was generally noisy and exciting. Students tended to go at their own pace,

sometimes ignoring the instructor, yet helping each other constantly. The instructor was always very busy, moving around the room and answering individual questions. There were often moments when students would "go off on a tangent", which made for a very interesting and creative environment.

Newton's Method.

How would you find the roots of the polynomial $p(x) = x^6 - 21x^5 + 13.4x^3 - 10$? Or how would you find the values of x for which $\cos(x) = x$? We will see in this assignment that Newton's Method is a simple, yet effective algorithm for approximating the roots to an equation of the form $f(x) = 0$. Consider the differentiable function $f(x)$ defined on the interval $[a,b]$, in which f has a root. Newton's Method begins by guessing an initial approximation to the root, x_1. The algorithm is based on the assumption that the x-intercept of the tangent line to the curve $y = f(x)$ at the point $(x_1, f(x_1))$ will be a better approximation to the root.

With some simple calculus, we can find this new (and hopefully more accurate!) approximation x_2 to the root. The equation of the tangent line is: $y - f(x_1) = f'(x_1)(x - x_1)$ Letting $y = 0$ and solving for $x = x_2$ as the new approximation, we get: $x_2 = x_1 - \dfrac{f(x_1)}{f'(x_1)}$. In general, Newton's algorithm is given by:

$$x_{n+1} = x_n - \frac{f(x_n)}{f'(x_n)}$$

It is often helpful to graph $y = f(x)$ in order to have a good first guess x_1 to the root you are trying to approximate.

Use the Newton program on your diskette to find both roots of: $x^2 - 2 = 0$.

Find the value of x for which $\cos x = x$.

Turn in these problems:

1. Graph the function $f(x) = x^3 + x + 1$. How many real roots does f have? Use Newton's method to approximate

the root(s) of f. Repeat this analysis for the function $f(x) = x^3 - 3x + 1$.

2. Find the first 10 positive roots to $\tan(x) = x$.

3. Try to use Newton's method to find the roots of $f(x) = x^{1/3}$. What happens? Why?

4. Consider the equation $f(x) = \sqrt{a}$ for finding the square root of the positive number a. Use Newton's method to derive the recurrence $x_{n+1} = x_n/2 + a/(2x_n)$ for finding this root, in which only addition, multiplication and division are used. Use a calculator or write a BASIC program to compute $\sqrt{10}$ to 6 digits of accuracy using this algorithm.

5. Find the real roots to the two polynomials:

$$p(x) = x^5 - 15x^4 + 85x^3 - 225x^2 + 274x - 120$$

$$q(x) = x^5 - 15.01x^4 + 85x^3 - 225x^2 + 274x - 120$$

Notice that even though these polynomials are similar, their roots are quite different.

6. The functions $f_1(x) = \ln x$ and $f_2(x) = e^x$ share two common tangent lines. Use Newton's method to find the equations of these lines. Graph the functions and their common tangents.

The Logarithm Function

The natural logarithm function is defined by the integral

$$\ln(x) = \int_1^x \frac{1}{t}\, dt$$

Use the program Riemann on the CALCAIDE diskette with 100 rectangles to approximate $\ln(2)$ and $\ln(8) = \ln(2^3)$. What relationship do you observe?

Find the approximate value of z such that $\ln(z) = 1$. What is this number called?

Plot the logarithm function $(f(x) = \ln(x))$ on the interval $[0,5]$. What is $\lim_{x \to \infty} \ln(x)$? What is $\lim_{x \to \infty} \ln(x)/x$? What is $\lim_{x \to \infty} \ln(x) - x$?

Which function grows faster, x or $\ln(x)$?

Turn in these problems:

1. Using the above definition of logarithm and program Riemann with 100 subintervals, verify that:

$\ln(ab) = \ln(a) + \ln(b)$, if $a=5$ and $b=10$.

2. Find the approximate value of z such that $\ln(z) = -1$.

3. Note that

$$\int_1^x t^{k-1}\, dt = \left.\frac{t^k}{k}\right|_1^x = \frac{x^k - 1}{k}$$

Graph $f(x) = (x^k - 1)/k$ for various values of k near 0. Carefully sketch a couple of these graphs on a sheet of graph paper along with the graph of $\ln(x)$. What do you observe?

4. Determine $\lim_{x \to 0} (x^k - 1)/k$.

5. The prime number theorem states that if $p(x)$ is the number of primes less than x, then the ratio of $p(x)$ to the integral $\int_2^x \frac{1}{\ln(t)}\, dt$ approaches 1 as x tends to infinity. Use program Riemann with 100 rectangles to approximate the number of primes between 2 and 100,000 (The actual number of primes in this range is 78498).

6. Other functions can be defined by integrals. For instance, the arctangent function is given by

$$\arctan(x) = \int_0^x \frac{1}{1 + t^2}\, dt$$

Use program Riemann with 50 subintervals to estimate $\arctan(1)$. Do you recognize this number?

Complex Numbers and the Fundamental Theorem of Algebra.

The Fundamental Theorem of Algebra states that every n-th degree polynomial has at least one zero in the set of complex numbers. This is equivalent to saying that every n-th degree polynomial has n roots, counting multiplicities. This theorem was first proven by Gauss (1777-1855) at age 22 in his doctoral dissertation.

Using a three-dimensional graphics package, we will show how to visualize the roots to a polynomial as the maximum points on a surface. This approach is based on the material in *Calculus* by Larson and Hostetler, third edition, Heath Publishers, 1986 (pp. 842-3).

Recall that the absolute value of the complex number $z = x + yi$ is $|x + yi| = \sqrt{x^2 + y^2}$. In order to graphically estimate the real and complex roots of the polynomial function $f(x)$, we consider the new real function

$g(x,y) = -|f(x+yi)|/2$. Since the absolute value of a complex number $a + bi$ is zero if and only if $a = b = 0$, we see that the graph of $g(x, y)$ is a surface lying below the xy plane whose maximum points correspond to zeros of f. Furthermore, the real zeros lie on the x axis.

Consider the polynomial $f(w) = w^2 + 1$. Show that the function $g(x,y)$ is equal to $-(x^2 - y^2 + 1)^2 - 4x^2y^2$. Graph g and verify visually that the roots are i and $-i$. You will get a better picture if you graph $\ln g(x,y)$ instead of $g(x,y)$, as this will emphasize the maximum values.

Turn in these problems.

1. Form the function $g(x,y)$, for estimating the roots to $f(w) = w^2 - 4$. Graph g and approximate the roots.

2. Graphically estimate the roots to $f(w) = w^4 - 1$.

Writing Project

This project will be done in teams of two or three students. You will have two weeks to write a 3-5 page report on the following problem. Neatness and clarity will count heavily.

Imagine you are a research team working for an engineering company. Your boss comes to you and says she needs an approximation for the arc length of one arch of the curve $f(x) = \sin(2x)$, $0 \le x \le \pi/2$, accurate to 0.001. Your boss has only taken calculus I (25 years ago), but wants to understand your solution method. Prepare a complete report for her, explaining all your steps and including all relevant formulas, graphs and equations.

You are to use the Trapezoidal rule for approximating the integral. Using the error formula, determine the minimum value of n, the number of subintervals to guarantee that the approximation is accurate to 0.001. You may use muMath

and either of our calculus programs to aid in the computations.

The project should be typewritten, except perhaps for formulas. Put your names on the cover page.

Conclusions

The student feedback about this experimental course was very positive. The second author writes:

"The weekly trips to the computer lab greatly enhanced the learning process. These hands-on explorations brought many of the principles alive. The iterative procedures are especially suited to automated calculations. For example, seeing a graphical representation of root-finding using Newton's Method gave far more insight into the process than repetitive grinding of the formula by hand. The session on the logarithm function sparked a lot of interest, leading to discussion of other transcendental functions and Euler's identity. The three dimensional graphing capabilities of today's computers can provide fascinating, and easy to understand representations of countless principles. I believe that the universal use of computers in calculus classrooms would bring many benefits, especially since so many analytical tools are now available and widely used for engineering and scientific computing in today's industry."

The success rate in this course was very high, and only two of 24 students dropped or failed. However, the students self-selected into this special section and were highly motivated. It will be important to see if this kind of limited experiment can be extended to the general calculus course. We hope so, as the computer laboratory definitely made calculus more fun for both the students and the instructor!

Revitalizing Engineering Calculus at Iowa State University

Elgin Johnston, Jerold Mathews, Clifford Bergman, Alan Heckenbach
Iowa State University

Introduction

Access to computing technology for undergraduate mathematics students at Iowa State began in 1970, when we received an NSF grant to build a lab containing 7 interactive (Teletype!) terminals. The language was BASIC. Following the lead of Warren Stenburg and a group at Florida State (CRICISAM) we taught several sections of calculus using CRICISAM materials. The aim was to teach calculus in an environment including

(1) A calculus text written with the view that computing technology may be relevant to calculus instruction, and

(2) Appropriate computing technology, easily accessible to students.

The hope was that fundamental ideas would be better understood and more realistic problems solved. Our experience was not unusual. The inertia of the calculus establishment, the time needed for students (and staff) to learn programming, and the state of computing technology combined to discourage the continuation of the combined course. At ISU the resources were rechanneled into an optional one credit course paralleling our three semesters of calculus. This laboratory course was taught from 1972 to 1985 to an average of 250 students per year.

Several of us attended the *Calculus for a New Century* meeting in Washington, D. C. in 1987. We were impressed with the interest in revitalizing calculus and with the exhortations to apply to NSF for an equipment grant. Although skeptical of our chances, we did apply and were fortunate to receive $80,000 from NSF, to be matched by institutional funds. This was in the early part of 1989. At about this time we learned about *Mathematica*. Initially we thought we could not afford *Mathematica* and the more expensive microcomputers it required. Two things combined to fund our lab of 44 Macintosh SE/30 micros, each with a copy of *Mathematica*. First, the director of our Computation Center saw S. Wolfram demonstrate *Mathematica*. Secondly, our new Provost is interested in increasing the amount of computing technology available to students and staff and has an inclination towards Apple computers. Something over $300,000 was invested to rebuild the 1970 lab.

Our new MathLab is in a 25 foot by 43 foot room containing some 58 carrels, 44 of which are used for student micros, 2 for SE/30 servers, and 4 for printers. One carrel is adapted for use by a handicapped person. There is space for a liquid crystal display device to project the screen display of a micro through an overhead projector. We are using Image Writers for student printing. A Laser Writer is present but not available to students. (Our Department of English ran through a large amount of money last year when they made a laser printer available to walk-ins. This moved the campus towards a policy of restricting access to laser printers to holders of special magnetically encoded plastic cards.)

Revitalizing Calculus

Two of us taught first semester engineering calculus during Spring, 1989, using a combination of *MuMath, Turbo Pascal,* and Flanders' *Microcalc*. A variety of problems were written, assigned, and graded. We made it clear to students that we wanted their written solutions to have good organization, thoughtful explanations, etc. Grading the projects turned out to require a major commitment of time from the instructor. One of the problems presented an empirical graph of velocity plotted against time and asked for approximations to distance and acceleration. This problem was assigned before much formal work had been done in the text on position, velocity, and acceleration. A second problem was a study of the function $f(x) = a^x$ and its derivative, all designed to lead the student to discover why e is a natural choice for a. After asking the students to graph f for several values of a using available hardware and software, we asked them to use paper and pencil showing that

$$f'(x) = f(x)f'(0)$$

(We made an error in notation here: we used $f_a(x)$ to denote a^x, which produced a fair amount of confusion.) Next, they were asked to compute $f'(0)$ for several values of a and to compute to two decimal places the value of a for which $f'(0)$ is 1. We have tried a variety of other problems. It has been our experience that finding, trying, and revising these problems is the most time-consuming part of our efforts. Our problems seem to fall into three kinds: problems chosen to help the students become familiar with our particular software and hardware, problems to accustom them to regard graphing as a relatively easy procedure to be used in other problems, and multiple step problems of the kind best solved with judicious use of a computer. We include a sample of the problems we have used.

At the present time we assign one problem project every two or three weeks, depending on midterms, vacations, and general stress level. We estimate the average student spends two to three hours per project in the lab. Several more hours are spent by the student in reading the problem and writing-up the results. Based on the experience.of others, we have begun encouraging students to form teams. Overall, this reduces the amount of student time in the lab and in preparing the results, it reduces the time needed to grade the finished reports, and it increases the capacity of the lab itself. The computer work counts about 25% of the total course grade. Students take the regular departmental midterm and final exams. They are doing substantially better on these exams. Part, but only part, of the explanation for the higher grades is that weak or lazy students drop the computer sections once they find out they are expected to do extra work.

In Spring Semester, 1990, we plan to offer eight sections of first semester and four sections of second semester computer-supplemented calculus (out of a total of 40 calculus sections). Over the next three years, we hope that 75-80% of the sections will be computer-supplemented. We plan to continue to offer traditional calculus. This seems wisest in view of the varying interests of our students and the varying views of the faculty members who teach calculus.

Although we are enthusiastic about the role microcomputers and computer algebra systems can play in calculus instruction, we emphasize that our course is first and primarily a calculus course. The microcomputer, like the handheld calculator, is a useful tool. Like all tools, it is not only important to know how to use it, but also when. We choose to teach these ideas through several challenging exercises, chosen to improve students' skills in problem solving and in communicating Mathematical ideas. These skills are extremely important to the science and engineering majors who form a majority of our calculus students. We believe the microcomputer is a valuable tool for revitalizing our calculus course. At the same time, this revitalization will not, in our opinion, transform calculus from a filter into a pump without a change in our student population.

EXERCISES IN PLOTTING
Mathematics 165

1. Consider the quadratic function $y = -3x^2 + bx - 2$ In this problem you will determine how the graph and the roots of the function change as b (the coefficient of x) changes.

   a) Graph this function for at least 5 different values of b (e.g., $b = 0, -2, 2, \ldots$) You may put the different graphs on the same set of axes, but be sure to label the different graphs clearly.

   b) If b starts at 0 and then increases, what happens to the graph of the function? What happens when b decreases from 0?

   c) For what values of b does the function have two real roots? One real root? No real roots? (You will not be able to answer this using the computer. Instead, examine the quadratic formula.)

2. Let f(x) = *some function*. We will study the effect of the absolute value function on the graph of the function. *Mathematica* uses Abs[x] for |x|.

   a) Graph the function. Describe the main features of the graph. For example, you should tell where the graph appears to be increasing, where it appears to be decreasing, and where the graph has asymptotes. (You should examine both the graph *and* the equation for the function in formulating your answer.)

   b) Now graph |f(x)| . What are the main features of this graph? How does this graph relate to the graph of f(x)? What is it about the absolute value that explains this relation?

   c) Graph f(|x|) and answer the questions in part b) for this new graph.

3. Suppose y = f(x) is an arbitrary function. How do the graphs of f($-x$) and $-$f(x) compare with the graph of f(x)? Experiment with at least four different functions before coming to a conclusion, and give reasons for your answers. (You will find it best to try "non-trivial" functions (e.g., f(x)=x^2 +sin$x$) and not just lines or simple polynomials.)

4. A right circular cylinder of height h is inscribed in a sphere of radius 1, as shown in the figure below.

   a) Find a formula for the volume of the cylinder. For what values of h is the formula meaningful?

   b) Let $v(h)$ be the volume of the cylinder when the height is h. Graph $v(h)$ for appropriate values of h.

c) By having *Mathematica* display coordinates of points on the graph, decide what the maximum possible volume of such a cylinder is. What does *h* appear to be when this maximum volume is attained?

d) **(Extra Credit.)** Suppose the sphere has radius R. What is the maximum volume for a cylinder inscribed in this sphere?

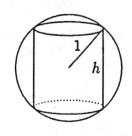

PORSCHE 911 CARRERA 4 TEST RESULTS

The graph in Fig. 1 is from an acceleration test of a 1989 base Porsche 911 Carrera 4. The data from this test are

(0, 0), (1.8, 30), (2.8, 40), (3.9, 50), (5.1, 60), (6.8, 70), (8.6, 80), (10.5, 90), (13.0, 100), (16.2, 110), (19.7, 120), (25.6, 130)

as reported in *Car and Driver* 35(1989), p43. A typical data point has the form (t_i, v_i), where v_i is the velocity of the Porsche at time t_i. The point (2.8, 40), for example, contains the information that 2.8 seconds after the acceleration test began, the Porsche was travelling at 40 miles per hour.

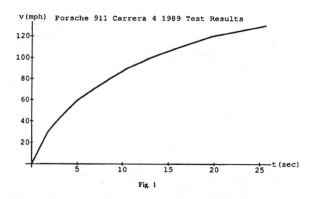

Fig. 1

Discussion

In Problem 1 you will be asked to find the position $x(t)$ of the Porsche at times t = 0, 1.8, . . ., 25.6. The discussion given here provides some background for these calculations. If the Porsche were moving at a constant velocity v mph at all times between $t = a$ and $t = b$ seconds, then the familiar formula $D = R \cdot T$, that is, "distance equals rate times time," would be applicable. The distance traveled in the interval [a, b] would be $D = v \cdot (b - a)/3600$ miles. We choose an x-axis such that the starting position of the Porsche is $x = 0$ and it accelerates in the positive direction. If $x(a)$ denotes the coordinate position of the Porsche at time a, then—still assuming uniform velocity v - the position $x(b)$ at time b would be

(1) $x(b) = x(a) + v(b - a)/3600$

Applying this argument to the time interval [6.8, 8.6] as an example and taking v as the velocity at 6.8 seconds, the increase Δx in the Porsche' position during this time interval would be

(2) $\Delta x = 70 \cdot (8.6 - 6.8)/3600$

If we had taken v as the velocity at 8.6 seconds the increase would have been

(3) $\Delta x = 80 \cdot (8.6 - 6.8)/3600$

It is clear that Δx as calculated in (2) is too small, and in (3) too large. It is reasonable to suppose that a better estimate for Δx might be found by averaging the two velocities. This gives (we use the symbol $\approx$ to mean "approximately equal to")

(4) $\Delta x \approx \dfrac{v(8.6) + v(6.8)}{2} \cdot (8.6 - 6.8)/3600$

Problem 1 Using the above ideas, calculate the position $x(t)$ of the Porsche at times 0, 1.8, 2.8, . . . ,19.7, 25.6 seconds. How far did the Porsche travel up to the time at which its velocity was 130 mph? Using the results of these calculations, prepare a sketch of the position $x(t)$ against time t. Use the same general format and scale of the above Figure. We suggest you do the numerical calculations with a hand-held calculator. You may also do the sketch by hand. If you would prefer to use *Mathematica* for the sketch, you may follow the procedure used in preparing the previous Figure.

(1) We entered the velocity data into *Mathematica* using the command

PorscheVelocityData = {{0, 0}, {1.8, 30}, . . ., {25.6, 130}}

Of course you will want to use different data and

change the name "PorscheVelocityData" to something more descriptive.

(2) After we had defined PorschVelocityData we used the command

ListPlot [PorscheVelocityData,PlotJoined->True, AxesLabel) {" t(sec)" ," v(mph)" }, Ticks->{ {5,10,15,20,25},{20,40,60,80, 100, 120} }, PlotLabel)"Porsche 911 Carrera 4 1989 Test Results"]

Problem 2 Give a geometric interpretation of the computations used in Problem 1. For this take another look at equation (4). The term

$$\frac{v(8.6) + v(6.8)}{2} \cdot (8.6 - 6.8)/3600$$

may be interpreted as the area of a certain rectangle in the above graph. For this it is helpful to think of the units on the t-axis as hours, so that the 10 second mark is actually the 10/3600 hour mark. Make a sketch (by hand) to explain your interpretation. Make a second sketch showing how this rectangle may be replaced by a trapezoid having the same area. Finally, relate the numerical value of $x(25.6)$ to an area on the graph.

In Problem 3 we wish to approximate the Porsche' acceleration as a function of time. Calculus (and physics) texts define average acceleration on a time interval $[t_1,t_2]$ as the quantity

$$\frac{\Delta V}{\Delta t} = \frac{v(t_2) - v(t_1)}{t_2 - t_1}$$

They define acceleration (some books use the term instantaneous acceleration) at any time t as

$$a(t) = \lim_{\Delta t \to 0} \frac{v(t + \Delta t) - v(t)}{\Delta t}$$

Problem 3 Compute the average acceleration of the Porsche on the time intervals [0,1.8], [1.8,2.8], . . . ,[19.7,25.6]. Using the above graph, give a geometric interpretation of these computed average accelerations.

Problem 4 For each of the intervals [0,1.8], [1.8,2.8], . . ., [19.7,25.6], find its midpoint. Letting t_{mid} denote a typical midpoint, use the results of Problem 3 in estimating $a(t_{mid})$. Do this for all of the midpoints. Use these results to sketch a graph of the acceleration function. What would be required to obtain a more accurate graph of the acceleration function?

DIFFERENTIAL APPROXIMATIONS

Problem 1.

a) Taking $\Delta x = .354$, use differentials to find an approximation to $(1.354)^2$

b) Now try approximating $(1.354)^2$ using two consecutive differential approximations with $\Delta x = .354/2 = .177$ in each case. (From 1^2, first approximate $(1.177)^2$ using $\Delta x = .177$, then using your approximation to $(1.177)^2$ and $\Delta x = .177$ again, get an approximation for $(1.354)^2$.)

c) Repeat b) but with three consecutive approximations and $\Delta x = .354/3 = .118$ in each case.

d) Draw a large graph of $y = x^2$ for $0 \le x \le 2$ and illustrate graphically the processes in a), b), and c). According to the graph, which of a), b), or c) gives the best approximation to $(1.354)^2$? What seems to happen to the error in the approximation when the step size (i.e Δx) decreases?

Problem 2. Suppose $f(x)$ is a function with

$$f'(x) = \frac{1}{1 + x^2} \quad \text{and} f(0) = 0.$$

a) Find an approximation to $f(1)$ using differentials and

　(i) one step　(ii) 5 steps　(iii) 100 steps
　(You may want to use *Mathematica* for part (iii). See below.)

b) Suppose that we wish to approximate $f(1)$ using differentials and n equal steps. Write a general formula for $f(1)$ in terms of f and n .

Mathematica **Discussion** The *Mathematica* **Sum** command can be used to add a sequence of numbers for which you have a formula. For example,

$$1 + \frac{1}{2} + \frac{1}{3} + \cdots + \frac{1}{100} = \sum_{k=1}^{100} \frac{1}{k}$$

can be summed using

$$\text{Sum[1./j, \{j,1,100,1\}].}$$

This command tells *Mathematica* to compute 1./j for j = 1, 2, . . . ,100, then to add these results. The expression {j,1,100,1 } says that j should start as 1 and end at 100, increasing by 1 at each step. The expression 1./j (instead of 1/j) tells *Mathematica* to work with decimals instead of fractions. (Try the above command with 1/j instead of 1./j and notice the difference in the form and convenience of the answers.) As a second example,

$$\sin(1 + 2(.01)) + \sin(1 + 4(.01)) + \sin(1 + 6(.01)) + \ldots + \sin(1 + 150(.01))$$

could be computed by entering

Sum[Sin[1+k(.01)], { k,2,150,2 }]

Sometimes the most convenient expressions for dy/dx involve both x and y (as in implicit differentiation). In such cases we can still use differential approximations to find a value $y(x_1)$ given the value of $y(x_0)$. The difference here is that we also need the y value (or an approximation of this value) in order to get the value of (or an approximation of) dy/dx.

Problem 3. Let $P = P(t)$ be the amount of Plutonium 241 present in a sample at time t. Plutonium 241 is radioactive and decays, with the passage of time, to Americium 241 (which itself decays to other elements). Laboratory experiments have shown that the quantity $P(t)$ satisfies the differential equation

$$\frac{dP}{dt} = -.0525\,P$$

where t is measured in years.

a) Suppose 50mg. of Plutonium 241 is present today ($P(0)=50$). Use differentials to approximate the amount present in 10 years. Do the problem with
(i) one step   (ii) three steps   (iii) 25 steps   (iv) 100 steps
b) How long will it take the 50 mg of Plutonium 241 to decay to half of this amount (i.e. 25 mg)? This time, t_h is called the half-life of the element. Experiment with various step sizes (up to 100 steps total) and times to solve this problem. Find the half life to one decimal place.

Mathematica **Discussion.** You can use *Mathematica* to handle the computations in problem 3 as follows. Let h be the step size, n the number of steps, and pl[k]=$P[hk]$ be the amount of Plutonium at time $t = kh$. First input pl[0], as given in the problem:

pl[0]=

Next write down the differential relation that allows you to approximate pl[k+1] from pl[k]. These computations can all be done using a **Do** command:

Do[pl[k+1]=   , { k,0,n-1,1 }]

In the above command, you are expected to fill in the blank space with the formula for pl[k+1] .

Problem 4. It is known from experimental evidence that in many cases the temperature of an object changes at a rate proportional to the difference between the temperature of the object and that of the surroundings. (This is sometimes known as Newton's Law of Cooling.) Thus, if $T(t)$ is the temperature of an object at time t and T_S is the constant temperature of the surroundings, then

$$\frac{dT}{dt} = k(T - T_S)$$

where k is a negative constant of proportionality. Suppose a cup of coffee is at 200°F when freshly poured. One minute later it has cooled to 190°F in a room at 70°F.
a) Find the value of k.
b) What will the temperature of the coffee be two minutes after it is poured?
c) How long will it take for the coffee to cool to 150°F?

Problem 5. (Extra Credit) Consider the expression

$$x^2 y + y + \frac{1}{3}y^3 - \frac{20}{3} = 0$$

a) Find $\dfrac{dy}{dx}$

b) Note that when $x = 1$, one acceptable value of y is 2 (so $y(1) = 2$). Using the expression you found in a), the fact that $y(1) = 2$, and differentials, find an approximate value for y when $x = 2$.

NEWTON'S METHOD

Recall that Newton's method is an algorithm for finding a solution to the equation $f(x) = 0$, in which f is a differentiable function. The method uses the recurrence formula

$$(*) \quad x_{n+1} = x_n - \frac{f(x_n)}{f'(x_n)} \quad \text{where } n = 0, 1, 2, \ldots$$

Thus, x_1 is computed from x_0, x_2 from x_1 etc. We must pick the value of x_0, which is called the *initial guess*. Newton's method is quite easy to implement in *Mathematica*. Since we will be applying the formula (*) many times, it is best to define a new function to represent the right hand side. For example
Newt[x_] := N[x−f[x]/f'[x]]

(The N[. . .] forces *Mathematica* to produce a numeric answer.) Suppose we wanted to solve problem 1 below.

We would first define our function f by f[x]:=$x^3 - 3x^2 - 3$, and make an initial guess, say $x_0 = 3$. We can compute x_1 with the command Newt[3]. *Mathematicaa* will print the value of x_1. To obtain x_2 type Newt[%]. (Remember, % yields the result of the last command.) By repeatedly typing Newt[%], we can compute $x_3, x_4, \ldots$, until we arrive at a satisfactory answer.

(1) The polynomial $x^3 - 3x^2 - 3$ has exactly one root. Find it, accurate to 3 digits. Explain how you decided when to stop.

(2) Find the (unique) solution to the equation $x^3 - 3x^2 - 3 = 15$. (Hint: consider the function g(x) $= x^3 - 3x^2 - 18$ Can you suggest a simple modification to (*) that will allow us to solve the equation f(x) = k, for a fixed real number *k?*

(3) How many solutions has the equation $e^x = 4x$? (Don't just write down a number - defend your answer!) Apply Newton's method first with : $x_0 = 0$, and then with $x_0 = 3$ What does this tell you about the significance of your initial guess?

(4) Let $f(x) = x - 4\sqrt{x} + 2$. What is the domain of f? Plot the graph of f as x ranges from 0 to 15. Observe that f has two roots. Notice that if you apply Newton's method with $x_0 = 2$, then $x_1 = -2$, which is not in the domain of f. (If you apply Newt again, you will get the complex number $.667 + 1.886i$ which is invalid in our context. You must stop, and try again with a new initial guess.) Find the two roots of f. Explain how you chose your initial guess for each root.

(5) Show that there are infinitely many solutions to the equation $\tan x = x$, one in each interval $[k\pi + \pi/2, k\pi + 3\pi/2]$ $k = 0, \pm1, \pm2, \ldots$ *(Extra credit: Does the distance between successive roots get larger or smaller as x ->∞?)* Find an approximation to the solution lying in the interval $[\pi/2, 3\pi/2]$ accurate to 3 decimal places. Describe the method that you used to choose your initial guess.

(6) The molar value V of a gas is related to its temperature T and pressure P by the ideal gas law $PV = RT$. A more realistic equation is van der Waal's equation

$$\left(P + \frac{a}{V^2}\right)(V - b) = RT$$

The constant $R = 0.08207$. For carbon dioxide, a $= 3.592$ and b $= 0.04267$. Find to 5 decimal places the volume V of 1 mole of carbon dioxide if P = 2.2 atmospheres and T = 320°K.

(7) Consider a mechanical device consisting (in part) of two wheels, one of radius 20 cm., the other of radius 4 cm., connected by a belt. Let u denote the distance between the centers of the wheels.

We wish to find the length of the belt as a function

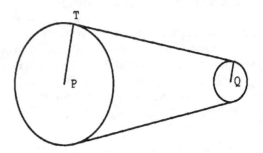

of u. First, find the belt length b in terms of u and ψ, where ψ = <*TPQ Show* that cos ψ = 16/u. Observe that, for any u > 16, there is a unique ψ satisfying this latter relation. We can rewrite this equation as ψ = arccos (16/u) (We will study the arccos function next semester.) Using the arccos function, write the belt length as an explicit function, f(u).

(8) Imagine that you are a designer for a company building the gadgets we analyzed in problem 7. Unfortunately for you, the purchasing department has already bought a large number of belts in three lengths: 150 cm., 165 cm., and 200 cm. It is up to you (as the only employee who still remembers calculus) to determine the appropriate spacing for the wheels. Do so, accurate to 5 microns. Explain the procedure you used to determine your initial guess in each case.

(9) In finding acceleration poles in planetary bevel gear trains the equation

$$\tan^3\theta - 8\tan^2\theta + 17\tan\theta - 8 = 0$$

must be solved. Find the three real roots. Report them in degrees. Note that it is not necessary to solve this equation in its present form.

RAINBOW PROBLEM - PART I

Rainbows occur when sunlight falls upon raindrops. Rainbows are visible to observers facing away from a relatively low sun and looking in the direction of the raindrops. A sketch is given in Fig. 1 showing the relationships between incoming sunlight, several raindrops, and an observer. We make the following assumptions:

(1) Raindrops are spherical.
(2) The speed of light is so large relative to the downward speed of the raindrops that the drops can be regarded as stationary as light passes through them.
(3) Rays of light from the sun are parallel to each other.

In this problem we ask you to show that the angle y made

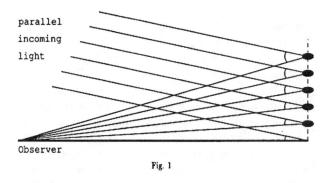

parallel
incoming
light

Observer

Fig. 1

between a ray entering and leaving a typical raindrop can not exceed a certain maximum value y_{max}. This is an explanation of why rainbows are limited in height. The analysis also gives reasons why rainbows have the form of a circular arc in the sky and show bands of colors. We leave to Part II an argument supporting the observed fact that rainbows are brightest in a relatively narrow arc. Some of the ideas we discuss are based on Descartes' work on refraction and rainbows, published in 1637.

For reflected light the angles of incidence and reflection are equal. For refracted light the angles are related by Snell's Law (discovered by Snell in 1621). The refraction of light as it passes from one medium to another is shown in Fig. 2 (at the top of the next column). Snell's laws relating the angles between the incident ray and the refracted or reflected ray are, respectively,

$$v_2 \sin x = v_1 \sin t \quad \text{and} \quad v_1 \sin x = v_1 \sin r$$

where v_1 is the velocity of light in medium 1 and v_2 is the velocity of light in medium 2.

A single raindrop of radius a is shown in Fig. 3, with a typical ray entering and leaving the drop. A portion of the incident light is refracted, reflected, and refracted as shown, finally entering the eye of the observer.

Problem 1 Give a careful argument showing that

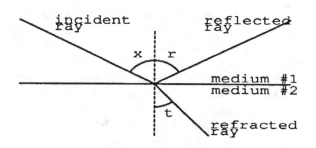

incident
ray

reflected
ray

medium #1
medium #2

refracted
ray

$$y = 4t - 2x = 4 \arcsin(k \sin x) - 2x \quad \text{where } k \text{ is a constant}$$

Include a diagram and give reasons for your conclusions. Note that y is independent of the radius a of the drop.

Problem 2 Use *Mathematica* to find y_{max} for both red and

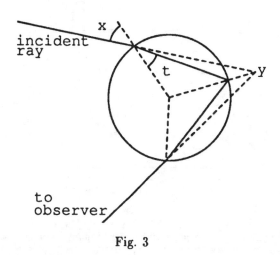

incident
ray

to
observer

Fig. 3

violet light. The ratio v_{air}/v_{water} varies with the color of light. For red and violet light

Red: $v_{air}/v_{water} = 1.332$
Violet: $v_{air}/v_{water} = 1\ 344$

Express your results in degrees. Check on the correctness of these results by using "exact" methods (which happen to work out well in this problem). You will need to know that

$$\frac{d}{dx} \arcsin x = \frac{1}{\sqrt{1 - x^2}}$$

In Fig. 4 we give a figure that is useful in showing that a rainbow is symmetric relative to the line L through the observer and parallel to the incoming rays. To understand this figure, imagine replacing a typical drop in Fig.1 by a small copy of the diagram in Fig. 3 and then rotating the resulting figure about the line L. It is not difficult to see that all angular relationships are unchanged. It follows that the rainbow light appears as an arc of a circle to the observer. This is part of an explanation of the rainbow. Given the results of Problem 2 and that the values of y_{max} for other colors of light are between the two values you have found, it follows that the bright arcs produced by the various colors are displaced relative to each other, thus producing a series of colored arcs. In Part II it will be shown that much of the reflected/refracted light reaching the observer has y values near y_{max}

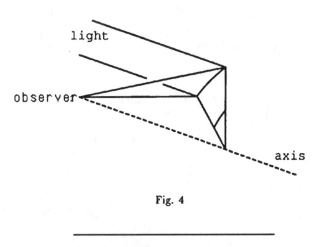

Fig. 4

BOSTON MTA PROBLEM

Several MIT students rode the Boston MTA subway from Kendall Square Station to Harvard Square, with an intermediate stop at Central Square. They were equipped with an accelerometer, a stop watch, and a notebook. They measured and recorded the acceleration of the subway car at intervals of one second. A graph of their data is shown below. They measured time in seconds and acceleration in miles per hour per second, for t= 0, 1, 2, ..., 230 seconds.

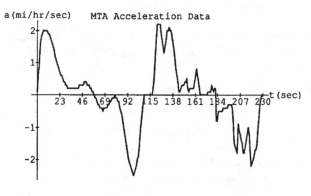

Fig. 1

Assuming that the car started from rest, use these data to

Problem 1 Prepare a graph of the velocity of the subway car as a function of t. Your graph should resemble the acceleration graph shown in Fig. 1, including appropriate "tick" marks and numbers on the axes, labels on the axes, and title.

Problem 2 Find the maximum velocity v_{max} (in miles per hour) of the subway car and the time t_{max} at which this occurs.

Problem 3 Find the distance x (in miles) from Kendall Square to Harvard Square.

Discussion

Letting $a(t)$ and $v(t)$ be the acceleration and velocity of the car at any time t and recalling that $a(t) = dv/dt$, it follows from the definition of derivative that the change Δv in the velocity during a "small" time interval Δt containing t is approximately equal to $a(t)\Delta t$. (You may assume throughout this problem that a one second time interval is small enough that all (reasonable) calculations are sufficiently accurate.) As an example, the change $\Delta v = v(21) - v(20)$ in velocity of the subway car in the time interval from $t = 20$ to 21 seconds is approximately equal to $a(t)(21-20)$. Using the Mean Value Theorem we may express this result in the form of an equality:

$$v(21) - v(20) = a(c)(21 - 20)$$

where c is some time between 20 and 21. Since we know the acceleration of the car only at 0, 1, 2, ... seconds, it is reasonable to replace the unknown $a(c)$ by the average of the values $a(20)$ and $a(21)$. It follows that
$$v(21) - v(20) \approx 0.5(a(20) + a(21))$$

This is the area of the trapezoid with vertices $(20,0)$, $(21,0)$, $(21,a(21))$, $(20,a(20))$. The area of this trapezoid approximates the area beneath the acceleration curve from $t = 20$ to 21 seconds.

It follows from these ideas that if we know the (initial) velocity at $t = 0$, then we may approximate the velocity at $t = 1$. Knowing the velocity at $t = 1$ we may then approximate the velocity at $t = 2$. And so on.

In what follows we outline the main steps needed to use the acceleration data in solving Problems 1, 2, and 3 listed above. The acceleration data is given in two forms. The first form is essentially a list of the raw data collected by the students. This appears in the *Mathematica* notebook MTA as the variable alist, that is,

alist = {{0, 0.1}, {1, 0.6}, {2, 0.9}, . . ., {228, –0.1}, {229, 0.0}, {230, 0.0}}

This form is useful in plotting the data. We used the command

ListPlot[alist,PlotJoined->True,PlotLabel->"MTA Acceleration Data",
    AxesLabel-> { "t(sec)","a(mi/hr/sec)" },
    Ticks-> {{23,46,69,92,115,138,161,184,207,230}, {-2,-1,1,2}}]

to produce the plot of the acceleration data. The first argument - alist - is the name of the list containing the data to be plotted. Specifying PlotJoined->True causes adjacent data points to be connected with a straight line. PlotLabel-> "MTA Acceleration Data" causes a title to appear on the graph. The effect of AxesLabel is evident from inspection of the graph. The effect of Ticks -> {{23,46, . . . ,230}, {−2, −1,1,2}} is to mark the axes at specified points. If you omit, say, ", {−2, −1,1, 2}" (the initial comma would be omitted as well), *Mathematica* will try to decide upon appropriate tick marks. You may accept those chosen by *Mathematica* or make your own choice.

The second form of the data is more convenient for use in calculations. We defined a function a on the domain {0, 1, 2, . . ., 230}, so that

$$a[0] = 0.1, \; a[1] = 0.6, \; a[2] = 0.9, \ldots, a[228] = -0.1,$$
$$a[229] = 0.0, \; a[230] = 0.0$$

This function may be used in your calculations. It is not necessary to use the variable alist in your calculations.

STEP 1 To calculate the subway distance x between Kendall Square and Harvard Square you may begin by computing approximate values of $v[0]$, $v[1]$, . . . , $v[230]$. You may use these values to plot the velocity function and to calculate x. You may wish to use your graph as a means for estimating the maximum velocity of the car and the time this occurs. One of these results can be inferred directly from the acceleration data. In your write-up of the problem, discuss the connection between the point on the graph of the velocity function at which the maximum velocity occurs and a point of the acceleration graph. It is best to compute the values of the function v succes-

sively, using values of v already defined. Use a "Do" command in calculating the values of v . First define $v[0]$ as a specific value and then use a Do loop. You may use the following format:

v[0] =

Do[v[i] =   ,{i,imin,imax,step}]

First you will need to choose a value for $v[0]$. Next, the Do command may be used to assign values to $v[1]$, $v[2]$,. . . The expression {i , imin, *imax, step*} specifies the name of the index (i in this case), the starting value of i, the ending value of i, and the step between successive values of i. If we were to execute the command
    Do[v[i] = 5*i+3, {i, 2, 8, 3}],
for example, the result would be $v[2] = 13$, $v[5] = 28$, and $v[8] = 43$. The effect of the Do command is to execute the expression in the first position of the Do command for the specified range of values of the index.

STEP 2 To graph the velocity function v you must convert your velocity data to the list form, as shown above in alist. Denote this list by vlist and execute the following command:

vlist = Table[v[i], {i,0,230}]

You may copy and adapt the ListPlot command given in STEP 1 to prepare a graph of the velocity data.

STEP 3 To compute the distance x (in miles) between Kendall Square and Harvard Square you may wish to compute $x[0], x[1], . . . , x[230]$ and, possibly, prepare a plot of this "distance data." Such a calculation would be quite similar to the calculation leading to the velocity data. However, note that Problem 3 asks for the distance $x = x[230]$ between Kendall Square and Harvard Square, and nothing else. In fact, x may be calculated without knowing $x[1],\ldots$. For this you may find it useful to use the Sum command. It has the form

Sum[f[i], {i, imin, imax}]

This form of the Sum command uses stepsize 1. We illustrate Sum with an example. Suppose that the function f[x] = x^2 has been defined. The command
    Sum[f[i], {i, 2, 5}]
would result in the output 54, that is,
    $2^2 + 3^2 + 4^2 + 5^2.$

mta.lab (display 1)

BOSTON MTA PROBLEM

The raw data are listed below. You won't need to use them in this form, but rather as provided by the function a defined below. The values a[i] of this function are defined for i = 0, 1, 2, ..., 230 and may be used when you need them. You may add your work to this notebook and turn in the results with appropriate arguments and supporting details. You may wish to use the ListPlot command given below. You may copy or move it to a new location, make appropriate changes and use it to plot the velocity data.

alist={{0,0.1}, {1,0.6}, {2,0.9}, {3,1.6}, {4,1.8}, {5,1.9}, {6,2.0}, {7,2.0},{8,2.0},{9,2.0}, {10,2.0}, {11,1.9}, {12,1.9},{13,1.8},{14,1.7},{15,1.6},{16,1.5}, {17,1.4}, {18,1.2},{19,1.1},{20,1.0},{21,0.9},{22,0.8}, {23,0.7},{24,0.6},{25,0.6},{26,0.5},{27,0.4},{28,0.4},{29,0.3}, {30,0.3},{31,0.2},{32,0.2},{33,0.2},{34,0.2},{35,0.2},{36,0.2},{37,0.2}, {38,0.2},{39,0.2},{40,0.2},{41,0.2},{42,0.2},{43,0.3},{44,0.3},{45,0.3}, {46,0.3},{47,0.3},{48,0.4},{49,0.4},{50,0.4},{51,0.3},{52,0.3},{53,0.3}, {54,0.2},{55,0.2},{56,0.1},{57,0.1},{58,0.0}, {59,-0.1}, {60,-0.1},{61,-0.2},{62,-0.3},{63,-0.3}, {64,-0.4},{65,-0.4},{66,-0.4},{67,-0.5}, {68,-0.4}, {69,-0.4},{70,-0.4},{71,-0.4},{72,-0.3},{73,-0.3}, {74,-0.2},{75,-0.2},{76,-0.1},{77,-0.1},{78,-0.1}, {79,0.0},{80,-0.1},{81,-0.1},{82,-0.2},{83,-0.2}, {84,-0.4},{85,-0.5}, {86,-0.6},{87,-0.8},{88,-1.0}, {89,-1.2},{90,-1.4},{91,-1.6}, {92,-1.8},{93,-2.0}, {94,-2.1},{95,-2.2},{96,-2.3},{97,-2.4}, {98,-2.5}, {99,-2.4},{100,-2.3},{101,-2.2},{102,-2.0}, {103,-1.9},{104,-1.6},{105,-1.1},{106,-0.8},{107,-0.5}, {108,-0.2},{109,0.0},{110,0.0},{111,0.0},{112,0.0}, {113,0.0}, {114,0.0},{115,0.0},{116,0.0},{117,0.0}, {118,0.2},{119,0.5},{120,1.1},{121,1.7},{122,2.2},{123,2.2}, {124,2.2},{125,2.2},{126,1.9},{127,1.7},{128,1.5},{129,1.3}, {130,1.5},{131,1.7},{132,2.0},{133,2.0},{134,2.1},{135,2.0}, {136,2.0},{137,1.8},{138,1.6},{139,1.4},{140,1.1},{141,0.8}, {142,0.6},{143,0.4},{144,0.1},{145,0.1},{146,0.2},{147,0.3}, {148,0.3},{149,0.3},{150,0.4},{151,0.4},{152,0.5},{153,0.2}, {154,0.1},{155,0.2},{156,0.2},{157,0.2},{158,0.2},{159,0.2}, {160,0.4},{161,0.6},{162,0.8},{163,0.6},{164,0.4},{165,0.2}, {166,0.0},{167,0.0},{168,0.0},{169,0.0},{170,0.0},{171,0.0}, {172,0.0},{173,0.0},{174,0.1},{175,0.1},{176,0.1},{177,0.1}, {178,0.3},{179,0.2},{180,0.1},{181,0.2},{182,0.2}, {183,-0.8},{184,-0.8},{185,-0.5}, {186,-0.5}, {187,-0.5},{188,-0.5},{189,-0.5},{190,-0.4}, {191,0.4}, {192,-0.4},{193,-0.4},{194,-0.4}, {195,-0.3}, {196,0.3}, {197,-0.3},{198,-0.3},{199,-0.4},{200,-0.8}, {201,-1.5} ,{202,-1.6},{203,-1.7},{204,-1.8},{205,-0.9}, 206,1.1}, {207,-1.2},{208,-1.4},{209,-1.5},{210,-1.8},{211,-1.8},

{212,-1.6},{213,-1.4},{214,-1.2},{215,-1.0}, {216,1.2}, {217,-1.8},{218,-2.2},{219,-2.1},{220,-2.0}, {221,-1.9}, {222,-1.7},{223,-1.7},{224,-1.5}, {225,-1.0}, {226,-0.7},{227,-0.3},{228,-0.1}, {229,0.0},{230,0.0}}

Do [a [i] =Last [alist [[i+l]]], {i, 0, 230,1}]

mta.lab (display 2)

ListPlot [alist, PlotJoined->True, PlotLabel ->"MTA Acceleration Data",AxesLabel->{"t(sec)","a(mi/hr/sec)"},Ticks->{{23,46,69,92,115,138,161,184,207,230}, {-2,-1,1,2}}]

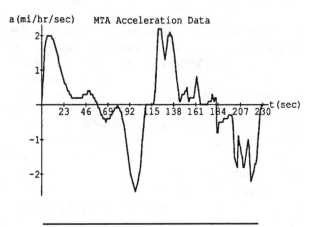

SMOOTHING THE ROAD AHEAD

A state highway is to be constructed in a hilly region. One stretch of the highway will run into and out of a "V-shaped" valley. The parts of the road running down the sides of the valley will be straight. One road will be at a 2% grade (the vertical elevation changes 2 feet for every 100 horizontal feet) and the other road is at a 3% grade. The stretches of road in the valley will be joined by a parabola 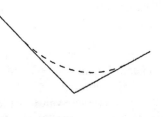 shaped section of roadway. This parabolic shaped section must meet the two straight sections so that there are no jumps or corners, i.e., smoothly. See the figure (not drawn to scale).

1. Assume that the parabolic portion of the road can be described by an equation of the form $y = x^2 - bx + c$, where

the units are measured in miles.

    a) What should b and c be if the road sections are to be joined smoothly? Where will the parabolic section of the road meet the straight sections? Be sure to tell where you are placing your x and y axes, and supply a good picture of the final configuration.

    b) Your boss decides that he would rather have the diagram from part a) with the units in feet (rather than miles). How do the labels change? How does the equation for the parabola change with these new units?

2. Special considerations also have to be made for night time safety. Suppose the speed limit on the road is to be 50 miles per hour. At this speed a car can stop in about 250 feet. Thus the road should be shaped so that headlights are always able to illuminate a point at least 250 feet ahead on the road. Hence the parabolic section of the road should not "curve" too much.

Now assume that the parabolic section can be described by the equation

$$y = ax^2 + bx + c$$

(different b and c than in problem 1.) What values would you advise for a, b, c to guarantee the necessary safety factor and still have the road be smooth? (You need not find the "best" values for a, b, c, but just values that you can demonstrate will work.) Assume that a car's headlights are three feet off the ground and that the beam is parallel to the tangent to the road at the point occupied by the car. Draw a compete diagram of the roadway with all necessary labels.

Hint: In class you discussed the error in using tangent line (or differentials) to approximate a function. The following error estimate was obtained:

$$\text{Error} = |f(x) - (f((x_0) + f'((x_0)(x - x_0))| \le M(x - x_0)^2$$

where M is the maximum value of $|f'''(t)|$ for t between x_0 and x. You can use this error bound to quickly find an acceptable value of a.

Mathematica **Discussion** With a little bit of work, you will be able to reduce the problem of finding b, c, and the 2 coordinates of the points where the road sections meet to a system of equations. You may wish to use *Mathematica* **Solve** command to solve this system. For example, to solve

the system of 3 equations in 3 unknowns

$$x^2 + xy - z = 2$$
$$x + y - 2z = 4$$
$$x^2 + y + 2z = -4$$

use the *Mathematica* command

Solve[{x^2+x*y−z==2, x+y−2z==4, x^2+y+2z==−4}, {x,y,z}]

OPTIMUM SPRAYER PROBLEM*

*This problem was given to us by Professor Thomas Tucker of Colgate University. It is based on a problem originally presented by Professor Jefferson Hertzler, Pennsylvania State University, Harrisburg Campus

A mobile field irrigation system is shown in Fig. 1. Pipe is supported on a framework attached to water-driven wheels

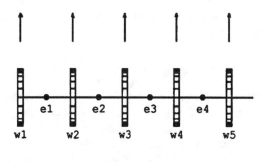

Fig. 1

w1, w2, w3, ... Water emitters el, e2, e3, ... are spaced evenly along the pipe. The entire apparatus moves through the field at a uniform speed. We assume that each emitter distributes water uniformly in a circular pattern. Ignoring the special cases of the first and last emitters, determine the spacing of the emitters to optimize the uniformity of coverage. We require that no point in the field may receive water from more than two emitters.

Discussion

Let the emitter spray radius be 1 unit. Since no point of the field receives water from more than two emitters, the problem reduces to determining the spacing parameter c. See Fig. 2 on the next page. Note that $1 \le c \le 2$. You will need to think about these comments.

Problem 1 Explain why it is useful to find an expression $w(x)$ whose value is proportional to the water received by points along the vertical line through $(x, 0)$, where $0 \le x \le c/2$. Show that $w(x)$ is given by

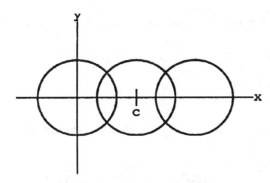

Fig. 2

$$w(x) = \begin{cases} \sqrt{1 - x^2} & 0 \le x \le 1 \\ \sqrt{1 - x^2} + \sqrt{1 - (x - c)^2} & c - 1 \le x \le c/2 \end{cases}$$

Why may the domain of w be restricted to [0, c/2]?

Problem 2 The total amount of water received on [0, c/2] is

$$\int_0^{c/2} w(x)\, dx$$

Use *Mathematica* in showing that (we assume here and elsewhere that the proportionality constant mentioned above is equal to l) the total amount of water received on [0, c/2] is π/4. Why should you be able to infer this result without calculation? The average amount of water received on [0, c/2] is the average value of the function w(x) on the interval [0, c/2].

Problem 3 From the above calculation of total water, the average value of w on [0, c/2] is π/(2c). Your instructor will lead a discussion in which it will become clear that we may optimize the uniformity of coverage by minimizing the "average variation" function

$$g(c) = (2/c) \int_0^{c/2} (w(x) - \pi/(2c))^2\, dx \qquad 1 \le c \le 2$$

Using a combination of *Mathematica* and hand work, show that g(c) may be reduced to a sum of three terms: $k_1(l/c)$, $k_2(1/c^2)$, where k_1 and k_2 are constants to be determined, and

$$\int_{c-1}^{c/2} \sqrt{1 - x^2} \sqrt{1 - (x - c)^2}\, dx$$

Problem 4 Use NIntegrate to calculate at a sufficient number of selected points so that the point at which g takes its minimum may be found, accurate to one decimal place. Why may the domain of w be restricted to [0, c/2]?

LENGTH OF ELLIPTICAL ORBITS

The ellipse shown in Fig. 1 may be described by either of equations (1) or (2). Corresponding formulas for the arc length L(a, b) of this ellipse are given in (3) and (4).

$$\frac{x^2}{a^2} + \frac{y^2}{b^2} = 1 \qquad (1)$$

$$\begin{cases} x = a \cos t \\ y = b \sin t \end{cases} \qquad 0 \le t \le 2\pi \qquad (2)$$

$$L(a,b) = 4a \int_0^{\pi/2} \sqrt{1 - \varepsilon^2 \sin^2\theta}\, d\theta \qquad (3)$$
$$\text{where } \varepsilon^2 = \frac{a^2 - b^2}{a^2} = \frac{c^2}{a^2}$$

$$L(a, b) = 4 \int_0^{\pi/2} \sqrt{a^2 \sin^2 t + b^2 \cos^2 t}\, dt \qquad (4)$$

Since Kepler's time, astronomers and mathematicians working on the "mutual disturbances of the planets" have sought approximations to these integrals. It was eventually shown that L(a,b) can not be expressed as a finite combination of elementary functions and therefore must be calculated separately.

Problem 1 Verify formulas (3) and (4) for the arc length

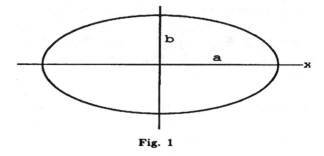

Fig. 1

L(a,b). For (3) start with (1) and the integrand

$$ds = \sqrt{1 + (dy/dx)^2}.$$

Make a trigonometric substitution. Show all of your work.

Gauss' hypergeometric series is given by

$$F(a, b; c, x) = \sum_{k=1}^{\infty} \frac{(a)_k (b)_k}{(c)_k k!} x^k \qquad |x| < 1 \quad (5)$$

where for any real number r and nonnegative integer k, the symbol $(r)_k = r(r+1)(r+2) \cdots (r+k-1)$.

Problem 2 Letting $a = b = -1/2$ and $c = 1$, verify that the first six terms of (5) are as given in

$$F\left(-\frac{1}{2}, -\frac{1}{2} ; 1; \lambda^2\right) \qquad (6)$$

$$= 1 + \frac{1}{4}\lambda^2 + \frac{1}{4^3}\lambda^4 + \frac{1}{4^5}\lambda^6 + \frac{1}{4^7}\lambda^8 + \frac{1}{4^9}\lambda^{10} + \cdots$$

A connection between the hypergeometric series and the length L(a,b) of the ellipse described by (2) was discovered in 1796 by J. Ivory. His result is

$$L(a, b) = \pi(a + b)\, F\left(-\frac{1}{2}, -\frac{1}{2}; 1; \lambda^2\right),$$

where $\lambda = \dfrac{a-b}{a+b}$ (7)

Problem 3 Use (6) and (7) in estimating the length of the orbit of Comet Halley (as astronomers are wont to call Halley's Comet). The data are given in astronomical units, where 1 AU = 149.6 x 10^6 km = length of the semi-major axis of the earth's orbit. For Comet Halley, a = 17.9435 and b = 4.55110. It is given that the error in using the first six terms of (6) to estimate the value of the series is less than 0.00001 for these data.

Problem 4 Compare the results of Problem 3 with the following approximations to the hypergeometric series in (7). These were found by Euler (in 1773), Ramanujan (in 1914), and Jacobsen and Waadeland (in 1985).

Euler $\qquad \sqrt{1 + \lambda^2}$

Ramanujan $\qquad 1 + \dfrac{3\lambda^2}{10 + \sqrt{4 - 3\lambda^2}}$

Jacobsen and Waadeland $\qquad \dfrac{256 - 48\lambda^2 - 21\lambda^4}{256 - 112\lambda^2 + 3\lambda^4}$

Ramanujan stated that his approximation was discovered empirically!

Problem 5 Use Simpson's Rule with 288 subintervals (289 points in the subdivision) to calculate $L(a, b)$ This

subdivision gives a result within 0.0000001 of the integral. Compare with the results of Problems 3 and 4.

HALLEY'S COMET AND EARTH - PART I

The planes containing the elliptical orbits of Earth and Halley's Comet are inclined to each other by approximately 18°. However, we suppose in this problem that the planes are coincident and ask for the the distance between Earth and Halley's Comet at the time the comet was at perihelion (the point in the orbit at which the comet is closest to the sun). The orbits are shown in Fig. 1. The sun is at the origin, which is the common focus of the two orbits. The major axis of the Earth's orbit lies on the y-axis. We have chosen the unit length to be the length of the semi-major axis of Earth's orbit.

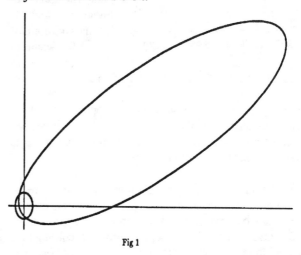

Fig 1

Equations describing the orbits are given in the following equations.

EARTH $\qquad x^2 + 0.033436x + 1.00028y^2 - 0.99972 = 0$

COMET $\quad 0.010095x^2 - 0.032323xy - 0.10223x + 0.042252y^2 - 0.046913y - 0.064331 = 0$

Problem 1 Use *Mathematica* in preparing an accurate graph of the orbits. It should resemble Fig.1, but should include axes with "tick" marks and numerical coordinates. In using PLOT include the option

$\qquad$ AspectRatio -> Automatic

so that the horizontal and vertical scales are equal.

Problem 2 Compute the distance between Earth and Halley's Comet at the time the comet was at perihelion. Halley's comet reached perihelion at time 1986.112, which is approximately February 9, 1986. At this time the coordinates of Earth were (−0.64706, 0.77621).

A MODEL FOR SETTING WHALING QUOTAS

This problem was adapted from *Applying Mathematics, a Course in Mathematical Modeling*, by D.N Burghes, I. Hentley, and J. McDonald.

In recent years, environmental groups like the Greenpeace organization have done a lot to publicize the plight of whales. Poor policy (or the lack thereof) by the whaling in-

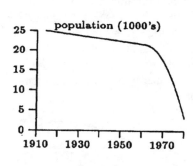

dustry has resulted in years of over hunting. As a result, many species of whales are at or near the point of extinction. The graph in Figure 1, which illustrates the population of Sei whale in the Indian Ocean in recent years, vividly illustrates this point.

In this project you will work with a proposed model for whale populations and see how such a model might be used to come up with a yearly quota for whalers that prevents crises in whale populations' level.)

When modeling animal populations with the intent of making predictions about future populations, it is usually necessary to have a means of predicting future populations based on the past and present populations. According to J. R. Beddington in an article in the *Report of the International Whaling Commision (1978)*, the following model has been used to model populations of Sei whales:

$$N_{t+1} = 0.94N_t + N_{t-8}\left[0.06 + 0.0567\left\{1 - \left(\frac{N_{t-8}}{N_0}\right)^{2.39}\right\}\right] - 0.94C_t$$

In the above equation, N_t is the whale population in year t, Ct_t is the number of whales harvested in year t, and N_0 is the population that would eventually be reached if the whale population were unexploited. In studying population dynamics it is a common practice to take N_0 as the unit of measurement and then to rescale all other quantities appropriately. For example, if an unexploited whale population were numbered at 12,000, then our unit of measurement would be in 12,000's of whales. For example, $C_t = .1$ would

be interpreted as saying that $.1 \cdot 12{,}000 = 1200$ whales were harvested in year t.

1. What do N_{t+1} and N_{t-8} represent in the above equation?

Equation (*) says that the population in year $t + 1$ is affected by three things: the number of whales surviving from the previous year $(.94N_t)$, the number of whales harvested in the previous year $(0.94C_t)$, and the birth of new whales from the breeding stock, the latter a function of the population eight years before year t.

2. Suppose that harvest levels are set so that the whale stock reaches some equilibrium level, N_e. If this equilibrium has been in effect for several years, then $N_{t+1} = N_t = N_{t-8} = N_e$. Explain why this is the case.

3. Suppose the equilibrium level N_e is reached. What is the corresponding harvest quota, c_t, that maintains this equilibrium? This harvest quota is called a *sustainable yield* for the equilibrium level N_e. Explain what is meant by this terminology.

4. In problem 3 you found c_t as a function of N_e. Sketch a graph of this function. (If you use *Mathematica* to sketch the graph, don't forget to set $N_0 = 1$ first.) What are the units for the horizontal and vertical axes of the graph? Suppose the point (N_1, C_1) is on the graph. Carefully explain the relationship between N_1 and C_1 in terms of harvest quotas, equilibrium populations, etc.

5. Using the equation found in problem 3, find N^* the population for which C_t is maximal. Does your value for this maximum seem consistent with the graph you produced in problem 4? This maximum value of C_t is called the *maximal sustainable yield* (MSY). Explain why this name is appropriate. What would happen if harvest quotas were set above MSY for several years?

Modeling populations of marine animals like whales is a very uncertain process. It is very hard to estimate N_0 (the unexploited stock level), and hard to test the validity of models like (*). Thus to be cautious it has been suggested that the harvest quota be set at some *fraction* of MSY, say 90%.

6. Assume that the harvest quota is set at .9MSY, where we are using the MSY value found in problem 5. Find the (approximate) population equilibrium level for this harvest quota. Be sure that you find all possible solutions. Explain the meaning of your answers.

7. Test the above solution with actual numbers. Looking at Figure 1, let us assume that the unexploited population level is the 1900 level; i.e., $N_0 \approx 25,000$. The graph also shows that from 1950 to 1960 the population stayed relatively constant at about 85% of N_0. Suppose that for the *next* ten years whales are harvested at .9MSY per year. Make a chart of the resulting whale populations for the years 1961-1970. From this chart, what seems to be happening to the population? Now suppose that the population from 1950-1960 were 40% of N_0. Again chart the whale population from 1961-1970 and discuss the results.

8. **Extra Credit** Extend the projections obtained in problem 7 a few hundred years into the future and discuss the long term effect of the 90% MSY harvest strategy on various initial populations. Doing this will require a more sophisticated use of *Mathematica*. Here are some hints that might be helpful.

(i) Define a function based on equation (1) to perform the year to year calculations. A function of the form f[n ,p] where n represents the population in the previous year and p the population eight years before is suggested.

(ii) Set up the 1952-1960 conditions with

pop=list[.85,.85,.85,.85,.85,.85,.85,.85,.85]

(iii) Use a Do loop and Append statement to add predictions about subsequent years to the list. for example

Do[w=Append[pop,f[pop[[k]],
        pop[[k-8]]]],k=9,100]

will give predictions for the years 1961-2052.

(iv) Plot the population predictions versus time by using the command

ListPlot[pop, PlotJoined->True]

Use the above commands or variations on them to experiment with different initial populations and see what the long term affects are. When feasible, do projections 100, 200, 300, and 400 years into the future. Try to do this for at least four different initial populations.

JUMPING OFF A BRIDGE, SAFELY

The Dangerous Sports Club was founded in late 1977 by a group of British thrill seekers anxious to add "excitement" to their lives. The club members periodically embark on expeditions on which they participate in unusual, exciting, and frequently life-threatening activities. Past activities of the club have included hang-gliding from the top of Mount Kilimanjaro, skiing an expert alpine slope on everything from a grand piano to a doubledecker bus, low altitude parachute-jumping, and formal dining on the lip of an active volcano. In this problem we will work through one of their more unusual activities. . . bridge-jumping.

In bridge-jumping, a participant attaches one end of a bungee cord to himself, attaches the other end to the bridge railing, and then jumps off of the bridge. (Large bungee cords (rubber ropes) are used to slow jet fighters when they land on the surface of an aircraft carrier.) One of the groups' jumps was aired on the television program "That's Incredible". This jump was made from the Royal Gorge Bridge, a suspension bridge that spans the 1053 foot deep Royal Gorge in Colorado. One of the jumpers used a 415 foot cord (hoping to touch the bottom of the canyon), two others used 240 foot ropes, and the final two used 120 foot cords.

In this problem, you will select one of these jumpers, and calculate how far he fell before the rope started to pull him back up, and estimate how far below the bridge he came to rest.

Discussion. A bungee cord acts much like a spring. If the cord is stretched past its natural (unstretched) position, it exerts a force that acts to restore the cord to its natural position. This force (hopefully) helps to slow the descent of the bridge jumper before he hits the canyon floor.

If the jumper uses a cord that is L feet long when unstretched, he will fall L feet before the chord starts to affect his fall. We will start the analysis of the problem at the instant when the jumper is L feet below the bridge. To do so, we will need to know how fast the jumper is travelling at this time. This information is provided in the following table. The velocities were computed with air resistance taken into account.

L=120 ft	v= 72 ft/sec
L=240 ft	v= 94 ft/sec
L=415 ft	v=113 ft/sec

Once the chord is stretched beyond its natural length, several forces that affect the jumpers' velocity come into play. We will consider 3 of these forces. The first and most familiar is the force of gravity. The force of gravity, F_g, acting on the jumper is equal to the jumpers weight, w. According to Newton's Law, this force is also equal to the mass m of the jumper times the acceleration of gravity, 32 ft/sec2. Hence we have

$$32m = F_g = w.$$

Thus the mass of the jumper is given by

$$m = \frac{w}{32}$$

We will assume that the jumper weighs 160 pounds, hence has a mass of $m = 5$. (The units of mass in the ft-lb-s system are called *slugs*, but that is not important for this problem.)

The second force to consider is the force F_c acting to restore the cord to its unstretched position. This force is proportional to the amount the rope is stretched *past* its unstretched position and acts in a direction that will restore the cord to this unstretched position. For the Navy Air Corps quality bungee chords, the constant of proportionality is .39. Hence our second force is

$$F_c = .39s$$

where s is the amount that the rope is stretched past its unstretched position.

The third force to consider is the force F_r of air resistance, which we assume to be proportional to speed and to act in a direction opposite the direction of motion. We will assume the constant of proportionality here is 1. Hence

$$F_r = -v,$$

the minus sign because the force is acting opposite to the direction of velocity.

So after the jumper falls L feet, we consider 3 forces acting to change the motion. As shown in the diagram, the force F_g always acts downward (which we take to be the negative direction), while F_r acts in a direction opposite the velocity

and F_c acts against any displacement that stretches the cord to a length longer than L.

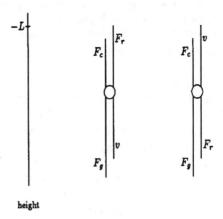

height

1. Let $h < 0$ be the position of the jumper (taking $h = 0$) to be bridge level. Tell why as long as $h \leq -L$, the total F of the 3 forces acting on the jumper is

$$F = -32m + .39(-L - h) - v.$$

Use Newton's Law to find the net acceleration of the jumper if his weight is 160 pounds (i.e. mass is $m = 5$).

We now choose to let t = 0 be the time at which the jumper is L feet below the bridge in his fall. Starting with this time we will calculate approximate values for his coordinate position (on the h axis shown in the figure), velocity, and acceleration at all later times. All we need to know to get started is the value of each of these 3 quantities at time t = 0.

2. For $t \geq 0$, let $h(t)$ be the position of the jumper at time t, $v(t)$ his velocity at time t, and $a(t)$ be his acceleration at time t. Based on the above discussion, what are $h(0)$, $v(0)$, and $a(0)$?

We will now approximate $h(t)$, $v(t)$, and $a(t)$ at each 0.1 second interval for the next 1 minute; i.e. at $t = 0.1, 0.2, 0.3, \ldots, 60$ seconds. For nonnegative integer j, we will let H[j] be our approximation to the position of the jumper at time $t = 0.1 *j$ seconds, and let V[j] and A[j] be defined analogously. Suppose that we have values for H[j], V[j], A[j] for some time nonnegative integer j. We can use these values to approximate the values H[j+l], V[j+l], A[j+l]; that is, to approximate position, velocity, and acceleration 0.1 seconds later. For example we can take

$$H[j+1]=H[j]+(0.1)\,V[j]. \qquad (1)$$

This seems reasonable since for the next 0.1 seconds, the velocity will be about V[j] ft/sec, so the jumper will fall (or rise) about V[j](0.1) feet in this short time. Adding this small change in position to the position H[j] gives us an approximation to H[j+1].

3. Write formulae for V[j+1] and A[j+1], given you know H[j], V[j] and A[j].

4. From problem 2 you know H[0], V[0], and H[]. Use these values and the results of problem 3 to find H[1], V[1], and A[2].

We now use *Mathematica* to calculate H[j], V[j], and A[j] for j = 1, 2, . . ., 600, which gives us values for these functions through the first 1 minute after the jumper is *L* feet below the bridge. We do this by means of the Do function. First initialize the three functions:
H[0]= ; V[0]= ; A[0]=
and then use the functions found in equation (1) and problem 4 to construct the Do command:
Do[{H[j+1]= ;V[j+1]= ;A[j+1]= },{j,0,599,1}] (You are expected to "fill in the blanks" in the above commands.)

We next put the values produced by the last command into lists suitable for plotting. We will produce three lists. The first will be called height and will consist of ordered pairs of the form {(0.1)j, H[t]} for j = 0, 1, 2, . . . 600. This list is easily constructed with the **Table** command:

height = Table[{ 0.1 * k, H[k]}, { k,0,600 }]

You should also construct analogous lists for the velocity and acceleration data.
We can now plot these three lists with the ListPlot command and from these plots get a good feel for the ups and downs of bridge-jumping. For example to plot the height function, use
ListPlot[height,PlotJoined—> True]

5. Plot the velocity and acceleration lists.

6. How far below the bridge does the jumper go before beginning to come back up? Did the jumper hit the bottom of the canyon? What was the maximum speed reached by the jumper during the jump? Maximum acceleration?

7. When the jumper finally stops bouncing, how far below the bridge will he be hanging?

8. **Extra Credit.** You will notice in analyzing your graphs that the jumper never rose to within *L* feet of the bottom of the bridge. If he had, our above calculations would be in error. Why?

Lightly Edited Samples of Student Writing in Calculus and *Mathematica*
D. Brown, H. Porta, and J. J. Uhl
University of Illinois

All claims made about a new course have to be backed up by student performance. Here is a selection of problems from the laboratory final project in first semester calculus turned in by Calculus & *Mathematica* students at Illinois. Note that student writing is a strong component of Calculus & *Mathematica*. Also, not everything the students said is correct or mathematically precise, but they did arrive at correct answers. Our goal is for the students to be able to solve the problem correctly and explain themselves to the best of their ability.

Example 1

Here is a standard optimization problem. The answer is by *Ginny Mark*, a first semester calculus student.

Problem:

Find the <u>highest and lowest</u> points on the graph of

$$f[x] = x^9 \, Exp[-x^2]$$

Recall Exp[x] is another way of writing e^x.

Student answer:

Simplify [Exp[x] - E^x]

0

Oh, okay, thanks.

f[x_] = x^9 Exp[-x^2]

$$\frac{x^9}{E^{x^2}}$$

**Plot [f[x],{x,-5,5}, PlotRange ->All,
        AspectRatio->Automatic]**

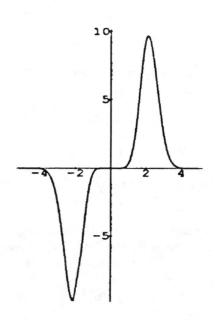

The highest point appears to be around (2.5, 9.5) and the lowest point (-2.5,-9.5).

**Plot[f'[x],{x,-5,5}, PlotRange->All,
        AspectRatio->Automatic]**

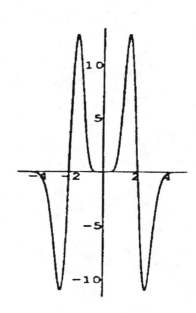

FindRoot[f'[x],{x,-2}]

{x -> -2.12132}

FindRoot[F'[x],{x,2}]

{x-> 2.12132}

f[x]/.x->2.12132

9.66343

Our high point is (2.12132, 9.66343) and our low point is (-212132, -9.66343).

Example 2

Here is a first class calculus problem suggested to us by Rod Smart of the University of Wisconsin. Answers to all parts by *Cody Buchman*, a student in first semester calculus.

-Problem: You own the Calculus & *Mathematica* Steel Plate Company. In comes an order for 750 square steel plates each measuring 12 feet wide and 12 feet long.

border = Graphics[Line[{{-6,0},{-6,12},{6,12},{6,0},{-6,0}}]]
Show[border, Axes ->(0,0),AspectRatio->Automatic, AxesLabel->("x","y")]

A drill (or router) is to be used to drill out a parabolic arch bounded by the parabola y = -4 (x - 3/2) (x + 3/2).

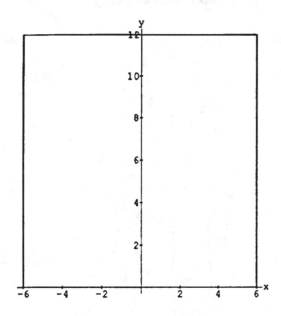

arch = Plot[{-4 (x - 3/2) (x + 3/2)},{x,-3/2,3/2}, DisplayFunction->Identity]
Show [border,arch,PlotRange->{{-6,6},{0,12}}, Axes->{0,0},AspectRatio->Automatic, AxesLabel ->{"x","y"},DisplayFunction->$DisplayFunction]

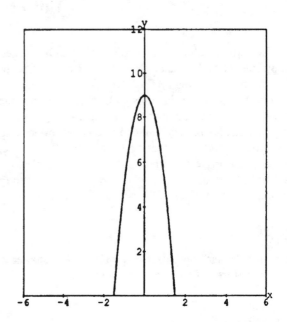

Part a:

You have a selection of router bits to do the job. The bits available to you come in the following cutting diamenters:

> 2 inch
> 3 inch
> 4 inch
> 5 inch
> 6 inch

Which of these bits can be used to cut out the arch? Discuss how you arrived at your answer.

Student answer:

First: to cut out such an arch, the bit must be no wider or narrower than the diameter of the smallest osculating circle. Why? We saw already that the smallest osculating circle of a parbola is always found tangent to the vertex of said parabola, and always fills the bottom of the parabola. If a bit larger than the diameter of this circle was used, it would cut a whole different, wider parabola, with possible ugly results (depending what the plates were being used for). Using a bit of smaller diameter would produce similar results. This will be illustrated later. For now, let's find the radius by setting the f[x] equal to -4 (x -3/2) (x + 3/2), and running it through the mill:

Clear[x,y,b,k,r,f,x0]

f[x_] := -4 (x - 3/2) (x + 3/2)

**circleeqn=((x - h)^2+(y[x]-k)^2
    == r^2):**

**eqn1= circleeqn/.{x->x0,y[x]->
    F[X0]}:**

**firstderiveqn=D[(x - b)^2+(y[x]-
    k)^2 == r^2,x]:**

**eqn2 = first deriveqn/.{x->x0,y[x]->
    f[x0], y'[x]->f'[x0]}:**

**seconderiveqn=D[(x-b)^2+(y[x]-
    k)^2 ==r^2, (x,2)]:**

**eqn3 = seconderiveqn/.{x->x0,y[x]->
    f[x0],y'[x]->f'[x0],y''[x]->
    f''[x0]}:**

**bkrsolved=Solve[{eqn1,eqn2, eqn3},
    {h,k,r}]**

$$\left\{\left\{r \to \frac{\text{Sqrt}[1+192\,x0^2+12288\,x0^4+262144\,x0^6]}{8},\right.\right.$$

$$h \to -64\,x0^2$$

$$\left. k \to \frac{71-96\,x0^2}{8}\right\},$$

$$r \to \frac{-\text{Sqrt}[1 + 192\,x0^2 + 12288\,x0^4 + 262144\,x0^6]}{8},$$

$$\left.\left. h \to -64\,x0^3 , k \to \frac{71-96\,x0^2}{8}\right\}\right\}$$

The first "r" is the correct one. Now let's solve for "r".

radius = bkrsolved[[1,1,2]]

$$\frac{\text{Sqrt}[1 + 192\,x0^2 + 12288\,x0^4 + 262144\,x0^6]}{8}$$

Solve[D[radius,x0 == 0,x0]

$$\left\{\{x0\to0\},\{x0\to-\tfrac{1}{8}\},\{x0\to-\tfrac{1}{8}\},\{x0 \to \tfrac{-1}{8}\},\{x0\to\tfrac{-1}{8}\}\right\}$$

This could also be seen from the illustration, but the numerical proof is here. Now, let's find the diameter by plugging back into r:

diameter =2 (N[radius/.x0->0])

0.25

Hmm.   The bit that I need must be 0.25 feet—3 inches—in diameter.

Part b:

-Choose a bit that will work and give a plot of the path of the center of the router.

-Student's Answer:

Let's start by solving for its slope:

y = - 4 (x - 3/2) (x + 3/2)

$$-4\left(-\left(-\tfrac{3}{2}\right)+x\right)\left(-\tfrac{3}{2}+x\right)$$

yprime = D[y,x]

$$-4\left(-\left(-\tfrac{3}{2}\right)+x\right)-4\left(-\tfrac{3}{2}+x\right)$$

This is the slope for the tangent line at any x along the parabola. Now, let's say a single point on the bitpath is (s,t). This point is on a line whose equation is given above, and which lies a distance a from a point on the parabola, {x,y[x]}. This gives us the equation (x,y[x]) + b{-1,1/y'[x]} = (s,t), where b is some multiple that increases as distance a increases. Now, we know that distance a = 0.125 (the radius of the osculating circle). Also, we can simlify the right term of the equation to {-b,b/y'[x]}, which represents the distance we must travel from the point {x,y[x]}, along the perpendicular line, to reach the point (s,t). Since we know this = 0.125, we get the equation. Sqrt[b² + b²/y'[x]²] = .125, and can solve for b:

Clear[b]

Solving for b in the above equation gives:

$b^2 = .1252/(1 + 1/y'[x]^2)$, or

b = .125 /Sqrt[1 + 1/yprime^2]

```
                   0.125
-------------------------------------------   -----------
                   3         3      -2
      Sqrt[1 + (-4 (-(-) + x) - 4 (- + x))   ]
                   2         2
```

Now, let's plug in "b" and plot:

pointt = {x,y} + Sign[x] b {-1,1/yprime}

```
                    0.125 Sign [x]
{x- ----------------------------------------------- ----,
                    3         3      -2
       Sqrt[1 + (-4 (-(-) + x) - 4 (- + x))   ]
                    2         2


       3         3
 -4 (-(-) + x) (- + x) + (0.125 Sign[x])/
       2         2


       3         3
((-4 (-(-) + x) - 4 (- + x))
       2         2
                   3         3      -2
 Sqrt[1 + (-4 (-(-) + x) - 4 (- + x))   ]}}
                   2         2
```

par = ParametricPlot [{x,y},{x,-3/2,3/2}, PlotPoints->75, DisplayFunction->Identity]

router = ParametricPlot[pointt,{x,3/2,3/2}, Plot Points-.75, DisplayFunction->Identity]

Show[par,router,
        DisplayFunction->$DisplayFunction]

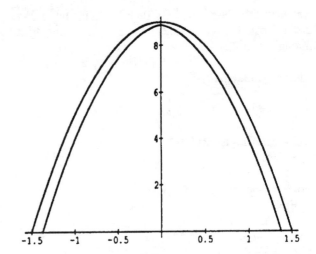

Note that glitch at the top. This is caused by a round-off error in the plotting mechanism. To solve this, let's go in for a close-up of the top:

Show[par,router,PlotRange->{{-.2,.2},{8.8,9}},
            DisplayFunction->$DisplayFunction]

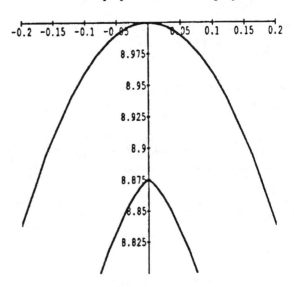

Note the peak at the top of the drill path. This is logical, for when the drill reaches the point of maximum curvature of the parabola, it must abruptly shift directions to maintain the cut's shape.

Example 3

Here is a problem dealing with accuracy of computation. Answer by *Anne Bierzychudek,* a first semester calculus student.

Problem:

If x is in [-12,12], then how many accurate decimals of x are needed to gbuarantee k accurate decimals of f[x] = x⁴ -3x³ + 5x² - 2x + 8?

-Student's answer:

Clear [f,x,y,z]
f[x_] = x^4 - 3x^3 +5x^2 - 2x + 8

$$8 - 2x + 5x^2 - 3x^3 + x^4$$

Plot[Abs[f'[z]],{z,-12,12}, PlotRange->All]

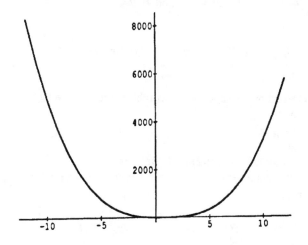

The plot makes it clear that for z in [-12,12]
        |f'(z) | < 10,000 = 10^4;
To get k accurate decimals of f[x] for x in [-12, 12], we need at least k + 4 accurate decimals of x.

Example 4

Here is a standard problem on tangent lines. The answer is by *Mable Chiu*, a first semester calculus student.

Problem:

Describe all constants a, b and c such that the parabola
    y = ax²+ b x + c
passes through {0,1} and is tangent to the line y = x. Plot several of these parabolas and the line on the same axes.

-Student's answer:

Clear [c,a,x,b,y]

Let's set the equation first:

y = a x^2 + b x + c

$$c + b x + a x^2$$

Since the equation y and y=x are tangent to each other they should share the same slope and touch at a certain point. So, the two equations are set equal to one another.

eqn1 = a x^2 + b x + c == x

$$c + b x + a x^2 == x$$

Since the two equations are tangent to each other their derivatives should be equal to one another. So the derivative of the two equations are set equal to each other.

eqn2 = D[c + b x + a x^2),x] == D[x,x]

$$b + 2 a x == 1$$

There are three unknowns that must be found a, b, and c. So one more equation must be generated to get what we want. We have three unknowns, so we need three equations in order to solve for a, b, and c

eqn3 = c == 1

$$c == 1$$

Now let's solve those three equations.

solved1=Solve[{eqn1,eqn2,eqn3}]

$$\{\{x \to \frac{-\sqrt{4 - 4\,(1 - 4\,a)}}{4\,a},$$
$$b \to \frac{2 + \sqrt{4 - 4\,(1 - 4\,a)}}{2},$$
$$c \to 1\}, \{x \to \frac{\sqrt{4 - 4\,(1 - 4\,a)}}{4\,a},$$
$$b \to \frac{2 - \sqrt{4 - 4\,(1 - 4\,a)}}{2}, c \to 1\}\}$$

Since the variable "a" is first in the alphabet the computer, chose   a   as the variable that everything else should be in terms of since it wasn't specified. So, x, b, c is solved in terms of "a".

y

$c + b\,x + a\,x^2$

Now, the solved values of x, b, c are plugged back into the equation "y" which is above. However, the plot will need the variable x to remain in the equation in order for it to be plotted since f(x) is dependent on the variable x. So, x remains to be x DO NOT SOLVE X IN TERMS OF A. It is not necessary, Look down. Only b and c is plugged into the equation.

alt1 = y/.{b -> (2 + 4*(1 - 4*a))^(1/2))/2,c-> 1}

$$1 + \frac{(2 + Sqrt[4 - 4\,(1 - 4\,a)])\,x}{2} + a\,x^2$$

alt2 = y/.{b -> (2 - (4 - 4*(1 - 4*a))^(1/2))/2,c -> 1}

$$1 + \frac{(2 - Sqrt[4 - 4\,(1 - 4\,a)])\,x}{2} + a\,x^2$$

There are solutions for this equation. The values b and c are part of one solution. They go together. Let's try to plot the solutions.

Plot[{alt1/.a->3,alt2/.a->3,alt1/.a-1,alt2/.a->1,
          alt1/.a->.4,alt2/.a->.4,x},{x,-2,2}]

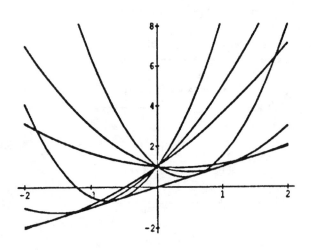

Answer: Here is the plot of the two solutions along with the tangent line y=x and random values for the variable a. All of the curves for equation "alt1" are on the left side of the graph. All of the curves for equation "alt2" are on the right side of the graph.

Example 5

Here is a problem on Fourier approximation. The answer is by *Rod Johnson*, a student in first semester calculus who had never heard of trigonometric polynomials.

Problem:

Find the constants a, b and c that make

$$\int_0^\pi (\,x(\pi - x) - (a\,Sin[x] + b\,Sin[2x] + c\,Sin[3x])^2\,dx$$

as small as possible. After you have found a, b and c, plot
     x(p- x)
and
     (a Sin[x] + b Sin[2x] + c Sin[3 x])
on the same axes for $0 \le x \le p$ and assess the quality of the fit.

-Student's answer

We will integrate this equation first.

Clear[a,b,c,x]

int =Integrate[(x(Pi -x)-(a Sin[x] + b Sin[2x] + c Sin[3 x]))^2,{x,0,Pi}]

$$\frac{Pi^5}{30} - 8\,a + \frac{Pi\,a^2}{2} + \frac{Pi\,b^2}{2} - \frac{8\,c}{27} + \frac{Pi\,c^2}{2}$$

All we have to do is take the derivative of the integral with respect to each constant, set it equal to 0 and solve for that constant. Let's start with a.

darva = D[int,a]

-8 + Pi a

Solve [{darva == 0},a]

8
{{a -> —}}
   Pi

Now let's find b.

darvb = D[int,b]

Pi b

Solve[{darvb == 0},b]

{{b-> 0}}

Finally we can get c.

darvc = D[int,c]

  8
-(—) + Pi c
 27

Solve[{darvc == 0 },c]
       8
{{c-> ———--}}
   27 Pi

To get

$$\int_0^\pi (x(\pi - x) - (a\,Sin[x] + b\,Sin[2x] + c\,Sin[3x])^2\,dx$$

as small as possible we will need to use

    a = 8/p
    b = 0
    c = 8/(27p).

Now let *Mathematica* know the values for a, b, and c:

a = 8/Pi; b = 0 , c = 8/(27 Pi);

We are going to plot the equation

    (a Sin[x] + b Sin{2x} + c Sin{3x})

with these values for a, b, and c so let's find out what that will be.

eqn2 = (a Sin[x] + b Sin [2x]+ c Sin[3 x])

$$\frac{8\,Sin[x]}{Pi} + \frac{8\,Sin[3\,x]}{27\,Pi}$$

Here is the plot of

    x(π - x)
and
  (8 Sin[x])/π + (8 Sin[3])/(27π)

from 0 to π.

Plot [{eqn2, x(Pi - x)},{x,0,Pi}]

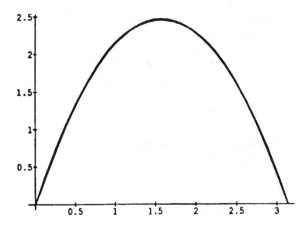

Now that is a hell of a fit. That is looking real sweet the whole way from 0 to π.

Example 6

Here is a problem from The Waterloo Maple Calculus Workbook. The answer is by *Ann Brzozkiewicz*, a first semester calculus student.

Problem:

The Waterloo Tile Co. is designing 1 foot by 1 foot ceramic tiles. On each tile, two fourth degree polynomial curves are running from the lower left hand corner to the upper right hand corner. They are positioned so that they trisect the square's right angles at the lower left hand and the upper right hand corners.   A shade of red paint is to be applied above the top curve, white paint is to be applied between the curves and a shade of blue is to be applied below the bottom curve. It is required that the red area = white area = blue area = 1/3 square feet. Determine the equations of the polynomial curves and plot.

- Student's answer:

There are two curves and therefore two equations. One I will call f[x] and the other will be g[x].

Clear[a,b,c,d,f,g,k,l,m,n]
f[x_] :- ax^4 + b x^3 + cx^2 + dx + e
g[x_] := kx^4 + 1x^3 + mx^2 + nx+0

There are five variables so I will need 5 equations to solve each one. The things I know are:

    a) The equations move from x = 0 to x = 1, so for x equal to zero the function equals zero, and for x equal to one, the function equals one.
    b) The slope of the tangent lines to the curves at 0 and 1 are given (because they trisect the corners of the tile.)
    c) The area between the curves must be equal to 1/3 sq. ft.

The following equations are therefore true.
$$f[0] = g[0] = 0$$
$$f[1] = g[1] = 1$$
$$f[0] = g'[1] = Tan[Pi/3]$$
$$f[1] = g'0[0] = Tan[Pi/6]$$
$$\int_0^1 f[x] - g[x]\, dx = 1/3$$

This last equation and the information given also indicate that the area under the lower curve will be equal to 1/3, and the total area under the higher curve will be 2/3. Keeping this in mind I start out to make f[x] the lower curve.

f1 = f[0] == 0

$$e == 0$$

f2 = f[1] == 1

$$a + b + c + d + e == 1$$

f3 = f'[0] == Tan[Pi/6]

$$d == Tan\left[\frac{Pi}{6}\right]$$

f4 = f'[1] == Tan[Pi/3]

$$4a + 3b + 2c + d == Tan\left[\frac{Pi}{3}\right]$$

f5 = Solve[Integrate[f[x], {x,0,1}] == 1/3]

$$\{\{a\text{->}\frac{5}{3} - \frac{5b}{4} - \frac{5c}{3} - \frac{5d}{2} - 5e\}\}$$

newf5 = a ==
$$a == \frac{5}{3} - \frac{5b}{4} - \frac{5c}{3} - \frac{5d}{2} - 5e$$

$$a == \frac{5}{3} - \frac{5b}{4} - \frac{5c}{3} - \frac{5d}{2} - 5e$$

N[Solve[{f1,f2,f3,f4,newf5},{a,b,c,d,e}]]

$$\{\{a\text{->}-2.11325,\ b\text{->}4.5359,\ c\text{->}-2.,\ d\text{->}0.57735,\ e\text{->}0.\}\}$$

These are all my variables, I just need to substitute them into my fourth power polynomial.

goodf = f[x] /. {a -> -2.11325, b-> 4.5359, c->, d-> 0.57735, e->0.}

$$0. + 0.57735\, x - 2x^2 + 4.5359\, x^3 - 2.11325\, x^4$$

Plot[goodf,{x,0,1}, PlotRange->{0,1}]

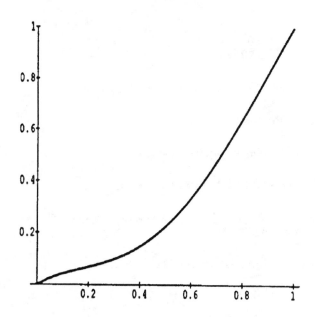

It looks like it might be ok, let's proceed to g[x].

g1 = g[0] == 0

0 == 0

g2 = g[1] == 1

k + 1 + m + n + 0 == 1

g3 = g'[0] == Tan[Pi/3]

$$n == Tan[\frac{Pi}{3}]$$

g4 = g'[1] == Tan[Pi/6]

$$4 k + 3 l + 2 m + n == Tan[\frac{Pi}{6}]$$

f5 = Solve[Integrate[g[x],{x,0,1}] ==2/3]

$$\{\{k-> \frac{10}{3} - \frac{5 l}{4} - \frac{5 m}{3} - \frac{5 n}{2} - 5 o\}\}$$

$$newg5 = k == \frac{10}{3} - \frac{5 l}{4} - \frac{5 m}{3} - \frac{5 n}{2} - 5 o$$

$$k == \frac{10}{3} - \frac{5 l}{4} - \frac{5 m}{3} - \frac{5 n}{2} - 5 o$$

N[Solve[{g1,g2,g3,g4,newg5},{k,l,m,n,o}]]

{{k-> 2.11325, l-> -3.9171, m-> 1.0718, n-> 1.73205,
o->0.}}

goodg = g[x]/.{k->2.11325,
    l->13.9171,m -> 1.0718,
    n-> 1.73205,   o -> 0.}

$$0. + 1.73205 x + 1.0718 x^2 - 3.9171 x^3 +$$

$$2.11325 x^4$$

Plot[{goodf,goodg},{x,0,1},PlotRange ->{0,1}, AspectRatio->Automatic]

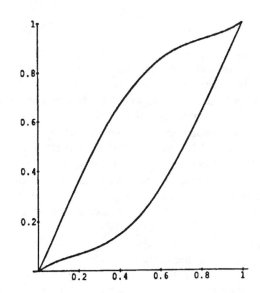

That's rather a funny looking tile, so to check my answers I will integrate the different areas to see if they meet the specifications. First, the area between the curves should be equal to 1/3.

Integrate[goodg - goodf,{x,0,1}

0.333333

Got it. Another thing to check is that the area from the top of the tile to the top function g[x], is also equal to 1/3.

Integrate[1 - goodg,{x,0,1}]

0.333333

Got it again! Lastly, the area under the bottom curve should be 1/3.

Integrate[goodf,{x,0,1}]

0.333333

Right on the money! Now the ratio of the three areas red to white to blue is 1:1:1 and the tangent lines to the curves trisect the right angles

Example 7

Here is a problem on chemical reactions. All answers are by *Maria Gonzalez*, a first semester calculus student She used the word processor to do interactive algebra with *Mathematica*.

Part i:

Two chemicals A and B react with each other and in the process, one molecule from A bonds with one molecule from B to form a new compound. Let a be the number of molecules of A present at the start and let b be the number of molecules of B present at the start. Let y[t] be the number of molecules of the new compound present t seconds after the rection starts. At anytime t, there are (a- y[t]) molecules of A available for reaction and there are (b- y[t]) molecules of B available for reaction. Since reaction requires collision of an A- molecule and a B- molecule, it makes sense to assume that y'[t] is proportional to both

(a - y[t])   and to   (b - y[t]).

What differential equation does this lead to?

-Student's answer:

y'[t] = K(a - y[t])(b - y[t]) with y[0]=0

Part ii:

Find a formula for y[t] in terms of a, b and K. What is the limiting behavior of y[t] if a > b?

-Student's answer: Rewrite the equation to get:

y'[t]/((A - y[t]) (B - y[t])) = K

Clear[y,t,x,K,a,b]

left = Integrate[y'[t]/((a - y[t])(b - y[t])), {t,0,t}]/.y[0]->0

$$-\left(\frac{Log[a]}{a-b}\right) + \frac{Log[b]}{a-b} + \frac{Log[a-y[t]]}{a-b} -$$

$$\frac{Log[b-y[t]]}{a-b}$$

right = Integrate[K,{t,0,t}]

K t

Simplify the left side of the equation to get:

(1/a-b))(Log[b(a-y[t])/(a(b-y[t])])

Now solve for y[t]:

(1/(a - b)) (Log[b (a (a - y[t])/(a (b - y[t])])] = K t
Log[b(a - y[t])/(a (b - y[t]))] = (a - b) K t

b (a - y[t])/(a (b -y[t]) = E^((a - b) K t)

Solve[(b (a - y[t]))/(a(b - y[t]))
**   == K^((a - b) K t), y[t]]**

$$\{\{y[t] \to -\left(\frac{-(E^{Kat}\, ab) + E^{Kbt}\, ab}{E^{Kat}\, a - E^{Kbt}\, b}\right)\}\}$$

To find the limiting behavior of y[t] if a > b then simplify y[t] by multiplying the top and bottom by E^(- K a t):

y[t]=((a b)(1-E^(Kbt))/(E^(Kat)))
**   )/(a-b(E^(K b t))/(E^(K a t)))**

$$\frac{(1 - E^{-(Kat)+Kbt})\, a b}{a - E^{-(Kat)+Kbt}\, b}$$

Now if a > b then the term
(E^(K b t))/(E^K a t))
will be 0 leaving:

y[t] = a b (1 -0)/a
y[t] = b (since a will cancel)

Therefore, the limiting behavior of y[t] if a > b will be approaching b.

Calculus, Concepts, and Computers:
Some Laboratory Projects for Differential Calculus
Ed Dubinsky and Keith Schwingendorf
Purdue University

Introduction

We give some examples of computer laboratory projects (assignments) we've used at Purdue University in our calculus project. In general, the tasks were one week projects which included two hours in the laboratory with undergraduate and graduate assistants in addition to the instructor. Students were encouraged to complete laboratory projects (or certain parts of them) prior to class discussions about the concepts. In general, grading of the assignments was not made for correct answers, but instead for a meaningful attempt to think about the mathematical concepts in the project. The projects included in this paper are revised versions of projects used during our NSF supported planning grant for AY 1988-89. Any significant changes in the original projects will be discussed and we will comment on what seemed to work and what didn't. Often, no significant changes were made in parts of the original assignments, but sometimes we shifted parts from one assignment to another in an attempt to smooth the length of the assignments or the instructor's preferences for the order of topics.

We should mention that ISETL was significantly enhanced prior to the Fall semester 1990, including changes with respect to the ease of use of ISETL on both the Macintosh and PC versions and also a 2-D graphics package was added. Future enhancements of ISETL's graphics package will probably include 3-D graphics, but there is no intention to include symbolic manipulation as part of ISETL. We believe that it is better to separate the construction of mathematical concepts with a mathematical programming language like ISETL and the use of a symbolic computer system like Maple, *Mathematica* or DERIVE. We have experimented with several 3-D graphics packages on the Macintosh, including 3-D Grapher, Macfunction, Funplot 3-D, Fields and Operators and combinations of these packages.

Student evaluations of individual projects were not made. However, evaluations were made by our students at the end of each semester and the students in our third semester course (MA 261a) wrote essays evaluating our approach to teaching calculus at the completion of our three semester sequence of courses. For samples of student comments from their essays see page 64 of this volume. In general, the students in our courses indicated appreciation of our prodding them to think about mathematical ideas and to construct mathematical concepts using ISETL. We should mention that this appreciation for our approach took time for our students to develop, probably due to previous courses in mathematics in which they primarily did instrumental learning (drill and practice). Our students also welcomed the sense of "community" gained in our courses which was transmitted via use of a team approach on laboratory projects and our cooperative problem solving approach used in class.

Our standard practice was *not* to discuss the details of the ISETL funcs our students were required to write until after they struggled with the constructions for at least one lab period. Class discussions after the first lab session of a week usually involved the use of a modified Socratic method to discuss the constructions to be made in the Lab and Homework Assignments.

In some cases however, we decided that a particular program was not worth the (often considerable) trouble for our students to write the program. In this case as an alternative we provided students with the program to use. On some occasions, a classroom activity we use consists of the class constructing the code with the instructor using a modified Socratic approach. We did this with Newton's method (in Lab 8) and the task of understanding a recursive program to help our students understand how the method works. On a subsequent examination they were able to decide and explain graphically situations in which Newton's method seemed to converge and situations in which it did not. On other occasions we have given our classes a piece of ISETL code to study and try to understand precisely how a mathematical algorithm is performed, followed later by class discussions.

One way in which we use a mathematical programming language and a symbolic computer system together is for the students to make ISETL constructions that represent mathematical notions, use them in simple cases, and then turn to Maple to make the same calculations in more complicated situations. This was done for finding (or approximating) zeros of functions. The students work with a bisection routine and Newton's Method in **ISETL**, and then use the **Maple** facility to find zeros automatically. Thus they develop some understanding of how the automatic procedure might be working. A similar approach is possible with limits where students use **ISETL** to construct long finite sequences that help them develop some intuition about approaching a limit and then use **Maple** to evaluate just about any limit that they run into. The same effect is obtained with an **ISETL** func ad (see Lab Assignment 4) that produces an approximate derivative that helps students deal with what the derivative means, and **Maple** calculations of derivatives with long lists of functions that help students discover some of the rules for differentiation.

Parts of any of the following homework assignments that assigned routine drill and practice problems are omitted here. Hence some homework assignments appear to be much shorter than they really were.

Example Laboratory Projects

Lab Assignment 4 of MA 161a

Summary

We begin with Lab Assignment 4 of our first course in calculus, Ma161a. In this assignment, students' investigations included: the δ–ε definition of limit at a point using **Maple** and what we refer to as δ–ε windows; pointwise "approximate derivatives" by constructing an **ISETL** func ad which the students were required to construct; the algebra of functions by writing **ISETL** funcs for addition, multiplication and composition; and the prediction of function values and limits by constructing **ISETL** tuples.

We asked our students to write a func in **ISETL** that will accept a function f (that is, a func that models it) and a small positive number ε, and then return a model of a function that is an approximation to the derivative of f. The solution looks like this.

```
ad := func(f,e);
      return func(x);
                 return (f(x+e)-f(x))/e;
              end;
      end;
```

Once the students have written this, they are asked to construct models of various functions f and to evaluate and

understand expressions like

```
ad(f ,0.00001) (3);
```

This is not easy for our students. Their trouble is not with the syntax (although they sometimes think that it is), it is with the idea (quite new and strange for them) that the answer to a problem is not necessarily a number, but could be a "whole function." Most of our students do get it, eventually. As a result, their concept image of functions is enriched and they are well positioned for understanding the idea of the derivative. They are able to work problems in which they are asked to compute the derivative by setting up the definition and taking the limit. In addition to the usual rational functions and roots, they are able to do this with what we shall call a function defined in parts, that is, a function given by different formulas for different parts of its domain.

In general, our students struggled with δ–ε windows as might be expected. However, as was the case with most of the parts of our projects, the students' work and hard thinking on the ideas helped a great deal in their understanding of class discussions. There were no significant changes in the parts of this assignment from the first time we used it in MA 161a.

MA161a Lab Assignment 4

1. For each of the functions described below, you are given a small positive number which we will call ε, a value c in the domain of definition of the function and a number L. Your task is to find another positive number δ having the property that if the independent variable is in the interval from $c - \delta$ to $c + \delta$, then the values of the function lie in the interval from $L-\varepsilon$ to $L + \varepsilon$.

One way to do these is to use **Maple** to sketch the graph and reason from the picture. When that is not possible, you will have to reason from your mind.

(a) The function is given by the expression $\frac{\sin(x - 0.5)}{x - 0.5}$, $\varepsilon = 0.01$, $c = 0.5$, and $L = 1$.
(b) The function is given by the expression $\frac{\sqrt{2+x} - \sqrt{2}}{x}$, $\varepsilon = 0.01$, $c = 0$, and $L = \frac{\sqrt{2}}{4}$.
(c) The function is my5, $\varepsilon = 1000$, $c = 0.01$, and $L = 10,000$.
(d) The function is my6, $\varepsilon = 1000$, $c = 0.01$, and $L = -9,999$.
(e) The function is my5+my6, $\varepsilon = 2000$, $c = 0.01$, and $L = 1$.

SUBMIT

(a) For each function, your value of δ

(b) How you got it and an explanation of why it works.

2. Write an **ISETL** func ad, i.e., ad(f ,a), that will accept a func representing a function f and a point a in the domain of f and return the value of the difference quotient,

$$\frac{f(a + 0.001) - f(a)}{0.001}$$

We call this expression the "approximate derivative of the function at a." Run your func on the following examples.

(a) The function is given by the expression, $\sqrt{x}$ and $a = 2$.

(b) The function is given by

$$r(u) = \begin{cases} u^3 + 3 & \text{if } u \leq -2 \\ -3u^2 + 7 & \text{if } -2 < u \leq 0 \\ -1 & \text{if } 0 < u \end{cases}$$

and $a = -2$.

(c) The same function r with a replaced by $a = 0$.

Change your func by replacing 0.001 by -0.001 (in both the numerator and denominator). Run your new func on the same three examples.

Use **Maple** or **Maple** and **ISETL** to produce a graph of the three functions with 3: in an interval of length about 0.001 and centered at a. Try to think up some way of indicating on the graph the meaning of the expression for the difference quotient in each of your two funcs. (Think about the meaning of the numerator, denominator and the whole difference quotient.)

SUBMIT

(a) A copy of your two funcs.

(b) Your six numerical answers.

(c) An explanation of how and why the three answers with the first func differed (or did not differ) from your three answers with the second func.

(d) Your three graphs with your interpretation of the approximate derivative, clearly labelled, on each graph.

It is possible for an **ISETL** func to return another func, which is a feature having many applications. For example, here is a func that will accept two funcs representing functions and return a func which represents the product of the two functions.

```
mult := func(f ,g);
    return func (x);
            return f (x) *g (x);
        end;
    end;
```

Using a similar structure, write two funcs, plus and comp that will give, respectively, the sum of two functions and the composition of two functions.

Let r be the function represented in problem 2(b) and let s be the function with the following representation

$$s(x) = \begin{cases} \dfrac{2}{x} & \text{if } x \leq -1 \\ |\sin(x)| & \text{if } -1 < x \end{cases}$$

Write out by hand as precise as possible an explanation of what would be the meaning of the following two **ISETL** expressions. This explanation should be in terms of the original formulas (expressions) used to define r and s.

```
plus (r ,s ) (x);
comp(r,s) (x);
```

SUBMIT

(a) A copy of your two funcs, plus and comp.

(b) Your explanations of the two **ISETL** expressions.

4. **ISETL** tuples can be used to represent finite sequences of numbers, and this can be used to "approximate" a sequence which approaches a number. Consider, for example, the following **ISETL** code where a represents a real number and f is an **ISETL** func representing a function.

```
a := 2;
s := [a + ((-l)**i)/(l000*i): i in [1. .60]];
s;
[f(x) : x in s];
```

Run this code for the following choices of f and a.

(a) The function h (from Lab 1 problem #4) given by

$$h(x) = \begin{cases} -(x+4)^2 - 2 & \text{if } x < -3 \\ (x+2)^2 - 4 & \text{if } -3 \le x < 1 \\ x + 4 & \text{if } 1 \le x < 2 \\ -2x + 5 & \text{if } x \ge 2 \end{cases}$$

with $a = -3$.

(b) The same function h as in part (a) with $a = 2$.

(c) The function given by the following expression,
$$f(h) = \frac{\sqrt{2+h} - \sqrt{2}}{h}$$
with $a = 0$.

SUBMIT

(a) A verbal description of the meaning of the code for s in the second line of code above. Include an explanation of the presence of $(-1)**i$ and of the meaning of s(2).

(b) A verbal description of the meaning of the fourth line of the code, [f (x) : x in s]; . Include a description of the effect of $(-1)**i$ and explain how it would be different if it were not there, and if it were replaced by -1.

(c) What do you think could be meant by "other ways of approaching a" and "other rates of approach to a"?

(d) Two variations of the **ISETL** code for s (in the second line) that implement your response to the previous question.

(e) Based on the result of running the code above for each function, what do you predict should be the value of each function at the indicated point a?

(f) How do the three examples differ with respect to the result of evaluating f at a?

Comments

In Homework Assignment 4 of MA 161a, our students further investigated limits of functions using **Maple**'s limit calculating facility for various limits of functions. Based on their experience with many limit situations we asked our students to guess limit rules for sums, differences, constant multiples, linear combinations, products, quotients and compositions of functions. In general, student performance was quite good.

Lab Assignment 5 of MA 161a

Summary

In Lab Assignment 5 of MA 161a, our students investigated the following topics: further predictions of function values and limits using **ISETL** tuples; "approximate tangents" to curves by constructing various **ISETL** funcs; the "approximate derivative" function Df for a function f by constructing an **ISETL** func ada which accepts an **ISETL** func f and returns the approximate derivative func Df; and the discovery of the graphs of derivatives of various functions from the graphs of the functions.

In Part 5 of this assignment the students were to graph the functions and their approximate derivatives obtained using their func ada from Part 4, and then match pairs of the original functions so that one was a function and the other was its approximate derivative. As a result of this part of the assignment, class discussion focused on the meaning of the values of the approximate derivative function in relation to the graph of the original function. On a subsequent exam, students were asked to compare two or three graphs and indicate which was the function and which was its derivative. Part 5 will be improved in the future by replacing the expressions for the functions listed in (a) to (h) by data functions (or what we refer to in these assignments as mystery functions). Since some of our students had calculus prior to our course, they focused on the formulas and derivative rules they knew, rather than the meaning of the values of the func Df returned by the func ada.

MA161a Lab Assignment 5

1. Recall the following **ISETL** code from Lab Assignment 4, #4:
```
a := 2;
s := [a + ((-l)**i)/(1000*i) : i in [1. .60]];
s;
[f (x) : x in s]; .
```

For each of the following functions and given point a, write a func to represent the function, run code similar to the above four lines and use the result of [f (x) : x in s] to predict what should be the value of the function at the given point.

(a) The point is 0 and the function is given by

$$k(u) = \frac{u}{\sin(2368u)}$$

(b) The point is 1 and the function is h given from Lab Assignment 1, #4.

(c) The point is 2.006 and the function is the one you defined by varying a in Homework Assignment 4, #1(b).

SUBMIT

1. Your prediction of the value for each function.

2. For each of the following functions and two points, use **Maple**, or **ISETL** and **Maple** (graph) to sketch the function and on the same graph sketch the line connecting the two points. (Of course, you will have to figure out the equation of the lines.) Use an interval for x that just includes the two points and only a little more on each side. This line is called the "approximate tangent" to the curve at the point.

(a) The function is given by the expression $|x - 15| + 83$ and the two points are [14.5, 83.5] and [16, 84].
(b) The function is given by the following expression.

$$U(t) = \begin{cases} t^2 & \text{if } t \le 1 \\ t & \text{if } 1 < t \end{cases}$$

and the two points are [0.9, 0.81], [1.1, 1.1].

SUBMIT

Your graph for each function and tangent line, clearly labelled, with the equation of the approximate tangent line written on each.

3. Now we will do an analytic version of the approximate tangent to the curve at a point. Write a func at that will accept a func say f that represents a function whose domain and range are real numbers, a point a in the domain of the function represented by f and a small positive number e. The func at is to return a func which represents a function f_1 given by the expression

$$f_1(x) = \frac{f(a + e) - f(e)}{e} (h-a) + f(a)$$

Run your func for each of the following functions and point a with an appropriate choice of e in each case. Use **Maple** to sketch a graph which has both a sketch of the original function f and a sketch of the function f_1 which your func at produces. Choose reasonable intervals (not too small) to display clearly what is going on.

(a) k(y) = 8.5y³–1.8y² + 0.026y + 3.2, a = 3.5,
(b) f(z)= |2z–1|+1,   a=1/2.

SUBMIT

(a) A copy of your func at.
(b) Your graphs.

4. In this problem we formalize the work you have done in defining a function whose process is obtained from the approximate derivative ad(f,a) with f fixed and a varying. Write a func ada that will accept a func, say f, that represents a function whose domain and range are real numbers and a small positive number e. The func ada is to return a func which represents the function Df given by the expression

$$Df(a) = \frac{f(a + e) - f(e)}{e}$$

Run your func ada for each of the following functions with an appropriate choice of e in each case. Use **Maple** to sketch a graph which has both a sketch of the original function f and a sketch of the function Df which your func ada produces. Choose reasonable intervals to display clearly what is going on.

(a) g(r) = 5r³ + 2r² – r + 4.
(b) f(z) = |2z – 1| + 1.

SUBMIT

(a) A copy of your func, ada.

(b) Your graphs.

5. In the following list of functions, there are some pairs f, g in which g is a reasonable approximation to the function Df of the previous problem. Your task is to discover as many of these pairs as you can by using **Maple** to sketch each function f as well as Df and then comparing the

graph of Df with the graphs of the other functions in the list.

(a) $a(b) = -\dfrac{1}{b}$

(b) $c(d) = \dfrac{\sin(d)}{1 - (\sin(d))^2}$

(c) $e(f) = -\dfrac{1}{f} + 0.01 * \sec(f)$

(d) $g(h) = -\dfrac{1}{h} + 0.01 * \sin(h)$

(e) $i(j) = \ln\left(\dfrac{1}{j}\right)$, assume $j > 0$

(f) $k(m) = \dfrac{1}{m^2}$

(g) $n(p) = p - p \ln(p)$, assume $p > 0$

(h) $q(r) = \sec(r)$

SUBMIT

For each pair f and g for which you think g is a reasonable approximation to Df, submit one sheet containing graphs of all three functions f, g and Df. Make sure you connect points so that your graphs are solid curves.

6. Use graph to sketch the function h from Lab Assignment 1, #4. Looking at the formula for the function, how many parts would you expect the graph to have? How many parts does the graph appear to have? Explain.

SUBMIT

Your responses to the questions and your explanation.

Comments

In Homework Assignment 5 of MA 161a, two parts of the assignment focused on the use of the **Maple** command diff to discover the usual derivative rules for sums, differences, constant multiples, products, quotients and compositions of functions. In the original assignment students were referred to a list of homework problems in a standard text to try and discover the derivative rules. This assignment was improved the second time it was used in Fall 1989 by telling the students to use the various combinations of the functions represented by the expressions : x^n , $\sin(x)$ and $\cos(x)$ to discover the derivative rules. Many students who had not had calculus before were able to successfully discover the basic rules with the exception of the chain

rule. However, some of these students did discover the chain rule.

Lab Assignment 6 of MA 161a

Summary

In Lab Assignment 6 of MA 161a, our students' investigations included: further investigation of derivatives using the **Maple** command diff; the use of an **ISETL** func print1 which we provided them to investigate the values of derivatives and approximate derivatives using their **ISETL** func ada; and various properties of the graphs of functions, including zeros of a function and its derivative, local extrema, the signs of function and derivative values, intervals where a function is increasing and decreasing and intervals where a derivative function is increasing and decreasing, and concavity. As usual these investigations are done prior to discussions of the related concepts in class.

We should note that student modifications of the func print1 were subsequently used in MA 161a, MA 162a and MA 261a for student investigations involving limiting values of various calculus concepts, including approximations of values of functions using linear and quadratic approximations (i.e., first and second degree Taylor polynomials, the values of definite integrals, the the Fundamental Theorem of Calculus and indeterminate forms.

MA161a Lab Assignment 6

1. For each of the following functions, write down what you think its derivative is. Then use the **Maple** operation diff to compute its derivative. *Note: You will not be graded on your first answer, only the one you get from* **Maple**. *The reason we need your attempt is to determine how much work you need to do on learning to compute derivatives.*

(a) The functions given by the expressions in the text, Page 134, #1, 3, 4, 8, 9, and 11.

(b) The functions given by the expressions in the text, Page 184, #2, 9, 10, 17, 18.

(c) $f(\phi) = \tan^2(3\phi)$

(d) $p(\theta) = \cos(\tan(3\theta^2 - 2))$

(e) $h(\gamma) = \dfrac{1 + \sec(\gamma^3 + \ln(\gamma))}{\gamma + \exp(\gamma)}$

(f) $\ln(\sin(\sqrt{u} + 1)$

SUBMIT

A list of two answers for each function - yours and **Maple**'s.

2. In the **ISETL** folder is a file containing the func print1 which accepts 3 functions, f, g, and h, the endpoints a, b of an interval contained in the domain of all three functions, a positive integer n and a filename fil. The func places in the file fil four columns of data. The first column gives the values of the division points of the partition of the interval from a to b into n equal subdivisions and the other columns give the corresponding values of the three functions at these points.

Place in a file the func ada that you did for Lab Assignment 5, Problem #4 to compute the approximate derivative, Df, given f. In the **ISETL** sessions for this problem, you will have to !include both print1l and ada. Print a hard copy of print1 to hand in and study it briefly.

For each of the following functions, write a func to represent it and one to represent its derivative (which you can find by hand or by using **Maple**). Apply your func ada with your own choice of e to obtain an approximate derivative (remember that ada(f,e) is a function). Give these three functions in that order to print1 along with the given endpoints, the value of n = 50 and a filename of your choice. That is, use the syntax for print1 to create your file. The correct syntax is print1 (f,g,ada(f,e),a,b,50,"fil");.

Produce one file and a one page hardcopy output for each function. In your final production, select a value for e that will make the last two columns agree up to three significant figures.

(a) The function given by the expression : x^3, and $a = 0, b = 2$.

(b) The function given by the expression $x^{1/3}$, and $a = 0.01, b = 0.2$.

(c) The function given by the expression $\sin^3(2x-\pi/3)$, and $a = -2\pi, x = 2\pi$. Note that in **ISETL** you may use $\pi = 3.14159$ and in **Maple** use $\pi = $ Pi.

SUBMIT

(a) A hard copy of the file containing print1.
(b) A copy of each func and its derivative func.

(c) Your output for each function, appropriately labelled. Indicate on the page the value that you used for e.

3. For each of the following functions, use **ISETL** and **Maple** to sketch a graph of the function and its derivative (on the same graph) in the intervals indicated. There are several ways in which you can compute its derivative. If the function has a name like mys1 then it is an **ISETL** mystery function represented by an **ISETL** func that you can call and use in your approximate derivative func ada from Lab Assignment 5 to get a graph of the approximate derivative (instead of the derivative).

(a) The mystery function mys1 in **ISETL** on the interval for x from -1.1 to 1.05 and for y from -1 to 1.

(b) The function is given by the expression

$$e^{0.063x}(x^3 - 2.14x^2 - 3.647x + 4.001)$$

and on the interval for x from -3 to 3.

(c) The function is given by the expression

$$2\sin(x) - \cos(3x)$$

and on interval for x from -0.1 to 7.

(d) The mystery function mys2 in **ISETL** on the interval for x from -1 to 4.

3. For each pair of a function and its derivative (which you can find by hand or by using **Maple**) or its approximate derivative, i.e., answer the following questions (approximately) about both the function and its derivative by inspecting the graph. In parts (a)-(e) answer each question by giving the appropriate values (approximations) of the domain variable.

(a) Where are the zeros of the function?

(b) Where are the zeros of the derivative?

(c) Where does the function have a local maximum value? a local minimum value?

(d) Where does the derivative have a local maximum value? a local minimum value?

(e) On which interval(s) is the function positive?

negative?

(f) On which interval(s) is the derivative positive? negative?

(g) On which interval(s) is the function increasing? decreasing?

(h) On which interval(s) is the derivative increasing? decreasing?

(i) On which interval(s) is the function concave up? concave down?

(j) On which interval(s) is the derivative concave up? concave down?

SUBMIT

(a) The answers to all of these questions for both of the function and its derivative (or approximate derivative) arranged neatly in a tabular form.

(b) Any interesting observations you can make from your answers (e.g., how does the sign of the derivative relate to the behavior of a function, the maximum of the function, etc.) together with guesses as to the explanation of your observations.

Lab Assignment 7 of MA161a

Summary

In this assignment, our students investigated higher derivatives and the computation of Taylor polynomials using **Maple**. However the "meat" of the Lab assignment was an investigation of implicit functions and the use of the **ISETL** func graphr which plots implicit functions represented by "approximate equations" in two variables. This investigation of implicit functions was very hard for students and an attempt was made to make the assignment easier for our students to "digest" by reordering some parts of the assignment and making it easier to read, although the essence of the assignment remained intact from the its first use in MA161a.

This lab combined the use of the **ISETL** graphics tool called graphr which we provide for our students to construct graphs of implicitly defined functions. The syntax of graphr is

graphr(eq,I,J,n, "filename");

where eq is is a func which approximately implements an equation in two variables, I and J are pairs of numbers representing endpoints of intervals, n is a positive integer giving the number of points to be plotted and filename is a name for a file.

The result of graphr is a file containing a set of pairs of points for which the equation is (approximately satisfied). The students leave **ISETL** and apply **Maple**'s graphing facility to produce a graph.

For example, given an equation like

$$x^3 + y^3 - 4.5xy = 0$$

the following func can be used with graphr to produce a reasonable approximate graph of the equation.

```
eq := func(x,y);
    e := 0.001;
    if x = 0 and y = 0 then return true;
      return (x**3 + y**3 - 4.5*x*y)/(abs(x)+abs(y)) < e;
end;
```

This is difficult, if not impossible, for most of our students, but for interesting reasons. Most of them at first simply use equality and the result is that the graph that they eventually produce is empty or almost empty. Working with small groups of students in the lab gives them the opportunity to begin to think about and understand about relative error, and some of the pitfalls of approximation. They also benefit from discovering that none of this difficulty occurs with an equation like $x^2 + y^2 = 1$.

The next thing that the student confronts is the choice of I and J which must represent intervals large enough so that their cartesian product is a box that contains the entire graph. The students begin to think about boxes made by products of intervals and they need to think seriously about approximation in order to make sure that their box is large enough.

Finally the students must deal with the choice of n and varying the error bound e. If n is large there will be many points, but it takes a long time. The value of e is more interesting. If n or e is too small, they will not get enough points to see what the graph looks like. If e is too large then the tool produces points that the students quickly decide do not belong on the graph. They play around with these quantities and eventually come up with a fair representation of the graph of the equation.

The next step is that the students are given the following code to enter into the computer. (The procedure bis is a func they have used earlier. It uses a bisection method to approximate a zero).

```
implicit := func(rel, I0, J0);
    return func (x);
         if not ((I0(I) <= x) and (x <= I0(2))) then return;
         end;
         f := func(y);
              return rel (x ,y); end;
         return bis(f, J0(I), J0(2), 0.0001);
         end;
    end;
```

The reason that we give them the code rather than just the tool (as in graph and graphr) is that they can go through it and understand what it is doing which, in this case, we feel has mathematical value. The code is discussed in class during the period that working with implicit is a lab assignment for them.

It is interesting to observe the variety of their choices (since one can put the box in many different places) which suggests that they are not locked into imitative behavior. They also benefit from realizing that not every reasonable way of cutting down a curve to get a function can be done by the local method of selecting a box. This opens their thinking to the possibility of discussing some global ideas from algebraic geometry. They struggle painfully but bravely with the meaning of implicit. It is a long time before they can look at implicit(rel, I0, J0) and realize that it is a function and that

$$implicit(rel, I0, J0)(3)$$

for example, stands for the value of this function at 3. They have a lot of trouble with the subsidiary function that implicit uses, and they must really understand the idea of a zero of a function, going back and forth between thinking of the function as a total entity and thinking about evaluating the function at several points. And of course they are putting together the use of several different computer systems which requires working carefully and not making errors, all of the time understanding what they are doing.

It is important to note here that these difficulties are part of the mathematics and so struggling to overcome them is a way (a very effective way, in our opinion) of learning mathematics. The students do get through it all and we feel

that it is one of the early triumphs of our approach. After going through implicit they can talk about implicitly defined functions and this helps later with things like related rates problems. Moreover, the students show on examinations that they were able to think about things like selecting intervals restricting the values of x and y so as to ensure that the function defined implicitly by a certain equation, and displayed on a graph, had certain pre-specified properties.

Also, in this assignment we introduced our first use of **Maple**'s *student package* which was designed to help students understand various topics in calculus. In Part 6 of this Lab, students used the **Maple** derivative operator D in **Maple**'s *student package* to help them find derivatives of implicitly defined functions.

Our students used the **ISETL** func graph together with **Maple**'s graphing facility in this and many other assignments. We now include the description of how to use graph that we gave to our students.

USING ISETL TO MAKE GRAPHS

The graphing facility that we are using in this course is a part of the **Maple** system. Some functions, however, are not easily expressed in **Maple**, but can easily be represented in **ISETL**. In order to draw graphs of functions that are represented in **ISETL**, it is necessary to use the two systems together. There are four steps in doing this. Read the following description as you perform the tasks on the indicated example.

1. Enter **ISETL** by double-clicking on the **ISETL** folder, double-click on the **ISETL** icon and then press return. Write a func to represent the function you wish to graph. For example, consider the function represented by the following code.

```
f : = func (y);
    if is_number(y) then if y < -3 then return -(y+4)**2 - 2;
    elseif y < 1 then return (y+2)**2 - 4;
    elseif y < 2 then return y+4;
    else return -2*y+5;
    end;
    end;
end;
```

2. Execute the following **ISETL** statement.

$$graph (f, a, b, n, "filename");$$

Here f is the variable whose value is the func you just defined, a and b are left and right endpoints of the domain interval for this function, n is the number of points (distributed uniformly) on the interval [a, b] for which you wish to sketch $(x, f(x))$, and f ilename is a string which is the name of a file in which you wish to hold the data for transferring to **Maple**. (DON'T FORGET TO ENCLOSE FILENAME IN DOUBLE QUOTES. ALSO, DON'T USE THE NAME OF ANY EXISTING FILE THAT YOU WISH TO KEEP). Thus, for example, you might execute,

```
graph(f , –10 , 10 , 100, "temp");
```

ISETL will return the message "Computation of data completed"

3. Leave **ISETL** by selecting quit from the File menu and be sure to put your file in the **Maple** folder.

4. Now enter **Maple** and execute the following two commands. (You will see many numbers on the screen after the first command is executed. Can you figure out what they are?)

```
read 'temp';
sketch(");
```

This will give you the graph of your function, provided you remembered to put your data file in the **Maple** folder. If you want to save it, select **Save ...** from the **File** menu.

MA161a Lab Assignment 7

1. Higher order derivatives (see Section 3.6 of your text) can be computed easily using **Maple**. If f is an expression in **Maple** which represents a function of a single variable x, then the following two **Maple** statements will give the second and third derivatives respectively (see your **Maple** flashcard for the syntax of **diff**).

```
diff (f, x, x);
diff (f, x, x, x);
```

Use **Maple** to find the first six derivatives of the functions represented by the following expressions. To save time, type the diff commands on the same line. For example, after typing a function f in **Maple** as an expression, you can find the first, second and third derivatives of f by typing the following on one line:

```
diff(f, x);   diff(f, x, x);   diff(f, x, x, x);
```

and then pressing **return**.

(a) The function given by the expression
$$\cos(2x) .$$

(b) The function given by the expression
$$\frac{1}{1 + x} .$$

(c) The function given by the expression
$$\frac{\cos(2x)}{1 + x}$$

SUBMIT

A hard copy of your **Maple** session.

2. It is possible to use **Maple** to compute Taylor polynomials (see Section 3.6 of your text.) If f is an expression in **Maple** which represents a function of a single variable x, the following **Maple** statement will compute the Taylor polynomial of the function, about the point c, up to order n.

```
taylorpol(f,c,n);
```

Use **Maple** to compute the following Taylor polynomials up to order 6, about the origin.

(a) The function given by the expression
$$\cos(2x)$$

(b) The function given by the expression
$$\frac{1}{1 + x}$$

(c) The function given by the expression
$$\frac{\cos(2x)}{1 + x} = \frac{1}{1 + x} \cos(2x)$$

SUBMIT

(a) A hard copy of your **Maple** session.

(b) An answer to the question: Why does the first one only have four terms?

(c) An answer to the question: Is there any way to get the third one directly from the first two?

The purpose of this problem is to write three funcs that approximately implement each of the following equations in two variables.

$$x^2 + y^2 = 1$$

$$|x|^{2/5} + |y|^{2/3} = 1$$

$$x^3 + y^3 = 4.5xy$$

Each of your funcs is to accept two real numbers x and y and return true or false indicating whether the pair is a reasonable approximation to a solution of the equation. (Hint: First rewrite each equation so that all the terms are on one side of the equation.)

There are two issues which you are to keep in mind at this point.

(a) Because most computer operations (like 5th root) are only approximations, you should not have your func base its decision on exact equality, but equality up to a particular tolerance, which we shall call tol. You must decide on how to use tol and choose a value for it.

(b) Your func should have exactly two parameters, for the two variables—no more and no less. Use an external constant for tol, i.e., assign a value to tol before you use your func. Hence, tol can be changed externally as needed. An external assignment of tol will be useful in parts 4 and 5 below.

SUBMIT

(a) A hard copy of the final version of your three funcs.

(b) A record of your **ISETL** session in which you apply your funcs for several test values. Make sure that you get some true results for each equation.

4. There is an **ISETL** operation which will draw the graph of an equation such as the ones you worked with in the previous problem. It has the following syntax (similar to that of graph),

    graphr(eq, I, J, n, "filename");

where eq is a func that approximately implements an equation (like the ones in the previous problem), I is a tuple of two numbers representing an interval on the x-axis (like [1,3], etc.), J is a tuple of two numbers representing an interval on the y-axis, n is positive integer, and filename is a string which represents a file name.

Here is how graphr works. First it partitions the intervals I and J each into n equal subintervals, and this gives a partition of the rectangle that I and J form into n^2 equal subrectangles. Next it runs through all pairs [x ,y] of points on the plane corresponding to the vertices of the subrectangles. For each point it applies the func, eq to decide if the point is a reasonable approximation to a solution of the equation. Finally, it places all points which eq decided were reasonable approximations into the file, filename. Note: if you use **ISETL**, then there is a file named graphr in the **ISETL** folder and you must ! include graphr to use it.

The result of all this is a file with data in a format that **Maple** can use to construct a graph. To graph the data file, named filename, convert the file to a readable file using Edit6.0 and put the file in the **Maple** folder. Then (as on the sheet entitled USING **ISETL** TO MAKE GRAPHS) type the following two lines in **Maple**:

    read 'filename';
    sketch(");

Do all of this for the *second* of the three equations that you worked with in the preceding problem. Again there are a number of issues that you will be concerned with.

(a) Your choice of tol in writing your original equation func will affect the graph in a number of interesting ways. For example, the larger the value of tol, the more points you will get for your (approximate) graph. You will learn a great deal by trying to figure out what these effects are and then finally choosing reasonable values for tol and n to obtain an approximate graph of the equation. For example, as you make tol smaller, you will want to make n larger. For your first try, take tol = 0.1 and n = 50.

(b) Your choice of the intervals I and J is also critical. You must understand enough about the equation so as to make a choice of a rectangle that will include the entire graph, but not be so large that you will only get a small number of points.

SUBMIT

A hard copy of your final (approximate) graph of the second equation from part 3.

5. In this problem you will use **ISETL** to construct and graph an implicitly defined function from an equation. The operation consists of the following steps.

(a) Take the *third* equation in part 3 above and use the method of the previous problem to sketch its (approximate) graph. Try I = [−2,3] and J = [−2,3], then adjust I and J as desired.

Keep the following in mind: A particular choice of tol may have very different effects if *x* and *y* are very small, than if they are very large. Try to figure out a way to take care of this automatically by modifying your func which approximately implements the equation. At the same time, make sure your func does not introduce a potential division by 0.

(b) In the **ISETL** folder there is a file containing the func implicit, which you will have to ! include to use it. The func implicit is as follows:

```
implicit := func(rel,I0,J0);
    return func (x);
    local f;
    if not ((I0(l) <= x) and (x<= I0(2)))
        then return; end;
f = func (y);
    return rel (x ,y);
    end;
    return bis(f, J0(l), J0(2), 0.001);
    end;
end;
```

Study it briefly. The func implicit accepts a func rel which represents a function of two variables, *x* and *y* that implements the left hand side of the equation that you took in part (a) (i.e., the expression that you got by setting everything equal to 0), and two tuples I0 and J0 (like I and J) that are the endpoints of intervals that restrict *x* and *y* respectively. The intervals I0 and J0 are assumed to be chosen so that in these intervals, for each 3: in I0 there is exactly one pair [x,y] which has rel return the value 0. The choice of I0, J0 is to be made by inspecting the graph you made in (a).

Note that implicit uses bis (from Homework Assignment 4), so you must ! include bis.

(c) Use the method described in **USING ISETL TO MAKE GRAPHS**, like you did in part 4 of this lab, to sketch the graph of the function represented by the func which implicit produces.

Apply this operation to the third of the three equations in part 3. Look at its graph to make choices of I0 and J0. Sketch the graph of the resulting implicitly defined function.

SUBMIT

(a) Your (approximate) graph of the third equation in part 3.
(b) A hard copy of the graph of the implicitly defined function you obtained using implicit and the operation described above.

6. In this problem you will use the **Maple** *student package* to differentiate functions defined implicitly by an equation in two variables.

The **Maple** *student package* can be called for use in **Maple** by typing the command **with(student);**. The result of typing this command will be a tuple containing a list of the topics covered in the package (see the attached list of topics and information about their use). In particular, the student package contains the derivative operator D which can be used to to find $\frac{dy}{dx}$ for *implicit* functions defined by equations in two variables x and y, like those in part 3. The operator D takes the derivative of what follows it, i.e., $D(x)$ and $D(y)$ represent the derivatives of x and y, respectively. If you want to find the derivative of y with respect to 3:, i.e., you want $\frac{dy}{dx}$, then $D(x) = 1$ (since the derivative of x with respect to x is 1) and $D(y) = \frac{dy}{dx}$.

Consider the equation

$$x^3 + y^3 = 4.5xy$$

We can think of y as a function of x (see your graph of the equation for part 5) by thinking of "pieces" of the curve that represent a function (and forgetting about the other pieces for the moment). Now type in **Maple** the following lines (be sure to press enter after each semicolon).

```
with(student);
 x**3+y**3 =4.5*x*y;
D(");
subs(D(x) = 1, ");
solve(", D(y));
```

Note that the third line above finds the derivative of each side of the equation, hence each term of the equation. Answer the following questions:

Can you explain why *two terms* appear on the right hand side after the operator D is applied to 4.5xy?

Can you explain why the term $3y^2D(y)$ appears after D is applied to y^3?

SUBMIT

    (a) A copy of the derivative $y' = \frac{dy}{dx}$ of the implicit functions defined by the equation

$$x^3 + y^3 = 4.5xy$$

    (b) Your answers to the two questions asked at the end of part 6.

Comments

In Homework 7, students investigated graphs of Taylor polynomials, with increasing degrees, of a given function as compared to the graph of the function in order to "visualize" approximations. They also investigated the intermediate value theorem via various graphs.

Lab Assignment 8 of MA161a

Summary

In Lab and Homework Assignments 8, the main focus was on an investigation of Newton's Method. The func newt was derived in class using the Socratic method. Students compared Newton's Method to the Bisection Method (which was also derived in class using the Socratic Method and used in earlier lab and homework assignments). No significant changes were made in this assignment from the first time it was used.

MA161a Lab Assignment 8

1. In the **ISETL** folder there is a file called newt which contains the func newt. The func newt is as follows:

```
newt := func(f,fp,x0,eps);
  local xl;
  if f(x0) = 0 then return x0; end;
  if fp(x0) = 0 then return "Method fails"; end;
  xl := x0 - f(x0)/fp(x0);
  if abs (xl-x0) < eps then return xl;
  else return next (f ,fp ,xl ,eps);
  end;
end;
```

Study it briefly. You will have to !include the file newt for this assignment. Newt takes four parameters f, fp, x0 and eps, where f is a func which represents a function, fp is a func which represents its derivative or an approximation to it, x0 is a number in the domain of f expected to be close to a zero of f and eps is a small positive number. The result of newt is an approximation to a zero of f which is obtained by running Newton's method until two successive approximations differ in absolute value by less than eps.

Run newt on the following functions with the given starting points and eps = 10^{-10} for tolerance, which can be entered in **ISETL** as 1.0e–010 (do not put a * between 1.0 and e) or 0.0000000001. For the representation of the function, you must write a func. Similarly, for the representation of the derivative, which you can either compute by hand or by using **Maple**, you must also write the appropriate derivative func.

Evaluate the function at your final estimate for the zero.

You might find it interesting to insert a statement like

        print x0;

in newt after the local statement. This will give you a picture of how the convergence is (or is not) going.

    (a) The function given by the expression

$$e - x + \frac{x^2}{2} - \frac{x^3}{6}$$

and x0 = 4. You can use exp(1) for the value of e.

    (b) The function given by the expression

$$\sin(2x) - \tan(x)$$

and x0 = – 9.5

SUBMIT

(a) A copy of your func.

(b) Your approximations for the zeros in each case.

(c) The value of the function at your approximation to the zero in each case.

2. Run your func newt for the function given by the following expression

with initial point x0 = 2. Try to figure out what caused the result you obtained. If necessary, use **Maple** to sketch the graph of the function.

SUBMIT
$$x\,e^{-x}$$
Your explanation of what happened.

3. Find approximations to all zeros of the following functions. Sketch the graph or use any other technique to roughly locate the zeros and then use newt to get your approximations. You may have to use your func ada from Lab Assignment 5 to get the approximate derivative *Df* (instead of the derivative) to use in newt.

(a) The function given by the expression

$$e^x - 3.62x^3 - 4.13x^2 + 13.24x + 13.11.$$

Note that for e^x you must use exp(x) in **ISETL**.

(b) The function whose domain is restricted to the interval from $-\pi$ to π and is given by

$$\sin(x) - \cos(x)$$

(c) The **ISETL** mystery function, mys1.

SUBMIT

(a) Your approximations for the zeros in each case.

(b) A description of your method in each case.

4. For each of the following functions, find all local maxima and minima on the given intervals by approximating the zeros of the derivatives and checking the graphs or nearby points to decide if they are a maximum or a minimum or neither.

(a) The function given by the expression

$$e^x - 3.62x^3 - 4.13x^2 + 13.24x + 13.11$$

on the interval from -3 to 7.5.

(b) The function whose domain is restricted to the interval from 0 to 2π and is given by

$$\sin(x) - \cos(x)$$

(c) The **ISETL** mystery function, mys1 on the interval from -1.1 to 1.1.

SUBMIT

(a) Your approximations for the maxima and minima in each case.

(b) A description of your method in each case.

MA161a Homework Assignment 8

1. The **ISETL** mystery function mys3 has a single zero and it lies between -6 and 6. Its derivative is represented by the **ISETL** mystery function mys3p so it is possible to use newt to approximate this zero to five correct decimal places. There is a difficulty however in that most choices for a starting point will lead to an infinite loop. (After quite a long run this builds up to a "stack overflow error.") Nevertheless there are choices that will lead to a very rapid convergence.

Explain why Newton's method leads to an infinite loop with most choices of a starting point. Find a starting point that will lead to convergence and approximate the zero.

SUBMIT

(a) Your explanation for the infinite loop.

(b) A choice of starting point for which the method converges, and the result.

(c) The reason(s) why you made your choice.

2. The purpose of this problem is to compare the performance of newt and bis in finding approximations to zeroes of functions. A copy of bis is attached for your reference. Of course, you will have to ! include bis to use it. In each of the

following cases use newt with the given starting value and also use bis for the same function with the given two endpoints at which the sign of the function is different. In all cases, use the tolerance eps = 10^{-10}

Time newt and bis in each case and also consider how the answers compare. Explain everything that happens as well as you can.

(a) The function given by the expression

$$e - x + \frac{x^2}{2} - \frac{x^3}{6}$$

with x0 = 4 for newt and the endpoints 2 and 4 for bis.

(b) The function given by the expression

$$\sin(2x) - \tan(x)$$

with x0 = 1.5 for newt and the endpoints 0.5 and 1.5 for bis.

(c) The function given by the expression

$$x^5 - x^3 + 2$$

with x0 = 4 for newt and the endpoints −4 and 4 for bis.

SUBMIT

(a) All approximations that you obtain.

(b) Your explanations.

Lab Assignment 9 of MA161a

Summary

In Lab Assignment 9 of MA 161a, our students investigated various "word" problem situations. They constructed **ISETL** funcs for the situations and then used **ISETL** and **Maple** to investigate extreme values of functions represented by **ISETL** funcs. No significant changes were made in this assignment from the first time it was used.

MA161a Lab Assignment 9

1. In each of the following, a situation is described and two variables, one "independent" and one "dependent", are specified. Your task in each situation is to do the following things.

(a) Write an **ISETL** func that accepts the independent variable as a parameter and, checking the domain which is determined by the the the realities of the situation and using appropriate formulas, returns an expression that gives the resulting value of the dependent variable.

(b) Use **ISETL** graph and **Maple** to sketch the graph of the function represented by this func using the restrictions on the independent variable determined by the problem.

(c) Read off approximate values of the independent variable at which the maximum and minimum values of the function occur.

(d) Read off approximate values for the maximum and minimum values of the function.

(e) Use calculus to find the actual values of the independent variable at which the maximum and minimum occur.

(f) Find the actual maximum and minimum values of the function.

The situations

(a) A man is in a boat, 3000 meters out from his house on the shore on a line perpendicular to the shore. A lighthouse is right at the shore 2000 meters down shore from his position. He can row 2 km/hr and walk 4 km/hr. He intends to row in a straight line to a point on the shore between his house and the lighthouse, and then walk to the light house. The independent variable is the amount of time he walks. The dependent variable is the total time it takes him.

(b) Same as first problem, except that the independent variable is the amount of time he rows and the dependent variable is the total time it takes him.

(c) A conical tank which is 12 meters across its circular top and 15 meters deep is partially filled with water. The independent variable is the volume of the water in the tank and the dependent variable is how far down you go before hitting water.

SUBMIT

For each situation, a single sheet which contains the graph of the function represented by your func, the function and a copy of your func which implements the function, and all values obtained in steps c, d, e and f above.

2. In each of the following, there is a situation which specifies a relationship between two or more variables. Write a func which takes the two variables as parameters and returns the value of the expression which is zero precisely when the relation is satisfied. For example, if the situation were,

the edge of a cube and its volume

your func might read (with an explanatory comment),

```
$ S is the length of an edge, v is the volume
f := func(s,v);
return v =s**3;
end;
```

The situations

(a) The radius of a circle and its area.

(b) Referring to page 245, # 7 of the text, the amount of water and the depth of the water.

(c) Referring to page 245, # 11 of the text, the length of her shadow and her distance from the lamp post .

(d) Referring to page 245, # 15 of the text, you decide.

(e) Referring to page 245, # 19 of the text, you decide.

Run each of your funcs a few times to see that they work.

SUBMIT

Copies of your funcs.

Computer Laboratory Experiments:
Introducing Applications into the Calculus Curriculum*

Michael J. Kallaher and Michael E. Moody
Washington State University

Introduction

The computer revolution has significantly altered the methods and scope of applied mathematics and has transformed the theory and methods of such fields as analysis and operations research. Furthermore, the widespread and inexpensive availability of powerful computer resources has quantified many formerly qualitative disciplines including business, psychology, and sociology. It has also significantly broadened the class of accessible problems in the traditionally quantitative fields of chemistry, engineering, and physics.

Recognizing the above, five years ago Washington State University established a Mathematics Computer Laboratory and started the development of a sequence of computer experiments for mathematics classes. This development rests upon six basic observations derived from our experience.

First, the effective and efficient use of computer hardware and software requires a new synthesis of curriculum, teaching methods, and instructional materials. The computational and graphical power now available, besides eliminating much drudgery, also offers the ability to explore concepts and applications in creative ways.

Second, computer experiments are designed to be performed out-of-class by the student. This avoids scheduling problems for the laboratory, and the student learns how to set up a mathematical problem and how to interpret computer output. Our calculus experiments are in workbook form, while those for later courses (for example, differential equations, linear algebra) are in an expanded word problem format and require more direction and insight from the student.

Third, experiments are independent of particular computer hardware and software, and they minimize the need for staff help. This simplifies the operation of the laboratory, and the normal turnover of laboratory staff (mathematics graduate students and seniors) is less worrisome.

Fourth, experiments involve real mathematics and realistic or actual applications. The computer is often a critical part of the problem-solving process, and thus the experiments emphasize problem-solving skills, not mechanical manipulation. Laboratory experiments help students exercise critical thinking about means for solving problems and the significance of answers.

Fifth, laboratory experiments are tailored to discipline: biological sciences, business, engineering and physical sciences, social sciences. Students realize more easily the importance of mathematics when they see it applied to problems in their own discipline.

Sixth, the experiments emphasize the mathematical aspects of the problem being solved, demand that the student do mathematics, and require that the student interpret the mathematical answer in the context of the original real problem.

The Present Environment

The Mathematics Department at Washington State operates a Mathematics Computer Laboratory containing 7 IBM XT clones and 13 Apple IIe computers and a wide assortment of software, which is maintained by a computer systems manager. The facility, supported by Departmental funds, is directed by a faculty member assisted by a graduate student and staffed by qualified undergraduate mathematics majors. It is open 75 hours per week and records over 4500 student "sign-ins" per semester; approximately 90% of its use is by students doing required exercises or projects.

Significant use of the computer is now required in the three semester engineering calculus course and the introductory linear algebra courses. Each semester four to six computer based assignments are marked and comprise at least 10% of the student's grade.

*This project was supported in part by NSF Grant USE-8814131

An extensive collection of computer exercises for calculus and linear algebra have been developed in order to overcome the lack of suitable materials. Exercises stress applications of the mathematics and usually involve sufficient complexity so that the utility of the computer is evident.

Other courses using computers in instruction include numerical analysis, differential equations, advanced linear algebra, and abstract algebra. In each course two or three computer projects are required; each project involves substantial mathematical work by the student and is more open-ended with respect to the use of the computer. Some projects have formats similar to that of an industrial project: a written report is required explaining the problem, how it was solved, and interpreting the mathematical results.

Comprehensive programs for the analysis—both quantitative and qualitative—of ordinary differential equations and difference equations have been prepared and used in our courses. Accompanying exercises, manuals, and student guides have also been prepared and used.

Students successfully completing our courses receive training in using mathematics and computers together to solve realistic problems in various disciplines. Graphical exercises give them better intuition for mathematical concepts and results, and they learn to use such software as MATLAB, GAUSS, EISPACK, LINPACK, and parts of IMSL. The laboratory experiments also encourage students to work and learn together.

Experience has taught us several interesting facts concerning the operation of the Laboratory. First, staff must be able to answer *mathematical* questions as well as questions concerning computer hardware and software; thus, all laboratory assistants are mathematics students. Second, our laboratory experiments require frequent printing of graphs, and thus it is imperative to have substantial printing devices to avoid time-consuming queues. Third, computer experiments must be nontrivial, go beyond the lecture, and relate to real world situations. Otherwise, students regard them as just "busy-work" assignments and do not appreciate the learning experience.

Sample Computer Experiments for Calculus

The following three experiments have been successfully used in the engineering calculus course. The first, *Graph-*

ing in 3-d, is a WARMUP. A WARMUP is a straightforward exercise for students to familiarize themselves with a pertinent software program—in this case, the 3-d graphing program—and is usually not graded. This type of experiment is also useful for helping the student visualize mathematical concepts.

The second experiment, *Newton's Law of Cooling,* is a sample of a more involved experiment. As is typical, there is an introductory discussion of the physical phenomenon and its representation as a mathematical statement. The student is given the appropriate mathematical model and is then given the solution[1], which must be verified. The main point is to direct the student in an exploration of the solution's behavior, in realistic contexts, and induce the student to think about how the mathematics relates to the physical phenomenon. Graphical analysis is frequently used to build intuition and geometrical insight about the mathematical objects being manipulated.

The third experiment, *Partial Derivatives*, is an example of an experiment requiring more initiative from the student. Note again that the student is being asked to interpret graphical output to determine the properties of a function.

These experiments demonstrate well the common elements. Each is preceded by a checklist that indicates required computer programs, prerequisite mathematical topics, and what must be turned in to the instructor. The student is then guided by and records his/her effort on a worksheet; this greatly facilitates independent work, minimizes questions, is easy to grade, and insures that most students are successful with reasonable economy of computer laboratory time. In addition to the worksheet and annotated graphical output, students are required to submit a one-page review of the experiment wherein they summarize their findings and explain the importance of the computer to the experiment.

Graphing in 3-d

As you learned in calculus class, we may think of the function $z = f(x, y)$ as representing a surface in three dimensions, standing above the x-y plane. It is difficult to study the graphs of these surfaces, since we are restricted to a two-dimensional representation (on paper and on your computer screen), and most of us are not artists skilled at drawing. The computer can be a great help in the study of surfaces, since they can be quickly drawn and manipulated

---

[1]Often realistic applications involve differential equations; as long as the solution is presented for exploration, students who have not yet discussed differential equations appear not to be bothered.

on the screen to see them from different points of view. We can study their shape and see interesting properties. Using the 3-D GRAPHING program of your software, draw graphs of the following surfaces. Print a hard copy if possible.

1. $z = \cos(x^2 + y^2)$, $|x|, |y| < 2.5$

2. $z = x(x^2 + y^2)^{-3/2}$, $|x|, |y| < 2$

3. $z = 1 - x^2 - \frac{y^2}{4}$, $|x| < 1, -2 < y < 1$

4. $z = e^{-xy^2} - e^{-yx^2}$, $|x|, |y| < 1.5$

5. $z = xy(y - x)$, $|x|, |y| < 2.5$

6. $z = e^{-x^2 - y^2}$, $|x|, |y| < 1.5$

7. $z = 1 - y^2$, $0 < x < 2.5, |y| < 1$

You should experiment with various points of view and plot the surface from different viewing angles. Observe and note any interesting features: relative maxima (peaks and ridges), relative minima (valley floors), saddle points, etc.

Newton's Law of Cooling
Exploring Exponential Decay

What you need to have:

Computer program: FUNCTION GRAPHER
Computer program: ROOT FINDER
WARMUP: Graphing Exponential Functions
WARMUP: Finding Roots
EXERCISE: Newton's Law of Cooling
WORKSHEET: Newton's Law of Cooling
Calculator

What you need to know:

Derivative of exponential functions

What you need to do:

O **Step 1:** Read the EXERCISE introductory material.
O **Step 2:** Read the EXERCISE Part 1.
   Fill in Part 1 of the WORKSHEET.
O **Step 3:** Do WARMUP: Graphing Exponential Functions
O **Step 4:** Read the EXERCISE Part 2.

Fill in Part 2 of the WORKSHEET.
Use the computer to graph $T(t)$ and $T'(t)$.
   • Label axes: Time, Temp.
   • Label curves: $T(t)$ and $T'(t)$.
   • Answer questions on graph.
O **Step 5:** DO WARMUP: Finding Roots
O **Step 6:** Read the EXERCISE Part 3.
   Make the estimate in Part 3 of the WORKSHEET.
   • Use the computer to graph $T(t)$ and $T'(t)$.
   • Label axes with time-of-day and temperatures.
   • Answer the question from the graph.
   Now use the ROOT FINDER program to find t when $T(t) = 98.6$
   Finish Part 3 of the WORKSHEET by hand.

O **Step 7:** Read and do the EXERCISE Part 4.

What you need to hand in:

WORKSHEET
Graph(s) for Part 2.
Graph(s) for Part 3.
Graph(s) or calculations for Part 4.

EXERCISE 8
Newton's Law of Cooling
Exploring Exponential Decay

In this exercise you will explore how objects cool using a simple principle from physics. The physical law you will use is an approximate relationship which under a wide range of circumstances reasonably describes the process of cooling.

When the temperature difference between an object and its surroundings is not too large, the combined rate of heat transfer to or from the object by the processes of conduction, convection, and radiation is approximately proportional to the temperature difference between the object and its surroundings. Since the rate of change of temperature of an object is proportional to the rate at which it gains or loses heat, it is then true that the rate of change of temperature of the object (as measured by a thermometer) is proportional to the temperature difference with its surrounding environment. This relationship was discovered by the great

English mathematician and physicist Sir Isaac Newton, and is known in his honor as *Newton's law of cooling*. The distinction between heat and temperature is subtle. In fact, most people assume that they are equivalent. However, scientists carefully distinguish between these two concepts.

If T is the temperature of the object and T_s is the temperature of its surroundings, then the proportionality of Newton's law is stated by the equation

$$\frac{dT}{dt} = -k(T - T_S) \qquad (1)$$

where t represents time and k is a constant. In Equation (1),

$\frac{dT}{dt}$ represents the rate of change of temperature at time

t. The right hand side of the equation has the constant of proportionality k, and the difference in temperature between the object and its surroundings, $T - T_s$. The minus sign in front of k is present because the object is cooling so that T is decreasing. The specific value of the proportionality constant k depends upon the composition and physical characteristics of the particular object that is cooling down. Some materials cool much faster than others and will have a larger value of k (like a T-bone steak), whereas other things cool more slowly and have a smaller value of k (like a ceramic pot).

We assume that T_s does not change as the object cools; this is equivalent to assuming that the environment near the object constitutes a very large "heat sink", so that transfer of heat from the object will not significantly change the temperature of the surroundings. For example, when a cup of hot coffee cools it usually does not significantly increase the temperature of the room it is in. Do the following exercises:

Part 1.
If T_0 is the initial temperature of the object (when t = 0), verify by substituting into both sides of Equation 1 that the function

$$T = T_s + (T_0 - T_s)e^{-kt} \qquad (2)$$

satisfies the equation and the initial temperature condition $T_0 = T(0)$.

Part 2.
While hiking in the mountains you come upon a snow-fed spring whose temperature is $T_s = 40°F$. You have a can of soda $(k = 2\ hr^{-1})$ at a temperature of 75°F and you wish to cool it down by immersing the can in the spring water. Using the appropriate graphing utilities of your software package, plot T and T' for $T_0 = 75°$ in order to observe the cooling process for your drink. Using your graphs:

• Indicate where the soda is cooling the fastest.
• Indicate where the soda is cooling the slowest (be careful!).
• Estimate how much time it takes for the soda to reach 62.5°F.
• Justify the significant figures of your computer estimates.

Part 3.
You are now a forensic pathologist (like Quincy, of TV fame) and have been summoned to investigate a homicide. The temperature of the room at the scene of the crime has been a constant 45°F for several hours, and at midnight you measure the victim's body temperature as 60°F.

• If we assume $k = 0.5\ hr^{-1}$ for a corpse, estimate from the appropriate graph of T vs. t [using Equation 2] when the victim met his unlawful demise. Recall that the temperature of the victim at the time of the incident would normally have been 98.6°F in the absence of a fever.

• If we wish to know at what time the body had a temperature of, say, 80°F, we must solve the equation $80 = T_s + (T_0 - T_s)e^{-kt}$ for t, using $T_s = 45°, T_0 = 98.6°$, and $k = .5$. You can do this analytically using logarithms, or with the ROOT FINDER program in your software package. At what time was the temperature 80°?

• Justify the significant figures of your computer estimates.

• This application of Newton's law is one of the means that pathologists actually use to determine time of death. There are, however, limitations to how useful this method is. Using whatever means you like, explain why this method will not work well to estimate the time of death if the body has been dead for too long a time.

Part 4.
If the victim had a temperature of $T_0 = 102°$ at the time of death, how much of an error would result in the estimation if you assume that the deceased had a normal temperature when he died?

WORKSHEET 8
Newton's Law of Cooling
Exploring Exponential Decay

[Part 1] Copy the expression for T from Eq. (2):
T = _____

Now find the derivative of this expression:
T' = _____

To verify Eq. (l) find $-k(T - T_s)$ and compare to T'

To verify the initial temperature condition,
$T(0) = T_0$ evaluate T when $t = 0$:
$T(0) =$

[Part 2] Write the general formula for temperature from Eq. (2):
T =

Substitute the special values for $T_0, T_s, k: T =$

Simplify and graph T:

Find derivative and graph: T' =

Record the initial conditions: $T(0) =$ _____,
T' (0) = _____.
Mark your graphs as requested.

[Part 3] HINT: The time variable *t* we are using is actually the "time since the object began to cool"; this is simply the elapsed time since we observe the object cooling from $T = T_0$. Therefore, we can estimate the time of death by (i) graphing the appropriate T vs. *t* curve for a long enough interval of time $t \geq 0$; (ii) ascertain how many hours *t* it takes the body to reach the observed temperature; (iii) label the time axis to correspond to "clock time" and estimate the time of death.

You must also choose the scale on the vertical axis. That depends on how warm you think the victim was at the time of death:

Normal body temperature: _____

Estimate the number of hours before midnight: _____

After graphing, estimating and solving by computer, derive the answer by hand for the time since death. Show your work:

Explain the limitations to using this method to estimate time of death:

[Part 4] Calculate the time of death in the manner of Part 3 assuming that $T_0 = 102°$, and use this result to determine the error. Do you think that whether or not the victim has a temperature is an important consideration?

EXERCISE 17
Partial Derivatives
Exploring Partial Derivatives and the Shape of Surfaces

This exercise will help develop your ability to visualize surfaces in three-dimensions. It will also improve your qualitative understanding of partial derivatives of the first and second order. In particular, you will learn how the sign of the various partial derivatives reveals much about the local shape of a surface.

You will use the program of your calculus software that is used to graph surfaces. This program should enable you to define a function of two variables $z = f(x, y)$, specify an arbitrary rectangular region of the plane $x_0 \leq x \leq x_1$, $y_0 \leq y \leq y_1$ over which the function $f(x, y)$ is plotted. There are usually various plotting options enabling the user to change his viewing position of the surface. Thus, by changing the point of view, you can see the surface from different perspectives. This will be important, since features of the surface can be hidden, depending on the view point. Graph each of the following two functions:

1. $z = e^{-xy^2} - e^{-yx^2}$, $-1.25 < x, y < 1.25$

2. $z = xy(y - x)$, $-2 < x, y < 2$.

On a printed graph of the corresponding surface, indicate points where the following relationships are likely to be true:

a. $\partial z/\partial y = 0$ b. $\partial z/\partial x = 0$ c. $\partial^2 z/\partial x \partial y > 0$

d. $\partial^2 z/\partial y \partial x < 0$ e. $\partial^2 z/\partial x^2 < 0$ f. $\partial^2 z/\partial y^2 > 0$

g. $\partial^2 z/\partial x \partial y = 0$ h. $\partial^2 z/\partial x^2 > 0$ i. $\partial^2 z/\partial x^2 < 0$

Please give a brief explanation for your answers that *does not involve computation*. In other words, justify your answers by the qualitative shape of the surface. For example, if the trace of a surface on a plane parallel to the y-axis is convex upwards, then it will be true that $\partial^2 z/\partial x^2 > 0$ there. A word of caution: for some surfaces, some of these conditions may have no points satisfying them; if so, indicate this on your graph. Lastly, you may find it helpful to look at a surface from different viewing positions.

A Computer Experiment for Linear Algebra

The following computer exercise is designed for students in a senior level linear algebra course. It is designed to be solved using a high level software program like MATLAB or GAUSS. Note that the student must do some mathematics before the computer can be fully utilized.

Math 420 Computer Project 1

This project involves modeling the flow of money among the eight cities of the small island nation of Maeeka. For background read the section entitled *Stabilization of Money Flow* on pages 2-5 of the class notes on reserve in the library. The following graph shows the eight cities of Maeeka and the highways connecting them.

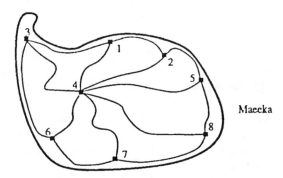

Studies have shown that money flows between two cities if and only if they are connected by a highway. Furthermore, these same studies have shown that if city i is connected to k cities (including itself) then the fraction of money in city i going to city j is given by the formula

$$p_{ij} = \begin{cases} \dfrac{1}{k} & \text{if city } i \text{ is connected to city } j \\ 0 & \text{otherwise} \end{cases}$$

for example, $p_{44} = 1/8$ and $p_{67} = 1/4$

Problem 1: Calculate the transition matrix $P = [p_{ij}]$.

For the last ten years the national government of Maeeka has been run by the Libertarian Party, whose members believe in the doctrine of noninterference by the national government. Thus, since 1977 the national government has refused to inject any money into the economy.

Problem 2: Assume that on January 1, 1977 the initial money supply for Maeeka is given by the 8-tuple
$$m = (1,1,1,1,1,1,1,1).$$
The unit of money used in this problem is millions of *disels*. A *disel* is worth approximately $0.20 in U.S. Currency.

• Calculate mP, mP^2, mP^5, mP^{10}, and mP^{20}.

• what is the limit of mP^n as n gets large?

• In how many years is this limit effectively reached?

• Which city (or cities) becomes the financial capital of Maeeka?

Let $s = (s_1, s_2, \ldots, s_8)$ be the 8-tuple obtained as follows: For each i, let s_i be the ith digit of your student ID number if that digit is nonzero; otherwise let $s_i = 1$. For example: if your student ID is 90161-629, then $s = (9,1, 1, 6,1, 6, 2, 9)$.

Problem 3: Redo Problem 2 with $m = s$. (Your s!)

Problem 4: What are some consequences—social, economical, political - of the results you obtained in Problems 2 and 3?

In September 1987 the Libertarian Party was voted out of office and replaced by the Swanduists who believe in active participation by the national government. The Swanduists are lead by Jerry Hart. Upon viewing the results of the Libertarian "benign neglect" of the past ten years, she decides to use government influence to obtain a more "equitable" distribution of money. But before acting, she calls upon an old friend, Profession Martin Krazo of Lowdie University, to do a study of the current situation and to make a recommendation.

Professor Krazo applies some basic Markov Chain theory. From this mathematical theory he knows that if P is a transition matrix with $p_{ii} > 0$ for at least one i, then

$$\lim_{n \to \infty} P^n = i^T * u$$

where i is the $1 x n$ vector all of whose components are 1's and u is a $1 x n$ vector satisfying

$$uP = u, \qquad u*i^T = 1.$$

Furthermore. Professor Krazo knows that the $n X n$ matrix

$$A = I_n - P + (i^T*u)$$

is invertible with inverse

$$A = B = I_n + \sum_{k=1}^{\infty} P^k - (i^T * u)$$

Recall that the $1 \times n$ vector q is the goal of Ms. Hart's money policy and a is the $1 \times n$ vector giving (by city) the amount of money supplied each year by the national government. Then, the $1 \times n$ vector

$$mP + \sum_{k=0}^{n} a P^k \qquad (*)$$

is the amount of money in circulation after n years. As in the notes, the vector m is the initial distribution of money at $t = 0$, the year in which Ms. Hart takes office. (We assume that no money leaves or enters Maeeka.)

In order for the vector (*) to converge to q the series $\sum_{k=0}^{\infty} a P^k$ must converge, and hence the nth term of the series must tend to 0 as n gets large.

Problem 5: Show that the inner product $a*i^T = 0$.

Problem 6: Interpret the result of Problem 5 in economic terms.
It follows from Problem 5 that the series in (*) equals

$$\sum_{k=0}^{\infty} a P^k = a B$$

Problem 7: Show that in order for (*) to converge to q the equation

$$q = (m*i^T) + aB$$

must hold.

Problem 8: Show that Problem 7 implies $q*i^T = m*i^T$

Problem 9: Using the above information show that in order to obtain her goal q Ms. Hart must choose $a = q(I_n - P)$.

Problem 10: For political reasons Ms. Hart wants the money to be evenly distributed among the eight cities of Maeeka. Assume that in 1977 (when the Libertarians took office) the initial money distribution was given in the vector **i**.

- What is m, the money distribution when Ms. Hart takes office?

- What is q, Ms. Hart's goal?

- Did Professor Krazo advise Ms. Hart that her goal was attainable?
- If he did so advise her, what is a?

Problem 11: Compute **u**, A, and B.

Problem 12: For the vector s, let $\alpha = 1/\beta$ where

$$\beta = \left[\sum_{k=1}^{8} s_k^2 \right]^{\frac{1}{2}}$$

Answer the last two parts of Problem 10 with $q = \alpha s$.

References

[1] Moody, M. E. and Shannon, K. *Microcomputer Exercises for Calculus: A Laboratory Manual.* West Publishing Company, St. Paul, 1988

[2] Kemeny, J.G. and Snell, J.L., *Finite Markov Chains*, Van Nostrand, New York, 1986

Using Computer Algebra as a Discovery Tool in Calculus

Richard Sours
Wilkes University

Introduction

Computer Algebra Systems (CAS's) have become effective teaching enhancements in undergraduate mathematics courses. They are being used, for instance, to process "real" data sets as opposed to contrived data sets in which the numbers are chosen to produce nice results. They are also being used to do "nontraditional" applications in the biological and social sciences. In the process they help demonstrate the applicability of mathematics to areas far beyond traditional areas like physics and engineering. Clearly, the computational capabilities of a good CAS allow consideration of classes of problems which have been inaccessible until now. There is yet another advantage of using a CAS which may exceed those listed above in importance in beginning courses. Whether the class meets in a computer lab or in a conventional classroom (with the computer used in a demonstration mode) the CAS is an excellent *discovery tool* .

Student discovery of mathematical truths is an important but elusive goal of undergraduate courses. Even though virtually all of the ideas being "discovered" are well known to more mature mathematicians, the process of uncovering mathematical ideas on one's own is exciting and stimulating. Indeed, *this is the very essence of the learning process*. The effective use of a CAS can make many mathematical concepts transparent enough so that even beginning students will reach the correct conclusions.

This paper concerns the use of a CAS, specifically as a tool for discovery, in first year calculus taught in a laboratory format. After some observations about helping students discover, two examples are presented. The first deals with the Product and Quotient Rules for derivatives, and the second deals with the error bounds for the Trapezoidal Rule or Simpson's Rule.

Three Problems Associated with Discovery in Calculus

The "Lecture/Response" style of teaching is one in which the instructor presents new material, works out examples and at any time will stop and respond to questions from the students. This technique is widespread and many teachers

are quite comfortable in this mode. Unfortunately, this technique is incompatible with a "Discovery" method. To encourage students to discover new ideas, the instructor must ask lots of questions, and answer very few. When he/she does get a question from a student, the best response is usually another question. This takes practice; the questions you ask must be well thought out to lead the students along the proper (thought) path. The biggest adjustment is learning to shut up, and learning to tolerate periods of silence in the class while the students are formulating their conjectures.

Clearly, if the students are supposed to discover a new idea, this discovery must occur before that topic is presented formally in a lecture. This leads immediately to two problems. First, the timing of "discovery classes," i.e., the scheduling of the calculus lab requires a great deal of flexibility. If the syllabus calls for presentation of the product rule on Monday, but the class doesn't meet in the computer lab until Thursday, the lab will certainly not be the first exposure to that topic. To use a CAS to present new topics as they come up, the instructor must have daily access to appropriate hardware/software; it must be made an integral part of the instructional process. One option is the availability of a demonstration computer for each class. Currently, the best arrangement seems to be a roll-around cart equipped with an overhead projector, a computer and an overhead projection device. If the school (or department) is willing to invest in this kind of hardware and if an easy-to-use CAS is available, then this equipment will surely see lots of use. A (pedagogically) better arrangement is to meet the class each day in a room equipped with enough microcomputers to have at most two students per machine. Obviously, this means dedicating a lab to a specific course when that course meets. This may not be practical in some cases.

There is yet one more problem which may interfere with using a CAS during the first presentation of a topic (so that some discovery might occur). Many students in the first-year calculus course have had some exposure to calculus before. The better ones will have placed into the next course, but some percentage of the class will certainly remember, say, the product rule. How can the true novices be given the opportunity to make an interesting discovery without someone who has been through it before blurting out the answer prematurely? This is a teaching challenge for which there is no easy answer. One approach is to have the students work in pairs or small groups which are chosen by the teacher. Alternately, the students might write their conjectures on paper and only the one called upon would give his/her ideas to the class. The issue of how to prevent the overly eager students from disrupting the thought processes of the rest of the class is a pedagogical challenge of long standing. The use of lab formats and/or CAS's does little to solve this problem.

"Discovering" Ideas in Calculus: Two Examples

1. The Product Rule and the Quotient Rule

Once derivatives have been covered and the students are comfortable taking the derivatives of polynomials, they are ready to *discover* the product rule and then the quotient rule. A CAS is the perfect tool for exhibiting to the students the patterns in these two rules. The students can make conjectures, test their hypotheses and eventually formulate for themselves a "Product Rule Theorem" (including the hypotheses). Having done this they will be anxious to see "their" theorem proved rigorously.

To make this discovery, first have the computer compute the derivatives of a number of functions to be used as building blocks. Sin x, cos x, tan x and even arctan x are good for this purpose. The students do not know the derivatives of any of these functions, but the functions are all familiar, and they are willing to believe that since the CAS knows how to differentiate polynomials, it probably knows how to differentiate these transcendental functions. Keep a catalogue of these functions and their derivatives visible on the chalkboard.

Now have the students enter $x^3\sin x$, and have the CAS differentiate. Repeat this with $x^3 \tan x$, and with $x^3 \arctan x$. Now enter $(\sin x)(\arctan x)$ or $(x^5+5x^2)\sin x$, and let the students guess at the derivatives before they have the CAS display the answer. Notice that the students do not have to know (thoroughly) that $(\tan x)'=\sec^2 x$, or that $(\arctan x)' = 1/(1 +x^2)$ to use them in this context. They just have to refer to the catalogue on the board. Finally, have one student offer a conjecture for the product rule. Test the student's conjecture with a couple more examples (with luck, this initial conjecture will be wrong!). Let the students try the (correct) conjecture on examples which they propose; they may propose some wild functions for which they must first have the CAS differentiate the building blocks.

After the class is convinced that they have formulated a correct product rule, have them work on the hypotheses "What would we have to require of f and g to be able to prove our conjecture?" In short order the students will state a complete theorem. Now the theorem *must* be proved. This is not an option; there is no reason why the students should believe the computer. The proof, however, will be believed because you are proving "their" theorem, not yours.

Although the quotient rule is more complicated, if the same list of easy functions is used to illustrate it, and since the class is now primed to look for a pattern involving the derivatives of these easy functions, they can discover this pattern also. In general, "discovering" these two formulas takes no longer than the traditional approach. The time spent working examples to illustrate the stated rule is replaced by letting the computer work out examples while the students look for patterns. The other advantages of this method, especially the joy of discovery, make it well worth using.

2. Error Bounds for the Trapezoidal Rule and Simpson's Rule

Most calculus texts present the Trapezoidal Rule and Simpson's Rule as numerical approximation methods for a definite integral. Some books present other methods (e.g., the left-hand endpoint, right-hand endpoint, etc.), but no first-year calculus text gives a thorough discussion of the error bounds for these methods. At best the book might give the appropriate inequality in an "...it can be shown that..." statement. Sometimes the error estimates are given as exercises, or not given at all.

Since the value of an approximating method is only as good as the size of its error, the existence of an (easy-to-compute) upper-bound on the error is essential. These formulas exist for both the Trapezoidal and Simpson's methods, they are relatively easy to compute (very easy with a CAS) and they should be discussed in much greater detail. This example serves to show that the students can "discover" all of the factors which contribute to the error. The Trapezoidal Rule is a little easier and it is presented thoroughly; the discussion for Simpson's Rule is similar.

Consider an easy example, $\int_0^3 (9 - x^2)\, dx = 18$

The Fundamental Theorem gives the exact value and we can use this to test the accuracy of the Trapezoidal method. In the table below, "Panels" gives the number of vertical strips into which [a,b] is divided. "Approx. Integral" is the approximation using the Trapezoidal method. "Error" is the true value minus Approx. Integral and "%-Error" is Error divided by the value of the integral, 18 in this case.

TABLE I

$$\int_0^3 (9 - x^2)\, dx$$

PANELS	APPROX.	ERROR	% ERROR
5	17.820000	.18	1%
10	17.955000	.045	.25%
20	17.988750	.01145	.0625%
50	17.998200	.0018	.01%
100	17.999550	.00045	.0025%
500	17.999982	.000018	.0001%

Clearly, as the number of panels increases, the error decreases. This is expected and students would be surprised if anything else occurred. A closer look at the Panels and the %-Error columns indicate exactly how this change occurs. As n = number of Panels increases from 5 to 10 and then from 10 to 20, the corresponding %Errors decrease to 1/4 of the preceding amount. As n increased from 10 to 100, the %-error is 1/100 of the preceding amount. Students can readily see that the %-Error is inversely proportional to the square of the number of panels. Therefore, part of an error bounds formula, E_T would be

$$E_T = \frac{c_1}{n^2}$$

Now consider $\int_0^3 |\, 3\pi \sin(\pi x)\,|\, dx\ = 18.$

The CAS can easily graph the integrand and also generate the following table.

TABLE 1

$$\int_0^3 |\, 3\pi \sin(\pi x)\,|\, dx$$

PANELS	APPROX.	ERROR	%ERROR
5	17.403890	.59611	3.31%
10	17.851511	.148289	.824%
20	17.962973	.037027	.204%
50	17.994077	.005923	.0329%
100	17.998519	.001481	.00822%
500	17.999940	.000060	.00033%

Notice first that the same inverse square relationship exists between n and the %-Error(i.e., as n doubles, the %-error changes by 1/4, etc.).

In addition, notice that the error (absolute or %-) is a little more than three times the corresponding error in the preceding example. The conjecture students usually make to explain this is that the second integral has three "humps." To pursue this idea, consider this example which is just the first hump of example 2.

TABLE 3.

$$\int_0^1 |\, 3\pi \sin(\pi x)\,|\, dx$$

PANELS	APPROX	ERROR	%-ERROR
5	5.8012967	.1987	3.31%
10	5.9505706	.0494294	.824%
20	5.9876579	.0123421	.206%
50	5.9980259	.0019741	.0329%
100	5.9995065	.0004935	.00822%
50	5.9999802	.0000198	.00033%

The integral is only 1/3 the size, but the %- errors are identical whether we are using one hump or three. So much for the "number of humps" theory. Also, it should be emphasized that the error changes by a factor larger than 3 (approximately 3.31) even though there are only 3 humps. The students will not discover the full reason for this increased error, but they know at this point that it has to do with the "humpiness," i.e., the curvature of the integrand.

The last factor contributing to the error is the length of the interval of integration. It is easy to illustrate that "b - a" affects error by an example like

$$\int_0^k |\, \pi \sin(\pi x)\,|\, dx\ = 2k,\ \text{for k} = 1,2,4\ .$$

Keep the number of panels constant while the length of the interval changes.

TABLE 4.
All integrals approximated using 20 panels.

INTEGRAL	TRAP. APPROX.	ERROR
$\int_0^1 |\, \pi \sin(\pi x)\,|\, dx\ = 2$	1.9958859	.0041141
$\int_0^2 |\, \pi \sin(\pi x)\,|\, dx\ = 4$	3.9670470	.03253
$\int_0^4 |\, \pi \sin(\pi x)\,|\, dx\ = 8$	7.7350623	..2649377

Now notice that when the length of the interval of integration doubles, the error increases by a factor of approximately 8 (i.e., $(.041141)/8 \approx .03253$ and $(.03253)/8 \approx .2649377$). A reasonable conjecture, which can be tested by further examples is that the error changes by $(b-a)^3$. This is in fact the right conclusion.

The error bound for the Trapezoidal Rule is

$$E_T^n \le \frac{(b-a)^3 M}{12\,n^2} \quad \text{where } M = \max |f''(x)|, \ x \in [a,b]$$

The students did not discover this formula, but they did discover the n^2 in the denominator. They also discovered that some factor related to the curvature must be present. Finally, they discovered that $b-a$ must be included. In a first course it is not necessary to go deeper than this, but a CAS used in this way can make this formula (and its companion for Simpson's Rule) come alive and be very believable. This kind of error analysis is not hard (with a CAS), and is a good example of the kind of insights students can have that makes the Trapezoidal Rule (and Simpson's Rule) much more interesting and relevant.

Observations

The two examples presented here have two important features in common. First, they both illustrate important ideas in calculus which can be *discovered* by the students. Since ideas which the students discover are much more likely to be retained than those which the teacher just states (with or without proof), it is worth going through the extra effort to help the students through this discovery process.

The second similarity in these examples is that each is impossible without a CAS. Some ideas in calculus, particularly the geometric ones such as the Mean Value Theorem or the Intermediate Value Theorem, can be discovered on a chalkboard. Neither of the ideas illustrated here can be done by hand. The moral here is that effective use of a CAS can help students learn traditional topics in calculus by helping them discover these ideas on their own. It can also open up new topics which were not previously accessible.

Evaluation

No formal evaluation of this program has been conducted. The difficulty is that the sections of calculus not involved are taught by different teachers than the sections using the CAS and a lab format. We are planning a test in which participating teachers will each teach two sections of randomly selected students in the same semester, one taught the "old way" and the other using the CAS and lab format. The sections will be given the same tests and final examination. Hopefully, these test results will provide some useful measures of the program.

In the absence of formal evaluation, two sources of informal feedback are in use. The first is regularly solicited student comments. Here is a sample:

> "...[the CAS] was mainly beneficial in showing relationships between a function and its derivative. The program also provided an alternative to the normal classroom environment."

> "The use of [the CAS] was helpful in that the use of the program made the graphing of functions and the finding of areas under and between curves much easier to do. But due to the time spent using the computer there was less emphasis on graphing by hand."

> "Very useful. I used [the CAS] in Calc II (even though it was not required in my class) and in Physics."

> "It helped us see the relationships between f, f ' and f "."

> "Using [the CAS] added nothing to the course."

The other source of informal evaluation is the author's opinion. Designing and preparing the lab assignments is very time consuming. Grading the lab reports is an added burden. Learning to adapt to a new teaching style can be unsettling (in the lab environment, the instructor must relinquish some control). In spite of these negatives, I endorse the program wholeheartedly. Using a CAS has renewed my enthusiasm for teaching Calculus. The tool that we call the computer can (and has!) become an addicting toy. I am convinced that my enthusiasm has shown through to my students. Finally, it is worth noting that DISCOVERY is not limited to the students. The CAS has helped me learn some of the subtleties of the subject that had eluded me for many years.

The Calculus Laboratory:
A Means to Increase Conceptual Understanding

Susan Hurley and Thomas H. Rousseau
Sienna College

Introduction

Two major conferences[1, 2] produced recommendations to improve the teaching of calculus. The recommendations focus on a need to increase conceptual understanding, to reduce time spent on mechanical skills, and to increase utilization of computer resources. At Siena College, we are addressing all these issues with the introduction of a laboratory for all first year calculus students. The unique feature of our laboratories is that they are primarily concerned with improving conceptual understanding, although we are also interested in the integration of computer skills into the calculus curriculum.

In this paper we present the philosophy, history, and logistical overview of our project, a brief description of the content of our laboratories including three selected exercises, and an evaluation of our first year's experience. The Appendix contains a description of the second semester's laboratories.

Laboratory Goals and Organization

Contemporary students of mathematics have several widely recognized deficiencies[1,2]. Their poor grasp of central concepts combined with weak analytical skills lead to an inability to complete or even attempt the solution of complicated problems. These shortcomings are amplified by the reluctance of students to use computers. The laboratory concept at Siena evolved through discussions among the Mathematics Department faculty because of our concern that students were indeed failing to develop quantitative thinking skills. The laboratory is an experience in directed discovery in which students complete a sequence of related exercises designed to introduce, amplify and/or apply the theory of calculus. While the ideas of calculus are central, the problems chosen for solution originate in diverse disciplines. Two to four students work in a group to solve the problems and their results are submitted in a laboratory report. The goal of the laboratories is to improve student performance in two main areas, general quantitative analysis skills and the understanding of the first-year calculus course. In addition, the laboratory offers a unique opportunity to introduce students to the computing resources available to them.

The laboratories became a required component of the first year calculus course during Academic Year 1988-89. After three years of debating alternative pedagogical philosophies, we proposed the laboratory concept to our Board of Instruction in September, 1987. Our concept was conditionally approved for two years so that we could assess its value as one method to enhance the mathematical thought processes of the modern college student. During the summer of 1988, we wrote the laboratory exercises. Each of the calculus courses is supported with ten laboratories. While following the traditional sequence of calculus instruction, the exercises are independent of the textbook used in the course.

Students meet for three contact hours of lecture and a two-hour laboratory each week. The laboratory is scheduled independently from the lecture and is generally taught by a different instructor. The course grade depends upon performance in both lecture and laboratory with four credit hours awarded for each semester's work. We conduct the laboratory in a classroom which has been modified to facilitate student working groups and to provide general access to computers. A faculty member instructs the laboratory and may be assisted by one or more laboratory assistants--third or fourth year undergraduate mathematics majors.

The laboratory period is long enough to allow the uninterrupted development of the topic for the day. While not leisurely, the two-hour period permits a deeper treatment of a concept than is afforded by the one-hour lecture. Our use of "laboratory groups" encourages peer teaching and a sharing of skills. This feature also develops cooperation among the students and is excellent preparation for the teamwork typical in modern industry. Group interaction with faculty provides the opportunity to direct students toward a solution without "solving" the problem. The modern scientist or engineer must be literate with computing devices and our guidance includes suggestions to use the immediately available computer resources when appropriate. Finally, the laboratories are written to include open-ended questions. We attempt to provide hints of further applications which are contingent upon a deeper understanding of the concepts of calculus.

Siena College provided several essential resources to allow the establishment of the Calculus Laboratories. We were given a classroom to be converted into a laboratory. A total of 24 students can be accommodated in groups of four. The college also provided five MacIntosh II computers plus an Image Writer printer. Funds for the computers came from a lab fee charged to the students. The number of faculty hours spent teaching freshman calculus increased by 25% because of the need to have five contact hours per week instead of four.

Lab Overview

We selected topics for the laboratories which explore the essence of calculus. An individual laboratory may amplify material covered in lecture, intuitively introduce concepts which will be discussed in the lecture, or expose the student to uses which are not normally covered in the calculus sequence. Titles of the laboratories appear in Table 1, following which we give a brief summary of the first semester laboratories. The labs on the concept of velocity, the definite integral, and volume transport are given in full.

Table 1--Titles of Laboratories

CALCULUS I

Concept of Velocity

Graphs, Limits and Continuity

Differentiation Rules

Applications of the Derivative

Linear Approximation

Optimization

The Definite Integral

The Fundamental Theorem of
Calculus

Differential Equations

Modelling River Flow

Volume Transport
(A continuation of River Flow)

Calculus II

Theory of Exponentials and Logarithms

Calculus of Exponentials and Logarithms

Exponential Growth and Decay

Waves in Nature

Applications of Trigonometric Integrals

Some Calculus Curiosities

Population Dynamics

Infinite Sequences

Complex Numbers and the Complex Exponential
Function

Differential equations

The Concept of Velocity

The first laboratory in Calculus I is designed to introduce the study of differential calculus. Students read and analyze data pertaining to a specific journey and make decisions on the physical meaning of the data. They are also asked to identify different graphs that will fit the given data and make inferences from the graphs. The concept of average velocity is reviewed and we introduce the geometric interpretation of average velocity as the slope of a secant line on the position/time graph. By computing average velocities over progressively shorter time intervals using measurements from the graph, students are led to the concept of instantaneous velocity and, simultaneously, to its link to the slope of the tangent line. We have included this laboratory in compressed format below.

Laboratory - Concept of Velocity

This laboratory is a sequence of exercises which help you to explore fundamental concepts from calculus through an example related to a journey from Albany to Buffalo. You will be making inferences about the velocity of the traveller based upon information related to the location of the traveller at a specified time.

The following table describes the distance s from Albany

at time t hours from the start of the journey.

Time = t (Hours)	0	1	2	3
DISTANCE= s (Miles)	0	60	90	140

Time	4	5	6	7
DISTANCE	150	200	270	300

Let us consider the graph relating distance travelled to time. The data given above in the table mean that the graph must pass through the eight points shown. There are many graphs which pass through these points. Examine the four shown below.

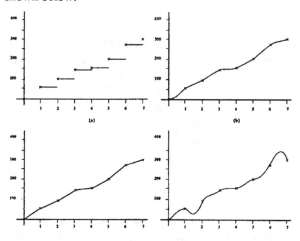

Which of the four graphs (a)-(d) could not possibly be the graph of this journey? Which seems to you to be the most plausible? Justify your answer.

We are interested in the relation between distance travelled and velocity. Compute the average velocity:

    a) For the entire journey.
    b) For the first four hours.
    c) For the last three hours.
    d) For the portion of the journey between Utica and Rochester.
    e) For each separate hour of the journey.
    f) Write the formula that you used to calculate the average velocity in terms of the function $s(t)$ and t.

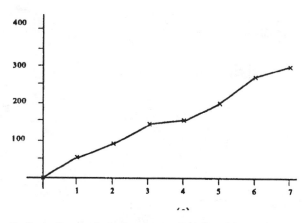

Refer to the graph shown above. You should be able to see a relation between your answers to e) and the straight line segments of the graph. What is that relation? Explain your answer. [Hint: Straight lines and line segments have slope.]

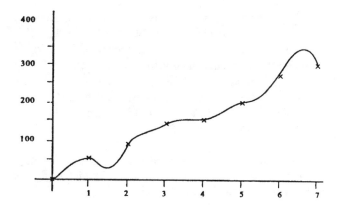

Consider the distance versus time graph shown above. Estimate as best you can the average velocity:

    a) During the time interval (1,3.5).
    b) During the time interval (1,1.5).
    c) During the time interval (6,6.5).

Are any of your answers peculiar? Explain the meaning of the answers to parts b) and c).

We wish to discover the instantaneous velocity when $t=4$. As a start we focus on shorter and shorter time periods. The graph on the next page shows the portion of the journey between $t=3$ and $t=5$.

On the expanded graph shown on the next page, draw the straight lines whose slope indicates the average velocity for the following time intervals: (3,4),(3.5,4),(3.75,4), (4,5),(4,4.5), and (4,4.25). You should have drawn six lines altogether.

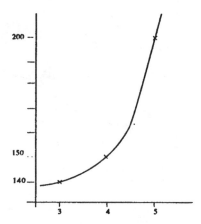

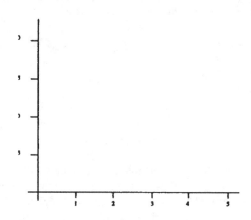

Compute roughly the average velocities for each of these time periods. Notice that all these time intervals begin or end at $t=4$.

What is happening at the precise moment $t=4$? How fast is our traveller going? How do we relate our knowledge of his velocity to the graph showing his position? How do we relate our knowledge of his velocity to his average velocity during time periods beginning or ending at $t=4$? The answers to these questions lie at the heart of that area of mathematics called Calculus. Let us attempt to infer the answers to these questions.

Do you think the velocity at $t=4$ will be close to the average velocity for any very short time interval which begins or ends at $t=4$? Justify your answer.

Do you think that the velocity at $t=4$ can be represented by the slope of a straight line? Justify your answer and sketch the straight line whose slope is the velocity at $t=4$.

Estimate the instantaneous velocity at $t=3$ and at $t=5$.

In the space below summarize what you have discovered about Average Velocity and Instantaneous Velocity.

This last exercise will test your understanding of today's lab. Draw a DISTANCE/TIME graph for a journey of 100 miles lasting 5 hours satisfying the following:

    1) The average velocity for the trip is 20 mph.
    2) The average velocity for the first 2 hours is 30 mph.
    3) The instantaneous velocity at $t=3$ is 15 mph.
    4) There is somewhere a 1/2 hour stop for lunch.
    5) During the last hour of the journey the speed is
       constant.

Graphs, Limits and Continuity

This Laboratory is designed to reinforce ideas that have proven difficult. There are two independent parts. The first part assumes that students have an intuitive grasp of the concepts of limit and continuity. They are led to the definition of continuity by analyzing graphs of functions. There is an opportunity to study the well known pathological functions. The second part of the lab focuses on the definition of limit.

Differentiation Rules

We choose to offer a different approach in this lab. For example, an alternate geometric proof of the product rule is developed from the area y of a rectangle with sides u and v. The increment Δy is easily seen to decompose as $\Delta y = u\Delta v + v\Delta u$ and the product rule follows in the limit. This approach reinforces the concept of the derivative as a rate of change. The lab also presents some practical situations where it makes sense to compute rates of change using the product or chain rules.

Applications of the Derivative

Several problems are approached in this lab. The students have to relate their analysis of a given function to questions about the physical situation that the function models. The computations involved are handled by the software, *Mathematica*. They then use calculus to solve a problem in analytic geometry. Finally they are asked to sketch the graph of a function.

Linear Approximation

The use of the tangent line in making linear approximations is introduced in this lab. Most text books introduce this as an application of the theory of the differential and students are misled into thinking that it is something new. The motivation for studying this topic is given as a question about how computers work and the promise that we will examine more accurate approximation methods in the next semester. The estimate of an error bound is given emphasis. We find this topic is hard for the students, and the lab is definitely a better forum than the lecture.

Optimization Problems

These problems are notoriously hard for most calculus students. Several problem sets of differing types and complexity are presented for group solution. Problem formulation is the focus of this laboratory.

The Definite Integral

The students are asked to compute Upper and Lower Sums and to compute The Definite Integral as a limit. They are also introduced to the Trapezoidal approximation. This lab was designed as a reinforcement of a fundamental concept.

The student worksheet follows.

The definite integral of a function has been defined as the limit of a certain sum when that limit exists. This laboratory explores approximations to this limit.

Requirements:

1. Define the function f by $f(x) = 2x^2$. Our goal is to evaluate

$$\int_0^1 f(x)\, dx = \int_0^1 2x^2\, dx$$

    a. Partition the interval [0,1] into n equal parts. Write equations for the upper sum S_n and the lower sum s_n for the given function over the interval [0,1].

    b. Evaluate $\int_0^1 2x^2\, dx$ as $\lim_{n\to\infty} S_n$ and $\lim_{n\to\infty} s_n$

2. Let $g(x) = x^3 + 4$. Evaluate $\int_0^1 g(x)\, dx$ by taking the limit of either the upper sum or the lower sum. Explain why you needed to evaluate only one of the sums.

3. There are many functions which are integrable, but the definite integral is difficult to evaluate as a limit.

    a. Evaluate the lower sum derived in (1 a) for n = 20, i.e., find s_{20}

    b. What is the error in using this lower sum as an approximation to the definite integral? How can you decrease the error?

4. Since the function $f(x) = 2x^2$ is not negative over the interval [0,1] (or any other interval for that matter), the definite integral

$$\int_0^1 2x^2\, dx$$

measures the area under the graph of f and above the x-axis. The lower sum 520 is a crude approximation to the definite integral. A better approximation to this integral is found by using trapezoids instead of rectangles for approximating elements.

    a. Find the formula T_n to approximate the integral using n trapezoids of equal width.

    b. What is T_{20}? Compare T_{20} to S_{20}. Which is more accurate?

There are many definite integrals which cannot be evaluated exactly. In such cases, we resort to numerical approximations such as the Trapezoidal method. Other, more accurate approximations are available and can be studied in courses covering numerical methods.

The Fundamental Theorem of Calculus

Here we focus on the meaning of The Fundamental Theorem of Calculus. Students tend to lose sight of the concept of definite integral once they learn the "easy" way to compute one. The graph of a function f is given and students have to answer questions about the related function

$$A(x) = \int_0^x f(t)\, dt$$

They identify zeros and extrema of A and find where A is increasing and decreasing. They use this information to answer questions about the derivative of A and compare their answers to the behavior of the function f.

Differential Equations

Students practice setting up and finding closed form solutions (when possible) to separable equations. Students are introduced to Euler's method to get an approximate solution to the problem dy/dx=y, y(0)=l. A computer program implementing Euler's Method is required.

Modeling River Flow

The last two laboratories in Calculus I introduce material not usually covered in a first course. *Modeling River Flow* is the first of these. A river 50 ft. wide is described and certain measurements of stream velocity are given. The students have to fit curves to the data, using a piecewise-linear, a quadratic and a trigonometric model. They then compare predictions from their models to an "actual" measurement of water speed to calculate error.

Volume Transport

We continue with *Volume Transport*. Students construct functions to fit given data on the depth of the river. Based on their choice of functions, they write a definite integral that represents the volume of water flowing in the river per unit time. They approximate the value of the integral using a computer and, if possible, compute the value using The Fundamental Theorem of Calculus. Since the choice of function was up to them, this may be impossible. This provides a useful lesson in the limitations of the Fundamental Theorem. The worksheet for the lab is given below.

Your last laboratory was an exercise in modelling the flow of water in a stream as a function of the position across the stream. This laboratory exercise explores the volume transport of the stream.

GOAL: To determine the volume of water transported through a cross-section of the stream every second.

ASSUMPTIONS:

1 . Water speed varies continuously across the stream.
2. Water speed does not vary with depth.

MEASUREMENTS:

1 . The river is 50 ft. wide.
2. The rate of flow at each bank is 2 ft/sec.
3. The flow rate is greatest at the center of the river and is 10 ft/sec there.
4. The flow rate at any distance from one bank equals that at the same distance from the other bank.

5. The water depth at each bank is zero.
6. The stream is 15 ft deep at its deepest point.

At the end of the last laboratory you found that information on water depth was needed in order to estimate the volume of water flowing in the river. You have now been given some additional data on water depth. Based on this information, the depth of the stream could, for example, be any of the models shown below.

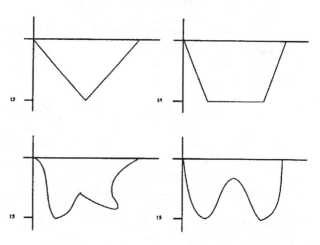

a) Make an assumption to describe the bottom configuration that you feel is reasonable. State your assumption and sketch your chosen bottom profile below.

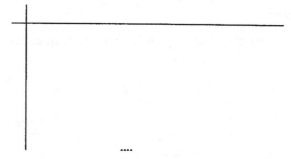

b) Write the function $h(x)$ which describes your model of the water depth.

3. Using your work from the last laboratory, model the water speed $f(x)$ with a parabola. Write that function below.

4. Describe the volume transported through a cross-section of the stream in terms of a definite integral. As a first step, think about the approximate volume transported through a strip of width Δx and depth $h(x)$ at a distance x from the bank.

5. Using your calculator, determine an approximation to the volume transport. Use $x = 5$.

6. The above approximation is very poor. Determine a more accurate value for the definite integral involved. Either

> a) Find an anti-derivative for your function.
> b) Write a computer program to find the upper sum for some large n.
> or
> c) Find a computer program or calculator that can evaluate the definite integral.

7. Discuss your model. What makes your model unrealistic? What additional information could you reasonably expect to get which would make your model better?

8. More sophisticated models of fluid flow consider variations of water speed with both depth and width of the stream and with time. You have not learned enough mathematics to work with all of these simultaneously. Are you able to model any of these variations? Describe some assumptions which you feel could improve your model.

The nature of our laboratories is apparent from the foregoing. A brief description of the content of the exercises for the second semester is in the Appendix.

Evaluation

We will describe the feedback from faculty, both lecturers and laboratory instructors, from academic advisors, from client disciplines, and from students. The general feeling is that the laboratory experience is beneficial. The topics chosen and the manner of presentation appear to satisfy our objectives--the development of quantitative thinking and sharpening of calculus skills. We then discuss the use of computers in the laboratories and give our evaluation of student performance during the first year.

The faculty laboratory instructors report that the laboratory is a great learning experience. We see student discussions, hear arguments, and watch as learning takes place. There is an enthusiasm among the students which most of us have never seen in a freshman class. The informal interaction between a group of students and the instructor develops confidence in the student and a willingness to question the propriety of problems and of concepts. The students quickly learn the value of group cooperation. Laboratory groups, once established, tend to remain un-

changed throughout the semester. We found that the development of problem-solving skills is a slow process. We observed and identified in our students conceptual shortcomings which had gone unnoticed in the past. As the students mature mathematically, the benefits of the laboratories become more pronounced. Many colleges currently teach calculus using three hours of lecture and a one hour recitation. Siena faculty conducting the three hours of lecture do not seem unduly hurried to complete the syllabus in the reduced time allocated. The principal observation is that there is less time to interact with the class. The lack of time to draw answers from the students does not appear to detract from performance.

Formal preparation for and grading of laboratory exercises is no different from a traditional lecture course. We found that the student lab assistant needs a review of the purposes of the exercise and of the mathematical concepts prior to the laboratory period to be effective. We discovered that students maintain a higher level of enthusiasm when lab assistants are present. We experimented with different methods to evaluate student performance. We tried individual lab reports, one report from each laboratory group, and, presently, are evaluating one report for each pair of students. The final grade for the laboratory is incorporated into the course grade; however, we are considering giving a separate lab grade in the future.

Academic advisors and faculty from client disciplines uniformly report that the laboratory experience appears beneficial. They enthusiastically support the laboratory concept as a tool to satisfy the long-term needs of students. We receive similar positive response from all groups who have worked with our laboratory exercises.

Comments from the students are enlightening and encouraging. We administered an opinion survey to all students in the laboratories. The primary result of this survey is that the students find the laboratory a worthwhile experience. Of course, comments range from "labs are great" to "labs were a complete waste of time". Students feel that the laboratories do help them to understand the concepts of calculus and they strongly favor the group solution atmosphere, yet are frustrated when the work in laboratories fails to improve their performance on tests in the lecture section. Many of the students enjoy the challenge of open-ended questions and the more difficult problems. Most of the students find the laboratories intellectually challenging and leave the laboratory period feeling drained.

During the first year, our students only had access to the main-frame computer of the college. They were hesitant to use this even when required to perform very tedious calculations. The majority were not enrolled in a Computer Science course and surprisingly many are not computer literate. This year, we find that students learn to use the Apple MacIntosh II without much formal instruction during the lab session. We give each lab group a short introduction to the PC and provide a tutorial covering the *Mathematica* commands useful in the exercise for the day. In this way the students' facility with the software develops naturally. We have implemented versions of some laboratories to be completed on personal computers. Some of these were written by two colleagues at Siena who have a National Science Foundation grant to develop curricular material in mathematics and physics using our symbolic manipulation program. The faculty has become sufficiently involved to write many more computer supplements to the labs.

The size of the college prevented our using a controlled experiment to discriminate between those students taking the laboratory and those taking the traditional lecture in calculus. Since all students taking a freshman calculus course were placed in the laboratory, our only objective means to compare different groups was the use of a final examination given to a similar group in the preceding year. The examination was a standard second-semester calculus final, emphasizing integration techniques and series convergence tests. The result of our comparison shows no significant difference in the scores achieved on the test between groups taking the laboratory and those who did not. We are concerned that we have not found a way to test students in a manner similar to the way we teach them to think in the laboratory. During the first year, examinations reflected the traditional emphasis on mechanical processes rather than our concept of problem analysis and quantitative thinking. During the Fall 1989 semester, we changed our testing procedure to include one laboratory-based problem on each of our full-period tests. In this manner, we hope to reinforce the thinking aspect of the laboratories. When we look at retention rates, we do see signs of success. Of approximately 200 students who enrolled in Calculus I during Fall 88, 83.4% continued to take the second semester of calculus in the spring with 72.9% receiving credit for the two semesters. The combined rates for the preceding two years were 75.8% and 61.9% respectively. This appears to be a significant improvement.

We anticipate that any major advances in the content of first-year college-level mathematics that result from the current curriculum development activity may cause changes and additions to the sequence of topics presented. We anticipate that our laboratory concept will be just as effective in that new environment.

Conclusion

In summary, the formal, two-hour calculus laboratory has been a required component of all freshman calculus courses at Siena College since the start of the 1988/89 academic year. The laboratories are designed to improve students' conceptual understanding by completion of a sequence of related exercises which reveal, amplify, and/or apply concepts from the theory of calculus. We are unable to report dramatic improvement of test scores during the first year; however, all who have read and used our manuals state that the quantitative skills being honed will benefit the students in the long run. We do see indications that we are retaining more students through the first year of calculus. We have modified the laboratories based on the first year's experience and have started writing computer-based versions of selected exercises. We believe that the calculus laboratories are an effective vehicle for improving the mathematics education of our students.

References

1. Douglas, Ronald G. (ed.).*Toward a Lean and Lively Calculus,* **MAA Lecture Notes,** Number 6, Mathematical Association of America, 1986.

2. Steen, Lynn Arthur (ed.). *Calculus for a New Century,* **MAA Lecture Notes,** Number 8, Mathematical Association of America, 1988.

3. Dubisch, Russell. National Science Foundation Grant CSI-8750276, "The Development of a Facility for Symbolically-Aided Instruction in Mathematics and Physics", 1987.

Appendix
Calculus II Labs

Exponential and Logarithm Review

The title speaks for itself. This Lab thoroughly reviews the concept of an inverse function and precalculus exposure to the general exponential and logarithmic functions.

Log-Log and Semi-Log Plots

The majority of calculus students will major in some science other than mathematics. The use of log-log and semi-log plots in the sciences led us to this topic which is now appearing in some calculus texts. The students are exposed to the theory and then asked to analyze two sets of data, using the appropriate technique.

Exponential Growth and Decay

This lab deals with traditional exponential growth and decay word problems. Problems modelled by the differential equations $dy/dx=ky$ and $dy/dx=ky+c$ are given.

The Wave Phenomenon

The motion of a piston attached to a rotating wheel provides a concrete introduction to simple harmonic motion. The students observe that the piston is moving over time according to a sine function and are asked the effect of altering the parameters involved. They then deduce the corresponding differential equation. The same equation appears in other physical phenomena. The second half of the lab deals with combinations of waves, both with equal and unequal periods.

Practice in Integration

This lab reinforces lectures on integration techniques. It involves a humorous episode with a water bomb. The students compute various areas and volumes which involve trigonometric integrals.

Some Calculus Curiosities

As a prelude to the work on series, and in order to reveal some mathematical curiosities, we decided to show the students that "A gallon of paint can coat an infinite surface" (Gabriel's Horn), that "An infinite curve can sit inside a dime" (Koch's Snowflake Curve) and that "A one dimensional curve can fill space" (A Peano Curve).

Population Dynamics

The Logistic equation is introduced as a model for inhibited population growth. Students work together to obtain the general solution and apply this result to several specific problems.

Taylor Polynomials and Series

The use of Taylor Polynomials to approximate functions is an excellent introduction to numerical approximation in general and is an especially appropriate topic for a math lab. It is, unfortunately, neglected in the traditional curriculum. We start with the observation that computers can only perform arithmetic, and ask how we can find sin(3) and ln(10). There is a brief review of the linear approximation lab from the first semester. Then students are led to discover how to do quadratic and cubic approximations, and how to find an error bound. They make the (by now) obvious generalization to nth degree polynomials and discover that a twelfth degree equation is needed to approximate e as accurately as their calculator does. Finally they are exposed to the Taylor Series for the exponential, sine, and cosine functions, and prove that these series converge to their respective functions.

Playing with Sequences

This is an amplification of the material presented in a traditional course.

Complex Numbers and The Exponential Function

The Complex Exponential function is needed in the study of ordinary differential equations. Its definition and properties use several concepts developed in Calculus II. The lab begins with a review of the algebra and geometry of complex numbers. Complex functions are discussed with an analysis of the squaring function. The exponential function is introduced as a power series and Euler's formula is deduced. The images of several sets are plotted and students develop a good "picture" of the behavior of the function.

Calculus I Computer Laboratory Assignments at Hesston College

Jean Alliman
Hesston College

Introduction

Hesston College is a private two year career and liberal arts college located in south central Kansas. The 550 students attending Hesston are enrolled in one or more of the 17 career programs or in the liberal arts program offered by the college. This article describes the computer laboratories which have been used in the Calculus I curriculum at Hesston College during the last two years.

Mechanics of the Laboratory Activity

The Calculus I classes at Hesston College consist of students who are planning to transfer to another college or university and for whom Calculus is required in their major field of study at the college or university to which they will transfer. In recent years half of the Calculus I students are business majors for whom this will be the last mathematics course. Most of the students have not used the software upon which the labs are based: True BASIC *Calculus* and *LOTUS 1-2-3*. Prior to taking this class some of the students have not used computers. Therefore, the directions included in each lab explicitly explain the procedures to be followed to run the necessary software.

Hesston College's Calculus I classes meet four times each week for one hour. The laboratory work is assigned in the place of textbook exercises over the weekend. Since the computer lab is open all day on Saturdays and on Sunday afternoons and evenings, there is ample time for all students to complete the laboratory assignments. The lecture which immediately precedes the weekend introduces the material to be covered by the laboratory to be done that weekend. Those labs which were originally designed to be exploratory and which were completed before an introductory lecture had been given were highly threatening to the students and the least satisfactory. The most successful labs reinforce material which has been introduced previously. Monday's class begins with a discussion of the laboratory activity. No labs are assigned at the beginning of a week during which there will be a test.

Students work in pairs on the parts of the labs which use the computer. This allows the instructors to pair a student with little previous computer experience with another student who has had more experience with computers.

After completing the exploration on the computer, each student is required to write a two to three page reaction which focuses on the two or three questions which are central to the concepts of the lab. These questions are posed twice in the lab: once at the beginning, in a section entitled PURPOSE, and once at the end of the lab where the students are directed to individually discuss the questions in the written report. The report is collected at the beginning of Monday's class period.

The Labs

The following list indicates the eleven labs which are assigned during the semester long calculus class. The software which each uses is indicated in parentheses.

Lab #1 - Exploring Functions and Their Limits (*Calculus*)

Lab #2 - Exploring Tangent Lines, Rates of Change, and the Derivative of a Function (*Calculus*)

Lab #3 - Further Exploration of Graphs of Functions (*Calculus*)

Lab #4 - Exploring Maximum, Minimum and Concavity of a Function (*Calculus*)

Lab #5 - Antiderivatives (*Calculus*)

Lab #6 - Exploring an Electronic Spreadsheet - *LOTUS 1-2-3*

Lab #7 - Areas Under Curves, Midpoint and Simpson's Rule (*LOTUS 1-2-3*)

Lab #8 - Volumes by Slicing, Disks and Washers (*LOTUS 1-2-3*)

Lab #9 - Inverse Functions, e, a^x, and e^x (*LOTUS 1-2-3*)

Lab #10 - Mathematical Models and Numerical Approximation (*LOTUS 1-2-3*)

Lab #11 - Newton's Method (*Calculus 2* and *LOTUS 1-2-3*)

A Typical Lab

Lab #11, which uses both *Calculus* and *LOTUS 1-2-3* to explore the approximation of roots of a function by means of the Intermediate Value Theorem and Newton's Method, is included here as an example of the structure and style of the lab assignments.

LAB #11 - Approximation of Roots of a Function, Intermediate Value Theorem and Newton's Method

Purpose:

In this lab you will explore two methods of finding a real root of a function. You will use the CALCULUS program to explore the Intermediate Value Theorem; and you will use the LOTUS 1-2-3 program to explore Newton's Method for approximating the roots of a function. Specifically, we will be exploring the following questions:

1. What characteristics of the graph of a function allow us to know that a real root is present?

2. How is this characteristic useful in finding a root by means of the Intermediate Value Theorem?

3. How does approximating the root of a function by means of the Intermediate Value Theorem compare with approximating it by means of Newton's Method in terms of accuracy and speed?

Hardware and Software:

The hardware requirement is a computer running under the MS/DOS operating system (version 2.1 or later) and having at least 256K of internal memory.

Remember to bring your data disk with you to the lab for this experiment. You have a copy of a spreadsheet called Newton on your data disk. You need that spreadsheet for this lab.

The Experiment: Finding the roots of a function

Part A: Intermediate Value Theorem

Use the CALCULUS program to obtain the graph of $x^3 - x + 1$ from -2 to 2. (If you have trouble with this, refer to LAB #1 and use $x^3 - x + 1$ as the function to be plotted.) When you have a graph of this function, you should note that it crosses the x-axis in only one place. This means that it has just one real root. The other two roots have imaginary values. This real root's value is somewhere between -2 and -1.

You will use the Table Generating Program (Fl, Evaluate) to help determine the value of this one real root. In response to the "From, to, step?" prompt, answer -2, -1, and 0.1 respectively. This will generate a table from -2 to -1 with a difference of 0.1 from one x value to the next value. The first column which is shown on the left side of the screen gives the values for x, while the second column gives the resulting $f(x)$ values.

Notice that $f(x)$ changes sign in this interval. In the chart below write the values for x and $f(x)$ which show where this sign change occurs.

Step = 0.1	x	$f(x)$
Negative value for $f(x)$		
Positive value for $f(x)$		

The values for x which you have written in the chart above should be adjacent to each other in the table on the screen.

The Intermediate Value Theorem says that since $f(x)$ went from a negative value to a positive value in the interval represented by the two values of x in the chart above, there is some value for x between those two values which would result in $f(x) = 0$. So we now know that the real root of the function is located somewhere between the two values for x that you listed in the chart above.

You can view the graph of $f(x)$ more closely by replotting it. Use the values for x in the chart above when you answer the "From, to?" prompt for plotting. Also, use the rescaling program. Press F2, Rescale, and then press the shift key and the Tab backward key simultaneously. Press the shift key and Tab backward simultaneously again. The cursor should be on the line which says, "x-from". Move backward over the -2 that is on that line and enter the first value for x which you listed in the chart above. Delete any part of the -2 which remains. Use the Tab forward key to move the cursor to the line with "x-to" on it. Backspace over the 2 which is already there and enter the second value for x which you listed in the chart above. If you still see the original 2 on the screen, delete it so that only your second x value appears. Tab forward to the "y-from:" prompt. Enter -0.5 in response to the "y-from:" prompt, and 0.5 in response to the "y-to:" prompt. Push the Enter key. This will give you a closer look at the graph along the y-axis. In other words, you have rescaled the graph so that you have zoomed in to see it up close.

If you have followed the preceding directions carefully, you should see a new graph in which the line for $f(x)$ crosses the x-axis between -1.4 and -1.3. In fact, it crosses the x-axis very close to -1.3.

Use a table again to help you find the root more closely. Use Fl and Evaluate to obtain a table generated from -1.4 to -1.3 with a step of 0.01. Again, look for the sign change. Put your results in the chart below.

Step = 0.01	x	$f(x)$
Negative value for $f(x)$		
Positive value for $f(x)$		

You can now analyze this table and realize that the real root of the function lies somewhere between the two values of x that you listed in the chart above. Use another table generated by the computer (Fl and Evaluate) with a step size of 0.001 to determine where the values for $f(x)$ change sign within the interval that you have indicated in the chart above. Copy your results into the chart below.

Step = 0.001	x	$f(x)$
Negative value for $f(x)$		
Positive value for $f(x)$		

You now know that the root of the function lies somewhere between_____and_____ (each number should have three decimal places).

Throughout this part of the experiment you have looked for the values of x which cause the sign of $f(x)$ to change. When you have found two values for x which cause the sign of $f(x)$

to change, you know that there is some value for x between those two values for which $f(x) = 0$. This value for x is a root of the equation. Continue the process until you reach the limits of this Calculus program and fill in the charts below.

	Step: 0.0001		Step 0.00001		Step: 0.000001	
	x	$f(x)$	x	$f(x)$	x	$f(x)$
$f(x) < 0$						
$f(x) > 0$						

Root of $f(x)$: $x =$_____
                (to five decimal places)

Use your calculator to check to see if the value for x that you found is in fact a root of $f(x)$. Remember that $f(x) = x^3 - x + 1$. You have found a value for x and want to see if this x value results in 0 when you put it into $f(x)$. Depending on the size of your calculator's memory and the care with which you did this part of the experiment, you will find that the result of putting this value for x into $f(x)$ will come close to 0, but will not equal 0 exactly. Write the difference here:

Follow the processes of this part of the lab to find the real roots of other functions. Be careful! Some functions will have more than one real root. You should choose your own functions, but some suggestions that you might try are listed below. Use the information that you find to fill in the chart below.

1. $f(x) = x^3 + x - 1$ 2. $f(x) = x^5 + x^4 - 5$

3. $f(x) = x^4 + x - 3$ 4. $f(x) = -x^5 - 5x^3$

Using Intermediate Value Theorem:

$f(x)$	Interval(s) in which Real Roots Occur	Real Root(s)	Number of Steps
$x^3 - x + 1$	(-2,-1)		
$x^3 + x - 1$			
$x^5 + x^4 - 5$			
$x^4 + x - 3$			
$-x^5 - 5x^3$			

Part B: Newton's Method

Background: Another method of approximating the root of a function is called Newton's Method. To use Newton's Method, you need to guess a value for a root and then use both the function evaluated at that guess and the derivative of the function evaluated at that guess to find a closer approximation for the root.

The experiment: For this part of the experiment, we will use both the CALCULUS program and LOTUS 1-2-3. To use the two programs efficiently without having to change frequently from one program to another, first use the CALCULUS program to fill in the chart which is located on the last page of this experiment. Then use the LOTUS 1-2-3 program to finish the experiment.

Let's go through the process with $f(x) = x^3 - x + 1$ as the first function. By now you can easily use the CALCULUS program to plot $f(x)$ and find that it has one real root between -2 and -1.

We can use either the -2 or -1 as our first guess (approximation) for the root of $f(x) = x^3 - x + 1$. So that you can learn the processes that are to be followed, choose -2 as the first approximation of this root. Let $-2 = x_1$. To find x_2, the second approximation for the root, we need the derivative of $f(x)$. We can use the CALCULUS program (F1, General, x^3-x+1, "From: -2", "To: -1,", F1, Derivative, F2, and Print Formula) to find the equation for $f'(x)$. On the other hand, you can easily find this derivative on a scrap piece of paper yourself!

$f'(x) =$_____. Enter this formula in the chart on the last page of this lab.

Before moving on to the next section of the lab, use the CALCULUS program to plot each of the other functions that you are exploring. Use these graphs to find the interval(s) in which their root(s) occur. Remember that some functions will have several real roots. Find the equation for each derivative. Put all of this information into the chart on the last page of this experiment.

After finding the information above, you are ready to use the LOTUS 1-2-3 program to facilitate the exploration of Newton's Method. Exit the CALCULUS program. When "C>" appears on the screen, type LOTUS, press Enter, 1 and Enter.

Insert your data disk into the 3 1/2 inch disk drive and load the data for this lab as you have loaded the previous

LOTUS spreadsheets (/, F, R, B:, highlight the word Newton, and press the Enter key).

On the screen you will see part of a spreadsheet entitled "Newton's Method." Column H is entitled $f(x)$. Move the cursor to cell H7. In the upper left corner of the screen you see that cell H7 contains "(F7)^3-F7+1." This means that the value which is in cell H7 is obtained by putting the value which is currently in cell F7 into the formula $x^3 - x + 1$. You will recognize this as $f(x)$, the function for which you want to find a root.

Use the arrow keys to move the cursor to cell J7. The top left corner of the screen tells you that J7 contains "3*(F7)^2-1." This means that the value which is in cell J7 is obtained by putting the value which is in cell F7 into the function $3x^2 - 1$. You will recognize this as $f'(x)$, the derivative of the function for which we want to find the root.

Use the arrow keys to move the cursor to F7. This value (-2) is the first approximation which we chose for the root of the function $f(x) = x^3 - x + 1$. Move the arrow down to F8. The top line of the screen tells you "+F7 - H7/J7." This means that the value that is seen in cell F8 is obtained by dividing the value found in H7 by the value in J7 and then subtracting the result from F7. Use your calculator and a scrap paper to verify that the value that you see in F8 is this result.

Another way of saying it is this: $x_2 = x_1 - f(x_1)/f'(x_1)$. This is the formula which is used in Newton's Method to find the second approximation for the root of the function. Use your calculator to check the resulting values that you see on the screen.

Move your cursor down column F. The value in F8 represents x_2, the second approximation of the root. F9 represents the third approximation of the root. As you move the cursor down column F, you will highlight successive approximations of the root of $f(x)$. Notice that within a few approximations, the value for the approximation of the root does not change. This value is _____. This value represents the best approximation for the root of $f(x) = x^3 - x + 1$ which is obtainable within the limits of this program. Check its exactness with your calculator by putting the value that you found as the root into the function $x^3 - x + 1$ and seeing how close the result comes to 0.

Which led to a smaller error, Newton's Method or the Intermediate Value Theorem?

Which required fewer trials?

To see the power of an electronic spreadsheet, suppose that we had picked -1 as the first approximation for the root of $x^3 - x + 1$. Move the cursor to F7 and type -1. Press Enter and watch the entire spreadsheet change as the computer calculates the various values which are affected by this change. The resulting value for the root should not change.

Some students wonder what would happen in Newton's Method if their first approximation for the root was not very close to the actual root. We can experiment with this very easily using this spreadsheet and the LOTUS software.

To do this, move the cursor to cell F7 and enter a negative value which is close to -1 and -2. Watch to see how the resulting value for the root changes. In fact, this program is able to calculate the root if the first guess is any negative number greater than -1000.

You can also see some of the difficulties that arise with Newton's Method. To do so, enter a positive value greater than 1 into F7. If you enter the value 2, for example, into F7, you will see how the values in this column come close to the value for the root and then jump farther away. In mathematical terms, we say that the results from Newton's Method sometimes <u>do not</u> converge toward the root of the function.

Use the Edit and Copy functions from LOTUS 1-2-3 to adapt the formula which is given on this spreadsheet to use Newton's Method to find the roots of the other functions for which you have already found roots using the Intermediate Value Theorem. Complete the following chart as you use Newton's Method to find the real roots of the functions.

Newton's Method

	$f(x)$	$f'(x)$	Real Roots(s)	Number of Steps
$x^3 - x + 1$				
$x^3 + x - 1$				
$x^5 + x^4 - 5$				
$x^4 + x - 3$				
$-x^5 - 5x^3$				

To leave this program, press /, Q (for Quit), Y (for Yes), E

(for Exit), Y (for Yes), and follow any other directions which appear on the screen before you leave the computer.

In writing your essay for this lab, be sure to include the charts that you completed during this lab and answer the following questions:

1. What characteristics of the graph of a function allow us to know that a real root is present?

2. How is this characteristic useful in finding a root by means of the Intermediate Value Theorem?

3. How does approximating a root of a function by means of the Intermediate Value Theorem compare with approximating the root by Newton's Method in terms of accuracy and speed?

Student Reactions

Following are some of the student reactions to the above laboratory as well as the laboratory exercises in general. These reactions were usually written in response to one of the questions on a critique page which accompanied each lab. Occasionally the reactions formed the introduction or conclusion to the student's written essay.

My lab partner and I felt that overall the labs were good because we gained a better understanding of calculus. Also, we gained more experience on the computer. Any experience on a computer in this highly advanced technological society is a bonus for that particular individual. The labs helped develop our minds not only in calculus and computers, but helped us reach into our brain and expanded our horizons.(DH)

I also found this lab to be interesting. It is great to see what computers can do in mathematics today. They have helped me very much.(AL)

I again enjoyed this lab, even though it wasn't dealing with a real life situation. I understood this lab very well which allowed me to enjoy it. All of these labs have been helpful for me because the paper at the end of each one helped me to think about what I did and made it clearer to me. Thank you.(TN)

This lab is another helpful aid in getting me more

acquainted with the computers. I thought it did a pretty good job of comparing the two different methods, both in presenting their similarities and also their differences. They were both very similar in their procedures, but the final result and the length turned out to be quite different.(CT)

Modifications Made as a Result of the First Year's Experience

Outstanding sophomore mathematics students at Hesston College become student assistants and help in various first year math classes. One sophomore student assistant is available in the computer lab during the time in which Calculus I students are usually working on their computer labs. This student has the experience of working through the labs as a freshman and is available to help current students as they struggle with the various aspects of the labs which may be difficult.

As a result of the first year's experience, it was clear that an additional lab was needed to precede those which rely on the LOTUS 1-2-3 spreadsheet software. Most of our students have no previous experience with an electronic spreadsheet and, therefore, had greatest difficulty with the first lab which used a spreadsheet. The additional lab is included as #6. It introduces them to the general idea of a spreadsheet and specific LOTUS 1-2-3 commands such as the Edit and Copy commands which are used frequently in later labs.

Conclusion

The laboratory experience was a very positive one for both the instructors of the Calculus I courses and the students who participated in them. While there is no scientific evidence to support the impression, the Calculus I instructors felt as though class interest was greater and questions were more focused on those Mondays which followed the laboratory experience than they had been in previous classes. Additionally, as the students realized that Friday's class period was providing them with some background information or ideas for the laboratory experience, they focused on the material more closely than previous students had.

The additional mode of teaching and learning – by means of the computer – provided a variety which many students appreciated. The visual orientation of the computer was advantageous to several students.

The entire experience was beneficial for both our students and faculty. We anticipate being able to expand upon the use of the computer throughout our entire Calculus curriculum.

Differentiability and Local Slope - A Lab Exercise
for the HP - 28S

Thomas Dick
Oregon State University

Objective

In this activity, students will explore derivatives in the graphical representation. The students will investigate the derivative of a function at a point as the "local slope" of the function. They will determine an approximation to the derivative at a point by finding the approximate local slope of the function at that point.

General Idea

A. Intuitive:

A function is differentiable at *a* if *f* is continuous at *a* and there exists a unique line tangent to the curve at $(a, f(a))$. If a function is differentiable on an interval, its graph has no "corners" or sharp turns, only smooth curves, on the interval. If one could zoom in on a graph at a point $(a, f(a))$ indefinitely, the graph at the point would appear to be a straight line. The slope of this "line" (the "local slope" of the curve) is the derivative of the function at *a*.

B. Mathematical:

The derivative of the function *f* is the function *f'* defined by

$$f'(x) = \lim_{h \to 0} \frac{f(x+h) - f(x)}{h}$$

for all *x* such that this limit exists.

If this limit exists at *x = a*, *f* is differentiable at *a*.

Programs

Two programs are used in this activity, ZOOM and PLUCK. The program listings are given below. The program PLUCK finds the exact point on the graph of a function given the coordinates of a pixel on the graph and, in this activity, is used to center the graph.

It is assumed that the graph of the function has been drawn previously. The input required by PLUCK is a point on level 1 of the stack.

Example:

Graph the equation sin(x)

1) Enter the equation 'SIN(X |ENTER|
2) Purge the plot parameters 'PPAR |PURGE|
3) Store the equation and draw the graph
    |SETQ| |DRAW|

The graph should appear on the screen looking something like this:

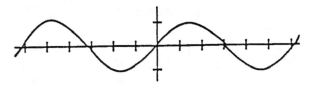

Digitize the pixel with x-coordinate equal to 1.

1) Use the arrow keys to move the crosshairs to the desired point.

2) Digitize the point and return to the stack. |INS| |ON|

The following should appear on level one of the stack: (1, .8).

Use PLUCK to find the exact value of the point on the graph. PLUCK The exact value of the point will appear on level one of the stack: (1, .841470984808).

Enter the following programs in your user menu:

```
ZOOM: << DUP
       *W
       *H
       DRAW
       DGTIZ
       |ENTER|
       'ZOOM'
       |STO|
```

PLUCK: << C->R   Places the x-coordinate of the point on level 2 and the y-coordinate on level 1 of the stack.

DROP   Drops the value on level 1 off of the stack.

DUP   Duplicates the value on level 1 of the stack.

'X' DUP   Stores the number on level 1 of the stack in the variable X.

RCEQ EVAL   Recalls the equation stored in the memory and evaluates it for the stored value of X.

R->C   Forms a point with the current value on level 2 of the stack as the x-coordinate and the current value on level 1 of the stack as the y-coordinate.

|ENTER| Places the program on level 1 of the stack.

'PLUCK' |STO| Stores the program under the name PLUCK.

Use of Calculator

The calculator will be used to find the approximate local slope of the functions. Each function will be graphed on the calculator and the graphs will be zoomed in on until the graph appears to be a straight line. Since it is impossible to zoom in on the graph indefinitely, the graph only appears to be a straight line because of the low resolution of the screen. All calculations give approximations to the derivative. The stress in this activity is for the students to understand the graphical interpretation of a derivative.

V. Lab Activity

Recall the equation for the slope of the line containing the points (x_1, y_1) and (x_2, y_2):

$$\text{Slope } m = \frac{y_2 - y_1}{x_2 - x_1} = \frac{\text{rise}}{\text{run}}$$

To find the approximate "local slope" of a function at a given x value, we need to zoom in on the graph until it appears linear.

Exercise 1: Determine the local slope of $2\sin(x)$ at 3.

1) Enter the function   '2*SIN(X |ENTER|

2) Purge the plot parameters before graphing a new function and then draw the graph

        'PPAR |PURGE| |STEQ| |DRAW|

3) Center the graph at the point corresponding to the desired x value of 3.
   a) Digitize the pixel with the desired x coordinate.
   b) Convert the pixel coordinates to the coordinates of the actual point on the graph.

        |ON| (to get back to the stack)
        |PLUCK| |CENTR| (in plot menu)

4) Zoom in on the graph until a straight line appears.
        .1 |ZOOM|
(If the graph is not a straight line, continue zooming in on the graph. Try different zoom factors.)

5) Find the slope.

   a) Digitize the endpoints of the line.
   b) Convert these pixel coordinates to the coordinates of the actual points on the graph.

        |ON| |PLUCK| |SWAP| |PLUCK|

   c) Calculate the slope by using
      $$m = \frac{y_2 - y_1}{x_2 - x_1}$$

Find the "local slope" of each of the following functions at the indicated value of x. Your answer should include: a) your calculation of the local slope, b) the zoom factor you used and c) a sketch of the line

1. x^2 at $x = 2$ 2. $\dfrac{1}{\sqrt{2x+3}}$ at $x = -1$

3. $\sqrt{1 + \sin \sqrt{x}}$ at $x=1$ 4. $\dfrac{x}{x-1}$ at $x = \frac{1}{2}$

5. $\tan(x+2)$ at $x = 2$ 6. $\sqrt[3]{2x - 7}$ at $x = 4$

7. $\sqrt{\dfrac{x^2 + 1}{x^2 - 1}}$ at $x=3$ 8. $\dfrac{\exp(x)}{x}$ at $x = 3$

9. $\dfrac{x+1}{x^2 - x - 2}$ at $x = 2.5$ 10. $\sqrt{1 - x}$ at $x = -3$

11. $\dfrac{\sin(x)}{x}$ at $x=3$ 12. $\dfrac{\cos(x)}{x+1}$ at $x=-2$

13. $\dfrac{1}{x^2}$ at $x=2$ 14. $2x^2-3x+5$ at $x=2$

15. $x^3\exp(\dfrac{-1}{x})$ at $x=1$ 16. $\sqrt{\dfrac{\sin^2(x)}{1+\cos(x)}}$ at $x=0$

17. $10-x^2$ at $x=3$ 18. $3x^2-10x+2$ at $x=3$

19. $\sqrt{\dfrac{x}{x^2+1}}$ at $x=5$ 20. $x(x-1)^{2/3}$ at $x=\dfrac{3}{2}$

Integrated Calculus and Physics Laboratory/Workshops

Joan R. Hundhausen and F. Richard Yeatts
Colorado School of Mines

Introduction

In our paper appearing in section two of this volume we have given a general description of the integrated course in calculus and physics at Colorado School of Mines. This course builds on the historical relationship between calculus and physics. These relationships are the basis for weekly Laboratory/Workshops sessions. During these sessions the students work together in informal groups that have a close interaction with the instructors and student assistants. The goal of these sessions is to promote transferability, i.e., to develop the facility for recognizing and applying mathematics to other contexts.

There are three general types of Laboratory/Workshop projects which we classify as Type A, Type B, and Type C. The Type A projects require the student to derive a mathematical formula describing a geometric or trigonometric situation. "Graphs and Modeling" is an example of this type of a Laboratory/Workshop. Type B projects use numerical methods (implemented by hand or with the HP 28S) to illustrate discrete versions of continuous processes. The "Initial Value Problem" example illustrates this type of project. Type C projects bridge the gap between calculus and physics. The "Linear Motion" project provides an example of this type.

We now present these projects in a form that is only slightly modified from that the students see in our course.

Laboratory/Workshop #2
Functions, Graphs and Models

OBJECTIVES: To study functions as they are used to model geometric and physical situations. To investigate the graphs of functions. To examine the role of parameters in functional relationships.

A. GRAPHS AND MODELING

1. A study in population growth:

A certain State experienced considerable economic turmoil over the decade from 1970 to 1980, yet showed a 15% increase in population for that period. Early in the decade growth was slow but steady. In the period 1973-75 the population actually declined below its 1970 level (call it Po) because of lay-offs in one of its major industries. Since 1975, the State has attracted new industries and the population surged in response. The population growth in the last year, 1979, however, was significantly less than that of the preceding three years.

a) Draw a graph of a function representing population versus time which might reasonably represent (i.e. model) the above situation. (Be sure to label your graph.)

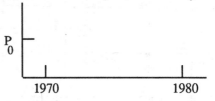

b) Describe briefly how the <u>average rate</u> of population growth compares during the periods: a) The decade. b) The period 1973-75. c) The final three years.

2. Geometric situations

In solving problems in physics and engineering as well as mathematics, it is often necessary to begin by determining a functional relationship (i.e. formula) between (or among) various geometrical objects. Here are two examples:

    a) Express the length b of the chord of a circle of radius 8 inches as a function of its distance p from the center. Provide a labeled sketch

$$b = b(p) = \underline{\hspace{2cm}}$$

What is the natural domain of b(p)? What is its range?

Is it an even or an odd function?

Graph your function b(p) on your calculator. (Adjust scales as necessary.) Sketch the graph below; label the axes. (Note that this function

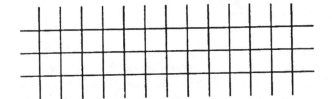

decreases over its entire domain.)

Over which part of the domain is b changing most rapidly (per unit change in p)? Answer this by indicating the appropriate interval on the p-axis.

b) A right circular cylinder of radius r is inscribed in a sphere of radius R. Find a formula for the

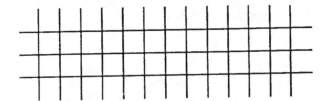

volume V of the cylinder in terms of r and R. Sketch and label a cross section .

V = V(r) = _____

Provide the following information about V(r):

Natural domain _____ Range _____ Even or odd? _____
Graph using R = 4. (You may again use your calculator.)

Note that the graph of this function is quite different from that of b(r), above. Explain this difference by describing clearly in complete sentences the behavior of the function over its domain.

There is one point on the graph which is very significant. Identify it and state why it is significant. (Your answer should involve r and V(r); write a complete and precise sentence.)

Find its approximate coordinates by using the

digitizing facility of the calculator. (We'll revisit this problem again using methods to be developed in the calculus.)

B. PARAMETERS

A parameter might be thought of as an unspecified constant in a functional relationship. Depending on the actual value of the parameter, the function may indeed have remarkably different properties.

1. A quadratic function

Consider a quadratic function f(x) which depends on the parameter p:

$$f(x) = px^2 - 3x + 2 \,.$$

Our objective is to see how the parameter p affects the shape of the graph of the function in general, and the value(s) of any real root(s) of the function in particular.

a) Before investigating the roots of this quadratic function, let us review the concept of the discriminant. for a general quadratic,

$$ax^2 + bx + c,$$

the roots are given by the quadratic formula

$$\frac{-b \pm \sqrt{b^2 - 4ac}}{2a}$$

The discriminant is just the term in the radical: $b^2 - 4ac$. Depending on the value of the discriminant the quadratic it can have two, one or no real roots. Indicate the values of the discriminant for which the quadratic has: -

Two real roots: _____ One real root:_____
No real roots:_____

b) For the quadratic function, f(x), given above, write out the discriminant (it will be a function of the parameter p, of course).

discriminant = _____

Indicate the values of p for which f(x) has:

Two real roots: _____ One real root: _____ No real roots:_____

c) Use the calculator to study the graphs of $f(x)$ for the seven strategically chosen values of p given below. Sketch your results on the seven grids provided. The values of p to be used are $\{-8, -2, -1/2, 0, 1/2, 9/8, 8\}$

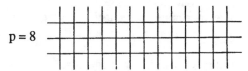

p = 8

Examine the graphs and describe as concisely as possible, in complete sentences, how the shape of $f(x)$ varies with the parameter p.

Are the graphs consistent with the condition on real roots established above? What about the case of one real root?

2. Another quadratic function: Determine precisely all the values of p for which $g(x)$ has one or two real roots. $g(x) = x^2 - 5px + 2p$

One real root(s): _____ Two real roots: _____

3. A practical problem using a parameter: *"Will he catch the train?"*

A passenger arrives at a railroad station just as his train is pulling away. The instant the train begins to move, he is at a distance D from its rear entrance. Running at top speed, he makes a desperate effort to catch the train. The question is: Under what conditions does he catch it?

To make a specific mathematical model of this situation, let us assume that the train pulls away with constant acceleration of 0.2 m/s^2 and that the man runs with a constant velocity of 6.0 m/s; also let D = 30 m. Let $x_T(t)$ give the position of the rear entrance of the train measured from its starting point (x = 0) as a function of time t; let $x_p(t)$ be the position of the passenger relative to x = 0:

$$x_T(t) = \tfrac{1}{2} (0.2 \text{ m/s}^2) t^2;$$

$$x_p(t) = (6.0 \text{ m/s}) t - (30 \text{ m}).$$

a) Mathematically, the question of whether the

passenger catches the train can be expressed as follows: *Does there exist a time t > 0 such that $x_p(t) = x_T(t)$?* In the space below, explain in words what this means.

b) Solve for t.   t = _____

c) Explain this result in words. In particular, answer the question as to whether the passenger catches the train. Why are there two answers?

d) Graph the functions $x_T(t)$ and $x_p(t)$ simultaneously on the calculator. Draw the result below. Be sure to label the axes. Show where the equality $x_T(t) = x_p(t)$ is satisfied.

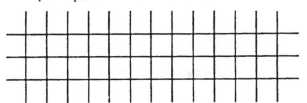

e) Consider now the more general question: For what range of initial distances D does there exist a time t when $x_p(t) = x_T(t)$? Replace 30m by the parameter D in the equation for $x_p(t)$ and solve for t as a function of D.   t = _____

For what values of D can the passenger catch the train? (Express as an inequality.)

f) For the *maximum* D-value, plot $x_p(t)$ <u>on the graph above</u>. What special relationship exists between the graphs of the two functions at the point where the curves meet ?

Laboratory/Workshop 7
The Initial Value Problem

OBJECTIVES: To use the Euler algorithm to solve the initial value problem for linear motion. To study the vertical motion of a body with and without air resistance.

A THE INITIAL VALUE PROBLEM

1. Statement of the Problem:

Given the position x_0 and velocity v_0 of a body at some initial time t_0, and its acceleration a for all later time $t > t_0$, find the position $x(t)$ and velocity $v(t)$ of the body at all later times.

This so-called "initial value problem" is very important to us because, in many important physical problems, the acceleration is a known quantity. This follows from Newton's second law which states that the acceleration of a body $a = F/m$, where F, the net force acting on the body, and m, the mass of the body, are often known quantities.

Since $a = dv/dt$, v can be found, in principle, by anti-differentiating (that is, integrating) the acceleration a. Furthermore, since $v = dx/dt$, the position x can be found, in turn, by anti-differentiating v. In certain special cases, these anti-differentiations can be carried out analytically using regular calculus, but in most "real world" applications, they can only be carried out approximately using some numerical techniques such as the Euler algorithm.

2. Application of the Euler Algorithm.

i. Let us assume first of all that the acceleration is a function of both velocity and time, thus

$$v' = dv/dt = a(v,t) .$$

The Euler algorithm can be applied directly to obtain velocity as a function of time:

$$v_{n+1} = v_n + a_n \Delta t,$$

where $v_n \approx v(t_n)$, $a_n = a(v_n, t_n)$ and $\Delta t = t_{n+1} - t_n$, and $v_0 = v(t_0)$ is the given initial velocity.

ii. Taking the next step,

$$x' = dx/dt = v(t);$$

thus the Euler algorithm yields

$$x_{n+1} = x_n + v_n \Delta t,$$

where $x_n \approx v(t_n)$ and $x_0 = x(t_0)$ is the given initial position. We have calculated v_n in a previous step, therefore the solution can be completed.

3. Vertical Motion without Air Resistance

A ball is thrown vertically upward from a height of 12.0 m with an initial velocity of 15.0 m/s. Calculate the velocity and position of the ball as functions of time. In particular, determine the maximum height achieved by the ball and its velocity as it hits the ground.

a) We can solve this problem exactly (which we'll do later as a check), but first let us obtain some approximate answers using the Euler algorithm. The given data:

$$x_0 = \rule{1.5cm}{0.4pt}, v_0 = \rule{1.5cm}{0.4pt}, a = \rule{1.5cm}{0.4pt}$$

b) Apply the Euler algorithm using $\Delta t = 0.1$ and $0 < t < 4.0s$, thus $N = \rule{2cm}{0.4pt}$

    i. Modify your Euler program to solve for v_n. Execute your program. (Attach a copy of your program and a print of your plot below.)

    ii. Modify your Euler program to solve for x (how will you introduce x_n into the loop?). Execute your program. (Attach a copy of this program and a print of your plot below.)

c) Now let's check on the work by comparison with the exact solution. To be specific, calculate the time of maximum height t_1 and the maximum height x_1; calculate the time of flight till the ball hits the ground t_2 and its velocity as it hits v_2. Record these values in the table below along with the corresponding values read off the Euler algorithm plots.

	t_1	x_1	t_2	v_2
EXACT				
APPROX.				

4. Vertical Motion with Linear Air Resistance

Repeat exercise 3, but with linear air resistance for which

$$a = -9.8 - 0.2v .$$

Attach your plots. Compare t_1, x_1, t_2 and v_2 with the case neglecting air resistance.

5. Vertical Motion with Quadratic Air Resistance (This is the best simple model of real air resistance.) Repeat exercise 4, but with

$$a = -9.8 - .01 |v| v .$$

(Why use $|v| v$ rather than just v^2 for the velocity dependence?)

Laboratory/Workshop 5
Linear Motion[1]

OBJECTIVES: To develop the concepts of velocity and acceleration both physically and quantitatively. To observe various examples of linear motion including constant velocity and constant acceleration; to represent these motions both graphically and analytically. To inter-relate the physical, graphical and analytical expressions of motion.

A. UNIFORM MOTION:

1. Constant Velocity.

Observe the motion of a body having constant velocity; consider the following definition:

A body moving with constant (i.e., uniform) linear velocity traverses equal distances in equal times.

a) The most general relationship between position x and time t of a body moving with constant velocity is the linear relationship

$$x = A + Bt,$$

where A and B are constants. Assuming that the body is moving with velocity v_0, and was at $x = x_0$ at time $t = 0$, evaluate A and B in terms of v_0 and x_0. (Use the difference quotient to find B)

A =_____

B =_____

b) Draw and label graphs of x vs. t, v = dx/dt vs. t, and acceleration a = dv/dt for this body. (Assume that x_0 and v_0 are positive.)

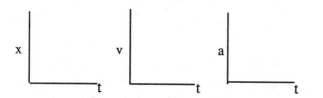

c) Discuss briefly, with the aid of a diagram, the statement:

Over a sufficiently short interval of time, the velocity of a body can be considered constant.

2. Constant Acceleration.

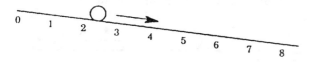

Observe the motion of an object (on a steady incline) having constant acceleration; consider the following definition:

A body moving with constant (i.e. uniform) acceleration traverses, in equal times, distances which follow the progression 1, 3, 5, 7, ..., as measured from the point of rest.

a) Show that this definition leads to the result that the total distance traversed, from rest, is proportional to the square of the time, i.e., $x = Ct^2$, where C is a constant.

b) The most general equation for the distance traveled (displacement) of a body moving with constant acceleration is the quadratic,

$$x = A + Bt + Ct^2.$$

Assuming that the body starts at point x_0 with velocity v_0 at time $t = 0$, evaluate the constants A, B and C in terms of x_0, v_0 and the constant acceleration a_0. (Use the difference quotient to find B and C.)

A =_____ B =_____ C =_____

c) Draw and label graphs of x vs. t, v = dx/dt vs. t, and acceleration a = dv/dt for this body. (Assume that x_0, v_0 and a_0 are all positive.)

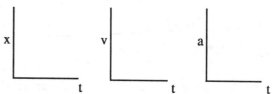

d) Observe the motion of the body projected up the incline.

Let x = 0 at t = 0 and take v_0 as positive; the acceleration is negative (down the incline). Call it a_0.

Draw and label graphs for x, v, and a versus time for the period from t = 0 until t = T, when the body returns to x = 0. (In this exercise, and in all those subsequent it is useful to watch the motion first with an eye to how position varies with time: then watch it again considering velocity; finally concentrate on acceleration.)

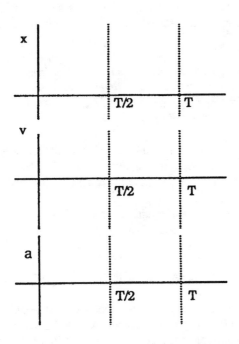

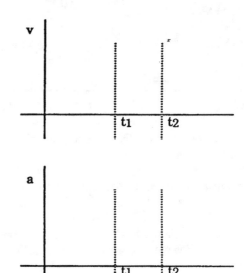

B. COMPOSITE MOTIONS:

1. Incline to Horizontal

Observe the motion of a body moving from rest ($v = 0$ at $t = 0$, $x = 0$) down an inclined track, then moving onto a horizontal track.

3. Velocity Comparison:

Body A moves with uniform velocity v_0. Body B moves on an adjacent track, slightly inclined, with constant acceleration a_0. Both bodies start at the same instant $t = 0$, with A at $x = 0$ and B at $x = x_0 > 0$. Observe that A passes B at a point we'll call x_1 at time t_1; later on, B passes A at a point x_2 at time t_2.

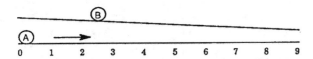

   i. Observe the motion and note the "number" giving the spot where the two bodies have the same velocity:_____

   ii. Draw and label graphs of x, v, and a for both bodies A and B.

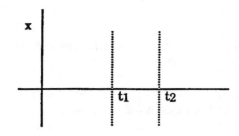

Let T be the time when the body moves onto the horizontal track. Draw graphs of x, v and a. (Be especially careful of how the graphs join at time T. Are x(t), v(t), and a(t) all continuous?)

2. The Bumper:

Observe the motion of a body moving with constant speed, then colliding against an elastic bumper so as to rebound with the same speed in the opposite direction.

Let T be the time the body hits the bumper. Draw graphs of x, v and a.

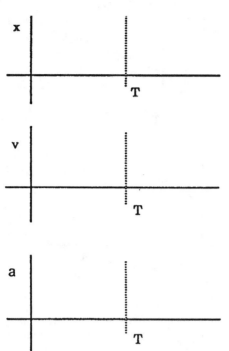

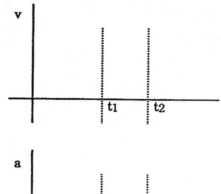

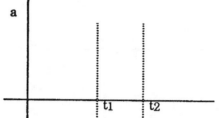

3. The Magnetic Brake:

Observe the motion of an object moving down an incline, through a magnetic field, then out again. Assume that the object starts from rest at x = 0 at time t = 0; it enters the magnetic field at time t_1 and leaves at time t_2.

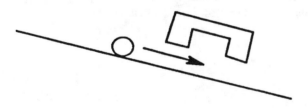

Draw graphs of x, v and a.

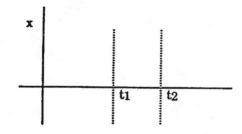

4. The Pendulum:

Observe the motion of a simple pendulum over one complete period of oscillation (forward and back).
Let T be the time for a complete oscillation. Draw graphs of x, v and a.

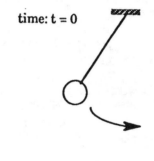

time: t = 0

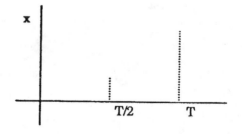

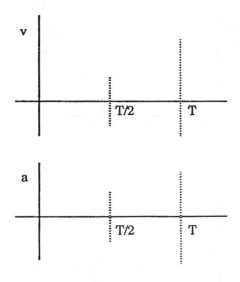

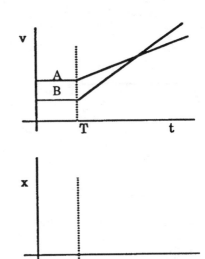

C. DISCUSSION TOPICS

1. Acceleration Comparison

Two objects, A and B, begin moving at the same time as shown in the velocity versus time graph.

   a) After time $t = T$, object _____ has the greater acceleration because_____

   b) Draw graphs of the positions of A and B, given that they are at the <u>same point</u> at time $t = 0$.

   c) Describe in words (complete sentences) a physical situation involving inclined and horizontal tracks which can produce motions consistent with the graphs.

2. The Continuity of the Functions: x(t), v(t), and a(t).

Discuss briefly whether it is necessary that the functions representing position $x(t)$, velocity $v(t)$ and/or acceleration of a body be continuous over their natural domains.

[1]NOTE: The "Linear Motion" project is based on the article: Renquist, Mark L. and Lillian C. McDermott, "A Conceptual Approach to Teaching Kinematics", <u>Am. J. Phys.</u>, <u>55</u>, (5), 407.

An Approach to the Basis of Calculus
Hervé Lehning

Introduction

In this paper, I would like to show how useful laboratory activities can be to acquire the basis of calculus. The most fundamental are the notions of number and function. I shall also add the notion of differential in order to show that my approach is not limited to only the most basic foundations.

Notion of number

Before studying calculus, the notion of number is rather vague but sufficient for precalculus needs. The only really known numbers are decimal and rational numbers.

Numerical presentation

The expansion of rational numbers in decimal form leads to the notion of decimal with an infinite number of digits. The use of computers allows students to achieve the calculations in all cases. Moreover, it is a very good programming exercise, so it is done in a laboratory. Through the examination of results, students easily conjecture the characterization of the expansion of rational numbers (which they generally succeed in demonstrating then). Therefore, they are convinced that some others exist. At that level, it seems useless to speak about the uniqueness of this expansion. I prefer to keep this question for a deeper study.

Geometrical presentation.

The historical example of the square root of two is excellent for this approach. It leads one to interpret numbers as points on the line. Other classical examples can and must be given (cube roots, the number π, etc). The connection with the numerical side is done in a natural way through showing and getting the bisection method computerized.

On that matter, I noticed that, on account of my previous presentation, at this level the most natural idea for students is the one of successive cut into ten parts to gain a decimal place at each step. Comparing these two algorithms is interesting as far as coding and efficiency are concerned. In terms of intuition, the cut into ten parts is dictated by numerical intuition while the cut in two parts is more natural in terms of geometry.

Error in a number

The previous example shows the difference between exact and approximated results. For this, and equally in order to demystify computers, I like having some calculations done. In the beginning, calculations of squares of square roots are not bad. Generally, calculations which depend on the representation mode of numbers are interesting to analyze (for example, try the real and extended Turbo Pascal modes on the calculations of the successive terms of the sequence defined by $u_0 = \frac{2}{3}$ and $u_{n+1} = 4u_n - 2$ for every n). Laboratory activities lead students to recognize the inevitable nature of obtaining approximated results and the importance of the notion of error, which allows a good understanding of the notion of real number.

Notion of function

The only notion of function which students have before calculus courses is the one of functions given by formula, a notion which I call algebraic in what follows.

Algebraic intuition.

This definition is the only one known by students at the precalculus level; nevertheless, they often fail by confusing a variable and its name (thus $x \to x^2$ and $t \to t^2$ are not always considered as the same function). Making explicit the corresponding calculation tree (see below) allows a better understanding of how a formal calculator or keyboard inputs of functions work. Coding this tree from keyboard inputs is a very good exercise but, at this level, it seems too difficult. It must be kept for a more profound study.

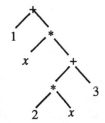

Calculation Tree of $1+x(2x+3)$

The notion of variable becomes obvious through this tree. The understanding of this notion is improved by learning a bit of coding (Basic or Pascal).

Numerical intuition

The effective use by students of little programs which they create themselves, such as the calculation of areas of circles, allows an effective understanding of the notion of variable in mathematics. So, a function, f, is conceived as something assigning a unique real value noted f(x) to each real value x. This notion strictly generalizes the previous one. For example, a function can be defined by its behavior on intervals.

At that level, specifying the meaning of the word "assign" seems useless because the geometrical interpretation gives it a very large intuitive meaning.

Geometrical intuition

We could move from a numerical to a geometrical interpretation through sketching the graphs of computed functions on the screen of a computer. The speed of that process greatly changes the way of considering functions. As a matter of fact, previously, students knew that each function possesses a graph but, for them, it was the result of long and hazardous calculations. So they never sketched a graph to get ideas! Now, this way of considering functions becomes usual: a function is an object which can be easily seen, and some properties fall into the domain of the visually obvious (for example monotony, the existence of a zero, a maximum, an asymptote, etc).

Eventually calculations must be used to specify a value. For example, the sketching of the function

$$x \to \frac{x^2 - 2x + 3}{x^2 + 2x + 3}$$

allows students to see that it has a maximum near the point −1.7; the use of a zoom can give a better precision, but only an exact calculation gives $-\sqrt{3}$. Of course, it can be done with the use of a formal symbolic calculator!

Notion of differential

One of the difficulties of the notion of differential for students is the simultaneous use of three intuitions. The definition is generally given numerically, the calculation is algebraic and the most important underlying intuition is geometrical.

Geometrical intuition

In order to distinguish between differentiable functions and the others, I make students use a curve sketcher with zoom possibilities. Students note that the curves they spontaneously define are linear near every point. They find it difficult to discover counterexamples. Of course, they meet the notion of tangent (see [2] and [5]). Moreover, the link between the sign of the derivative and the monotony of the function becomes obvious through this interpretation.

Numerical consequence

Then, reading this definition numerically and noting the regularity in the sequences of function values are interesting. The resulting approximation at the neighborhood of a point leads to the limit calculations we know and to the use of differentials in physics.

Through discretization, we obtain the notion of finite difference and so approximate calculations of derivatives. These questions are seen again at the time of the study of duality in linear algebra and of the study of Taylor's formula in analysis. These questions, also used for the integral, are the occasion of a great number of laboratory activities in the studies of calculus and linear algebra.

Formal calculation

One of the classical obstacles to a good understanding of the notion of differential is the important part of formal calculations of derivatives. This part must now be reduced and partly replaced by the use of a formal calculator when needed. My experience shows that this does not lead to a loss of meaning for students. This fact has already been observed in elementary arithmetic. It is not the knowledge of calculation algorithms, but the understanding of their meaning which allows a pupil to use the right operation in order to solve a problem. At the second level, coding a formal calculator by using the writing of a tree is interesting.

References

[1] Hervé Lehning, "From experimentation to proof", **American Mathematical Monthly** 96 (August 1989).

[2] David Tall and Beverly West, "Graphic insight into calculus and differential equations, The influence of computers and informatics on mathematics and its teaching," **ICMI Studies**, Cambridge University Press 1986.

[3] Hervé Lehning, "Mathematics in a Computer Age," **Educational Computing in Mathematics** 87, North Holland.

[4] Harley Flanders, "Teaching Calculus as a Laboratory Course," **Educational Computing in Mathematics** 87, North Holland.

[5] Hervé Lehning, *Mathematiques superieures et speciales*, Masson 1985.

Calculus Lab Projects Without Computer Components

Michael D. Hvidsten
Gustavus Adolphus College

Introduction

In May of 1988 I received a Bush Foundation grant, administered through Gustavus Adolphus College, to develop laboratory projects to be used as an integral part of a new Calculus I course at Gustavus Adolphus College. During that summer I carried out the development of this course, and have now completed teaching six sections of the new Calculus I.

The subject matter of each Calculus laboratory project was chosen from the standard syllabus for Calculus I. Nine topics were chosen as follows: Functions and graphing, Limits and continuity, Derivatives and slope, Error analysis, Maxima and minima, Newton's Method, Areas, The natural logarithm, and Growth and decay.

Most of these projects are done in a computer lab setting, using True BASIC Calculus software. However, the two labs on Error Analysis and Areas are done by the students at their desks in the classroom. The complete text of these two labs is included in this paper.

Error Analysis

Discussion

In the lab on Error Analysis, lab #4, students are given the task of constructing a cylindrical container (without a lid or bottom) where the height equals the radius. This task is set in the practical context of making cat food tins for the Big Kitty Cat Food Company, where the student is employed. This fictitious, fanciful setting is less threatening and more accessible to the students than the more abstract "x-y axis" setting.

On page 2 of the lab, the students are given their main task, that of analyzing the difference in volume between their container and the ideal. This is accomplished by analyzing the error in the volume of the cylinder. For a fixed value of error in r, the radius, they find the maximum error possible in V.

They are then asked to determine a *general formula* for the error in V, ΔV, as the error in r, Δr, changes. The students are then asked a series of questions designed to help them understand the derivation of the linear approximation formula $\Delta V \approx V'(r)\Delta r$.

Finally, the lab ends with an analysis of relative errors and percentage errors.

The overall goal of this lab is to show the students how the tangent line, as the best linear approximation to a function, can be used to give practical information about the possible errors generated during manufacturing processes and other "real-world" situations.

Overall, the lab has been successful. I encourage students to write down comments and conjectures at the end of each lab. For lab #4, I have received comments such as "could come in handy some day," "Interesting lab, the results surprised us," and "This lab (is) enlightening in that it bridged a gap I had long recognized between the sciences and math. Math is the theoretical, science is the pragmatic." On the other hand, I also received "we found this lab to be confusing and difficult." For many the connection between tangent lines and approximations to functions was very hard to grasp.

Title: Error Analysis and the Derivative

Purpose: Derivatives are useful in many situations. In Lab #3 we saw how the derivative can measure slope and velocity. In this lab we will use the derivative to estimate errors that occur when doing measurements. All practical measurements are inexact to some degree, and it is important to know how errors in one quantity affect the computation of other quantities. To analyze such errors we will use linear approximations to functions, based on the tangent line.

Materials: You will need to bring a ruler, scissors, and a calculator with you to class.

The Experiment: You are employed by the Big Kitty Cat Food company and the boss has asked you to start production of new cat food cans that should be cylindrical and have a height of 2 inches and a base radius of 2 in. These cans will hold the company's exciting new "munchy mouse" flavor of cat food.

Your job will be to oversee the production of these cans and also to determine a simple and fast way to analyze the maximum error in volume that could occur during production.

The volume of a cylinder of height h and base radius r is:

$$V = \pi r^2 h$$

In our case r=h=2, so the formula reads $V = \pi r^2 = 8\pi$. Thus, the volume of "munchy mouse" our can should hold is 8π cubic inches.

To build these containers, we will cut out a long strip of material and roll it into a cylinder. The width of this strip will become the height of our container and thus should be 2 inches.

Q1: What should the length of our strip be?

Using your ruler and scissors, cut out a strip of these dimensions from the legal size paper attached to this lab, but extend its length about 1/4 in. Then, roll the paper into a cylinder and tape it together across the 1/4 in. tab.

Error Analysis

How close to the desired volume will our container be? Most rulers have subdivisions marked down to 1/16th of an inch, or 0.0625 in. Also, rounding off $2\pi r$ when measuring out the length introduces another error of about 0.06. Thus, we can be confident that our measurements were accurate to within ±0.1225 in., or about ±1/8 of an inch. Using your prototype container, estimate the radius (1/2 the diameter) and check that your radius is close to this degree of accuracy.

Q2: Using the fact that $V = \pi r^3$ determine the greatest error in V that could occur, given the accuracy of your measurement of r. That is, when r=2+.1225 or 2−.1225, how much error could occur in the volume calculation of $V=8\pi$?

Linear Approximation to $V=\pi r^3$

Suppose that the boss asks you to give him/her a general idea of how the error in V changes as the accuracy in the

measurement of r gets better or worse. To get a good formula for the error in V we will use tangent lines.

Sketch in a tangent line for $V=\pi r^3$ at r=2in. on the graph below, and record this on your report.

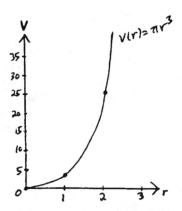

Q3: For values of r very near r=2, the tangent line to $V=\pi r^3$ at r=2 will closely approximate the values on the graph of V. Does this tangent line closely approximate the values of V near r=0.5? How would you construct a "closest Line" to V at r=0.5?

We call the tangent line to V at r=2 the linear approximation to the graph of V at r=2.

Q4: To calculate a tangent line to the curve V(r) we need to know the slope at each point on V. Find V'(r) and then find the equation of the tangent line to V at r=2 inches, $V=8\pi \text{ in}^3$.

Since the tangent line closely matches the curve V(r) near r=2, then the slope of the tangent line, V'(2), should closely match the secant line slope between V(2) and V(2+Δr) when Δr is small.

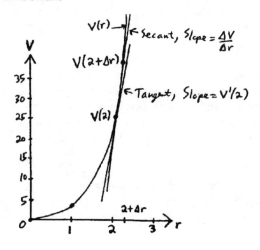